Introduction to Scientific Computing

Introduction to Scientific Computing

Brigitte Lucquin and **Olivier Pironneau**
Université Pierre et Marie Curie.
Paris, France

translated by

Michel Kern
INRIA
Rocquencourt, France

JOHN WILEY & SONS
Chichester · New York · Weinheim · Brisbane · Singapore · Toronto

National 01243 779777
International (+44) 1243 779777

e-mail (for orders and customer service enquiries): cs-books@wiley.co.uk

Visit our Home Page on http://www.wiley.co.uk or http://www.wiley.com

Reprinted May 2001

Other Wiley Editorial Offices

John Wiley & Sons, Inc., 605 Third Avenue,
New York, NY 10158-0012, USA

Wiley-VCH Verlag GmbH,
Papellallee 3, D-69469, Weinheim, Germany

Jacaranda Wiley Ltd, 33 Park Road, Milton,
Queensland 4064, Australia

John Wiley & Sons (Asia) Pte Ltd, 2 Clementi Loop #02-01,
Jin Xing Distripark, Singapore 0512

John Wiley & Sons (Canada) Ltd, 22 Worcester Road,
Rexdale, Ontario M9W 1L1, Canada

Library of Congress Cataloging-in-Publication Data

Lucquin, B.
[Introduction au calcul scientifique, English]
Introduction to scientific computing: / B. Lucquin. O. Pironneau.
p. cm.
Includes bibliographical references and index.
ISBN 0-471-979341 (PPC) 0-471-97266-5 (Paper)
1. Differential equations. Partial-Numerical solutions—Data processing. 2. Engineering mathematics—Data processing. 3. FORTRAN 77 (Computer program language) 4. Mathematical physics—Data processing. I. Pironneau, Oliver. II. Title.
QC20.7.D5L8313 1998 97-22466
620′.001′515353—dc21 CIP

British Library Cataloguing in Publication Data

A catalogue record for this book is available from the British Library

ISBN 0-471-979341 (PPC) 0-471-97266-5 (Paper)

Contents

Foreword

'A little point revealed the infinitudes'
Sri Aurobindo's Savitri

In this book we present some basic Scientific Computing methods for the solution of partial differential equations. Applications are mainly in engineering and physics, but some examples are also applicable in finance, economics, biology, etc.

The contents of this book are taught in Mathematical Engineering at the BS level at Pierre et Marie Curie University (Paris VI) in courses that have as their goal to train students to program at a high level. Thus it is above all a practical course; student evaluation is by means of a programming assignment, in C or in Fortran.

We chose Fortran 77 as our programming language because it is (still) an industry standard, and because it is well adapted to intensive computations and to vectorisation. But it is also recommended to study the C language, which is better suited to pre-and post processing. As it would have been confusing to give all programs in both languages, we only give a few hints on C techniques.

As a companion to this book we also give an integrated software package for solving partial differential equations with finite elements: FreeFEM. This software is accessible via ftp on the Internet(*) and it is strongly advised to obtain it, particularly to illustrate Chapter 1. This software uses a dedicated language (and thus a compiler): ^{G}fem. Moreover, the software illustrates the aim of all the projects, since the goal of this book is to provide techniques to let the reader write products like ^{G}fem.

We were fortunate to benefit from the assistance of two colleagues whom we wish to thank highly: D. Bernardi and F. Hecht. We also wish to thank Ph. Ciarlet and J.L. Lions for allowing us to publish the original version of this book, as well as the students of University Pierre et Marie Curie who took part in setting up the computer programs.

(*) Mosaic or Netscape server: //http:/www.ann.jussieu.fr/
ftp server: ftp://ftp.ann.jussieu.fr/pub/soft/freefem.tar.gz

Introduction

1 *WHAT DO WE MEAN BY SCIENTIFIC COMPUTING?*

The words 'Scientific Computing' denote any computation useful for Science; actually in this book we shall limit ourselves to applications from physics or engineering. We could also have given this book the title 'Computer Aided Engineering', or CAE. In practice, CAE has now become synonymous with CAM (*'Computer Aided Manufacturing'*) and CAD (*'Computer Aided Design'*) applied to engineering. The phrase 'Scientific Computing' usually refers to all the numerical simulations that are part of an engineer's work (shape optimisation for aircraft, nuclear safety questions,...) or that of a physicist (weather forecasting, pollution...).

Engineering covers such domains as fluid or structural mechanics, electromagnetism, as well as many other domains; a listing of all the applications of scientific computing would be too long to include here.

Nowadays, most industrial companies have at their disposal large computer codes allowing them to solve this type of problem. This book is an introduction to the construction of such programs.

2 *OUR APPROACH FOR A SIMPLE EXAMPLE*

In order to help understand what scientific computing is about, let us illustrate it by a simple example.

We wish to know the temperature of a metallic rod heated at both ends and placed in a room itself at a given temperature. This rod is considered equivalent to a line segment, whose extremities are denoted by a and b, and our goal is to compute the temperature at any point x in the interval $[a, b]$: we shall denote it by $\theta(x)$. The data of the problem are: the room temperature, denoted by θ_0, and the temperature at both extremities, denoted by θ_a and θ_b respectively.

In this problem, there is a loss of heat due to air convection, which can be modeled by a function α; the temperature θ solves the ordinary differential equation

$$-\kappa \frac{\mathrm{d}^2\theta}{\mathrm{d}x^2} + \alpha(x)(\theta - \theta_0) = 0, \quad a < x < b, \tag{1}$$

where κ denotes the thermal diffusion coefficient of the rod. To this equation we add the boundary conditions

$$\theta(a) = \theta_a, \quad \theta(b) = \theta_b. \tag{2}$$

Mathematicians tell us that if κ is strictly positive, and if α is a positive valued function, then problem (2.1)–(2.2) is *well posed*, which means in particular that it has a unique solution $(x \rightarrow \theta(x))$. This is doubly an important information: we can hope (approximately) to compute this solution, and if the program we write does not work, the fault will necessarily lie with the programmer!

It is not very difficult to set up a computer program for this simple problem. We start by discretising the differential equations with *finite differences*. This method, which we shall explain at greater length in Chapter 6, consists in approximating the derivatives by difference quotients, starting from the definition of the derivative that lets us write, for small h:

$$\frac{\mathrm{d}f}{\mathrm{d}x}(x) \simeq \frac{f(x+h) - f(x)}{h}.$$

The second order terms are approximated by:

$$\frac{\mathrm{d}^2 f}{\mathrm{d}x^2}(x) \simeq \frac{f(x+h) - 2f(x) + f(x-h)}{h^2}.$$

The interval $[a, b]$ is decomposed into $M - 1$ intervals of length $h = \delta x = (b - a)/(M - 1)$ (cf. Figure 1), and the solution is sought at each of the subdivision points $x_m = a + (m - 1)h$, $m \in \{1, ..., M\}$; let us denote this value by $\theta_m \simeq \theta(x_m)$. The approximate problem is equivalent to the solution of the following linear system

$$-\frac{\kappa}{h^2}[\theta_{m+1} - 2\theta_m + \theta_{m-1}] + \alpha_m(\theta_m - \theta_0) = 0, \quad m = 2, .., M - 1, \tag{3}$$

$$\theta_1 = \theta_a, \quad \theta_M = \theta_b. \tag{4}$$

Theory tells us that the finer the mesh (i.e. the smaller h), the better the approximation: the error $\theta_m - \theta(x_m)$ goes to 0 as h goes to 0.

A possible algorithm to solve this system is to use the Gauss–Seidel iterative method. Let us briefly explain the basic principle. To solve the linear system $Ax = b$, the Gauss–Seidel method consists in defining inductively a sequence of values x^k as solutions to

$$(D - E)x^{k+1} = Fx^k + b,$$

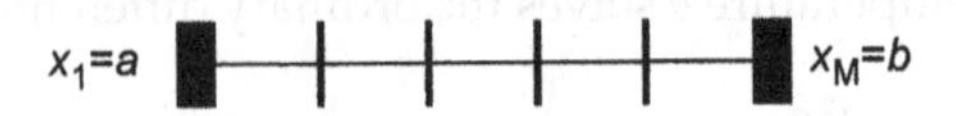

Figure 1 A subdivision of the rod into $M - 1$ intervals

where we have used the following splitting of A: $A = D - E - F$, with D a diagonal matrix, whose entries are those of the diagonal of A, $-E$ a strictly lower triangular matrix, whose non-zero entries are those of the strictly lower triangular part of A, and $-F = A - D + E$ the strictly upper triangular part of A. Theory tells us that if the matrix A is symmetric and positive definite (which is precisely the case here), this method is convergent, meaning that, as the number of iterations k becomes very large, the error between the exact solution x and the approximate solution x^k goes to 0.

Programming this algorithm for the solution of system (3)–(4) consists in extracting the value of θ_m from equation m, which gives:

$$\forall m \in \{2,..,M-1\}, \quad \theta_m = \left\{\frac{\kappa}{h^2}[\theta_{m+1} + \theta_{m-1}] + \alpha_m\theta_0\right\} \Big/ \left[\alpha_m + 2\frac{\kappa}{h^2}\right], \tag{5}$$

$$\theta_1 = \theta_a, \quad \theta_M = \theta_b. \tag{6}$$

Let us note that the Gauss–Seidel iteration index k has disappeared in this notation. Indeed, there is no need to store in memory the vectors corresponding to two successive iterations. As the computation proceeds (i.e. when m increases), the value of θ_m at iteration k gets replaced by its new value at iteration $k+1$.

This algorithm is iterated as many times as is necessary, i.e. until values of θ_m at two successive iterations are almost the same.

Let us summarize the algorithm:

1. *Read the data.*
2. *Initialise the Gauss–Seidel iterations,* for example by linear interpolation between boundary values:

$$\theta_m = \theta_a\frac{(M-m)}{(M-1)} + \theta_b\frac{(m-1)}{(M-1)}.$$

3. *Iterations*
   for k from 1 to kMax do the following:
   for m varying from 2 to $M-1$ replace θ_m by

$$\theta_m := \left\{\frac{\kappa}{h^2}[\theta_{m+1} + \theta_{m-1}] + \alpha_m\theta_0\right\} \Big/ \left[\alpha_m + 2\frac{\kappa}{h^2}\right]$$

4. *Visualise the results.*

The Fortran program is the following (to avoid confusion between M and m, m has been replaced by im in the program).

```
PROGRAM Heat
PARAMETER (M=10)         !      { number of x points}
PARAMETER (dx=0.1)       !      { mesh size}
PARAMETER (kmax=20)      !    { nb. Gauss-Seidel iters}
DIMENSION teta(M), alpha(M)
CALL datainput (xkappa,tetaa,tetab,teta0,alpha,M)
CALL init (teta,M, tetaa,tetab)
```

```
      DO 1 k=1, kmax
1     CALL relax (teta,alpha, M,dx,xkappa,teta0)
      write( * , * ) 'plot'
      pause
      CALL plotresult(teta,M)
      pause
      END

      SUBROUTINE relax ( teta,alpha, M,dx,xkappa,teta0)
      DIMENSION teta( * ),alpha( * ) ! { Gauss-Seidel iters}
      xkappadx2=xkappa/dx/dx
      DO 1 im=2, M-1
1       teta(im)=(xkappadx2 * (teta(im+1)+teta(im-1))
    >   +alpha(im) * teta0) / (alpha(im)+ 2 * xkappadx2)
      END

      SUBROUTINE init (teta,M, tetaa,tetab)
      DIMENSION teta( * )
      DO 1 im=1 , M
1       teta(im)=tetaa * (M-im)/(M-1.)+tetab * (im-1.)/ (M-1.)
      END


      SUBROUTINE datainput (xkappa,tetaa,tetab, teta0,alpha,M)
      DIMENSION alpha( * )
      write( * , * ) 'kappa?'
      read( * , * ) xkappa
      write( * , * ) 'teta0?'
      read( * , * ) teta0
      write( * , * ) 'tetaa?'
      read( * , * ) tetaa
      write( * , * ) 'tetab?'
      read( * , * ) tetab
      DO 1 im=1,M
        write( * , * ) 'alpha[', im, ']?'
1       read( * , * ) alpha(im)
      END

      SUBROUTINE plotresult(teta,M)
      DIMENSION teta( * )
      iteta=teta(1) * 10              !      { changes scale}
      CALL moveto(100, iteta)    !       ( moves cursor}
      DO 1 im=2,M
        iteta =teta(im) * 10
        im10=100+im * 10
1     CALL lineto(im10, iteta)      !  ( plot and move cursor}
      END

      include grafic.for     ! { contains moveto and lineto}
```

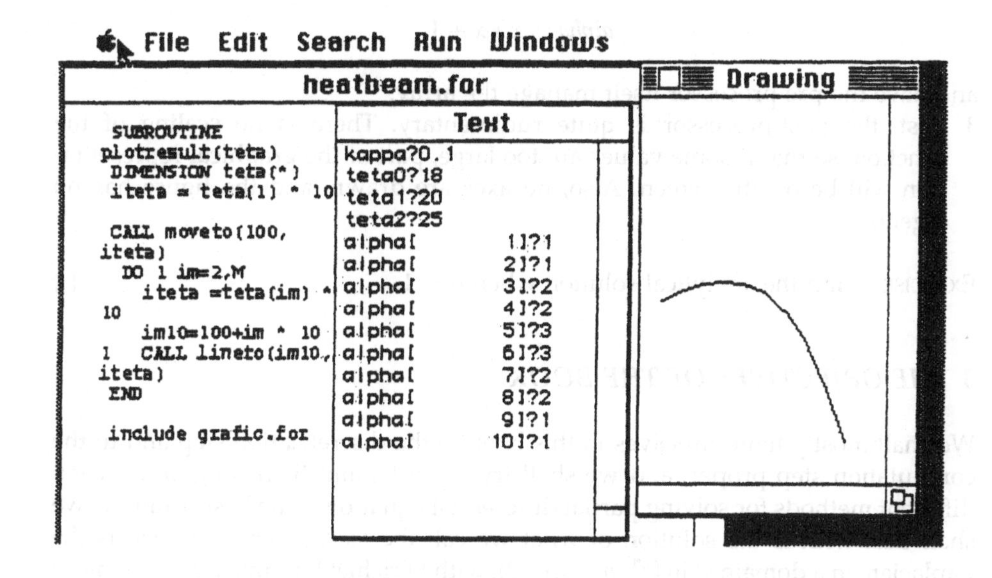

Figure 2 Results; screen dump after executing the program

This simple example shows us the four steps in a scientific computing software:

1. Write up the equations, choose the discretisation algorithm
2. Input the data (*pre-processor*)
3. Compute
4. Visualise (*post-processor*).

Programming steps are steps 2–4, step 1 being a modelling and numerical analysis step. Even on this simple example, the number of instructions in the pre- and post-processor parts is much larger than the number of instructions in the computational part. This means that the programmer spent much more time writing the pre- and post-processors. This state of affairs is rather general in scientific computing, and this is the reason programmers would rather use commercial software than write their own data input or graphical visualisation programs.

Last, let us note that the program we wrote is not very 'user-friendly', for several reasons:

1. Data input must be done in a fixed order, and there is no possibility of correcting errors. Nowadays, more user-friendly programs have menus, and the user tells the program (with the help of the mouse) which data should be input first.
2. The way data is input is not optimal. Indeed, let us assume that we that change $M = 10$ to $M = 100$ so as to increase accuracy. The user will then need to enter 100 values by hand in order to define the array α. To avoid this tedious task, it is better to read the values from a file, or to define them via a function, for instance by writing:

$$alpha = x * x + 1,$$

and have the pre-processor itself manage the array.

3. Last, the post-processor is quite rudimentary. There is no scaling of the function, so that if some values are too large, part of the graphical representation will be off the screen. Also, no axes are drawn, and the figure has no legend...

EXERCISE Find the analytical solution when $\alpha = 1$. □

3 THE OBJECTIVES OF THE BOOK

We shall mostly limit ourselves in this book to the discretisation step and to the computation step proper, and we shall try to write, and then program, several different methods for solving partial differential equations. The basic example we shall deal with is the solution of what we call the *model problem*, which is the Laplacian on a domain Ω in $\mathbb{R}^n$ ($n = 2$ or 3), with Dirichlet boundary conditions on the boundary Γ of Ω

$$-\Delta\varphi(x) = f(x), \quad \forall x \in \Omega, \tag{7}$$

$$\varphi(x) = g(x), \quad \forall x \in \Gamma, \tag{8}$$

where the Laplacian is defined by:

$$(\Delta\varphi)(x) = \sum_{i=1}^{i=n} \frac{\partial^2 \varphi}{\partial x_i^2}(x), \quad x = \begin{pmatrix} x_1 \\ x_2 \\ \dots \\ x_n \end{pmatrix}. \tag{9}$$

Indeed, we shall see in Chapter 1 that several algorithms for solving other partial differential equations require the solution of this simple equation. We shall also see in Chapters 5 and 6 that extending the methods to more general partial differential equations (PDE for short) is easy.

Even though we do not deal with pre-and post-processors, we shall try to give programs that are easy to port from one system to another (Unix, DOS-Windows, Mac-OS), and programs in which we have tried to provide a minimum level of user-friendliness.

At the present time, when the size and complexity of computations is growing rapidly, one might wish to run programs on a different 'host' computer, with a vector or parallel architecture. Some programming or algorithmic changes will then be necessary: we shall discuss them in the last chapter.

The organisation of the book is as follows. In the first chapter, we give numerous examples showing how partial differential equations occur in real life, and propose a programming environment designed for their solution. Then the book is divided into three parts.

In the first part, we present *the finite element method* for solving the model problem in dimension $n = 2$. We handle in detail several different formulations and solution methods, for which we systematically give programs in an appendix.

We complete this study in a second part, where we deal with more general partial differential equations, both at the level of the differential operator (leading to systems that are not necessarily symmetric), and at the level of the boundary conditions. As equations from physics are mostly evolution equations in time, we shall propose in Chapter 7 of this part, a *finite difference in time* discretisation for these equations. The discretisation with respect to the space variable can be of finite difference, finite element, or *finite volume* type: we shall briefly introduce this latter method.

The last part will include some numerical methods for problems in unbounded domains (Chapter 8), as well as programming for multiprocessor machines (Chapter 9). We have chosen to talk in Chapter 8 of *boundary integral methods,* as these methods are commonly used for solving problems in unbounded domains (more accurately in domains that are the complement of a bounded domain) in three space dimensions. Also, since they reduce to the solution of a numerical problem in two dimensions, they only require surface discretisations.

In a single book, we could not deal with all the numerical methods in use at the present time. We have chosen to present the finite difference method in one dimension (the time variable), the finite element method in two dimensions, and the boundary integral equation method in three dimensions. We have thus chosen to ignore other methods, such as particle methods, spectral methods, wavelets,...

We have also tried to emphasize numerical solutions and programming against a theoretical study of the problems. Some of the programs given in this book can be integrated at the heart of larger numerical algorithms; or we hope at the very least that the methods presented here will help shed light and/or improve existing programs.

APPENDIX A

The Program for Solving the Heat Equation, Translated into C++

```
#include <iostream.h>
#include <stdlib.h>
#include <math.h>

const int M=100;
const float dx=0.1;
const int kmax=20;

class data
    {
      public:
      float kappa, theta0, theta1, theta2;
      float alpha[M];
      void read();
    } ;
```

```
data d;

void data::read()
{
cout << "enter kappa:";
cin >> kappa;
cout << "enter theta0:";
cin >> theta0;
cout << "enter theta1:";
cin >> theta1;
cout << "enter theta2:";
cin >> theta2;
for(int i=0; i< M ; i++) alpha[i]=i * i/(M-1.)/(M-1.);
}

void init( float * theta, data& d)
{ for(int i=0; i< M; i++)
theta[i]=d.theta1 * (M-1-i)/(M-1.)+d.theta2 * i/(M-1.);
}

void relax( float * theta, data& d)
{
    const float kappadx2=d.kappa / dx / dx;
    for(int i=1; i< M-1; i++)
      theta[i]=(kappadx2 * (theta[i+1] +theta[i-1])
      + d.alpha[i] * d.theta0) / (d.alpha[i]+ 2 *
      kappadx2);
}

void plotresult( float * theta)
{
    MoveTo( 0, (int)(theta[0]));
    for(int i=1; i< M; i++)
      LineTo( i, (int)( theta[i]));
}

void main()
{
float * theta=new float[M];
data d;
d.read();
init(theta, d);
for (int k=0; k < kmax; k++)
    relax(theta, d);
plotresult(theta);
}
```

1 Some Partial Differential Equations

Summary

Numerical analysis of partial differential equations constitutes a major part of scientific computing. Indeed, in this chapter we illustrate how numerous problems from physics can be modelled by partial differential equations (PDE). We also give their numerical solution with the ^{G}fem language, a programming environment for the finite element solution of PDEs. The aim of the following chapters will be to explain the theoretical basis of such a program, the other possible choices, and how to build a program library such as ^{G}fem.

1.1 MEMBRANES

1.1.1 The Problem

An elastic membrane Ω is glued to a planar rigid support Γ, but a force $f(x)\mathrm{d}x$ presses on each surface element $\mathrm{d}x = \mathrm{d}x_1\mathrm{d}x_2$. The vertical membrane displacement, $\varphi(x)$, solves Laplace's equation:

$$-\Delta\varphi = f \text{ in } \Omega. \tag{1.1}$$

As the membrane is glued to its planar support, one has:

$$\varphi|_{\Gamma} = 0. \tag{1.2}$$

This is the homogeneous Dirichlet problem for the Laplace operator.

Let us recall that the Laplace operator Δ is defined by:

$$\Delta\varphi = \frac{\partial^2\varphi}{\partial x_1^2} + \frac{\partial^2\varphi}{\partial x_2^2}. \tag{1.3}$$

Problem (1.1)–(1.2) admits a unique solution; we refer the reader for instance, to Brezis (1987), Dautray and Lions (1990), Strang (1986) and Brenner and Scott (1994).

1.1.2 Example

Let an ellipse have principal radius $a = 2$, and secondary radius $b = 1$, and a surface force equal to 1. Programming this case with ^{G}fem* gives:

```
border(1,0,2*pi,40) /* The number 1 boundary            */
 begin                /* parametrised from 0 to 2 pi      */
   x := 2 * cos(t);   /* has 40 vertices. We recognise */
   y := sin(t);       /* the equation of an ellipse       */
 end;

buildmesh(800);       /* call the mesh generator          */
                      /* with maximum of 800 vertices     */
solve(v) begin        /* PDE: definition + solution       */
 onbdy(1) v = 0;
 pde(v) -laplace(v) =1
end;
plot(v); save('v.dta',v);
```

A triangulation is built by the keyword *buildmesh*. The keyword calls a triangulation subroutine based on the Delaunay test, which first triangulates with only the boundary points, then adds internal points by subdividing the edges, without ever stepping over the indicated maximum, here 800. How fine the triangulation becomes is controlled by the size of the closest boundary edges, except when the maximum number of vertices is reached.

The PDE is then discretised using the triangular first order finite element method on the triangulation; as will be shown in Chapter 3, one obtains a linear system whose size is the number of vertices in the triangulation. The system is

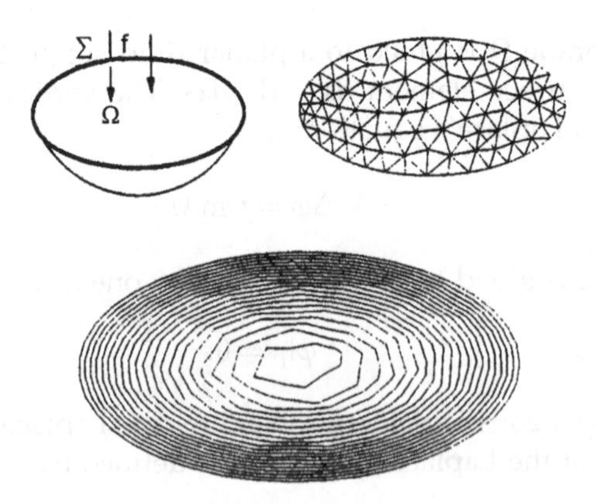

Figure 1.1 Membrane deflected by a vertical force

* ^{G}fem is a programming environment with a language dedicated to PDEs; its kernel, Fr$_{ee}$FEM, is machine independent, and is freely available. The following program is read and executed by Fr$_{ee}$FEM, and the results are visualised and stored on disk. The graphical part used to produce the figures in this chapter (essentially a screen dump in Postscript) is not provided; Fr$_{ee}$FEM uses the techniques explained in this book to solve PDEs

solved by a Gauss LU factorisation (preferred over a Choleski factorisation since it also enables us to solve non-symmetric systems).

We show in Figure 1.1 the triangulation and the iso-φ lines obtained for a membrane deflected by a vertical force.

1.2 ELECTROSTATICS

1.2.1 The Problem

Let $\{C_i\}_{1...N}$, be N conductors within an enclosure C_0. Each one is held at an electrostatic potential φ_i. We assume that the enclosure C_0 is held at potential 0.

In order to know $\varphi(x)$ at any point x of the domain Ω, we must solve

$$\Delta\varphi = 0 \quad \text{in} \quad \Omega, \quad \varphi_{|\Gamma} = g, \tag{1.4}$$

where Ω is the interior of C_0 minus the conductors C_i, and Γ is the boundary of Ω, that is $\cup_0^N C_i$. Here g is any function of x equal to φ_i on C_i and to 0 on C_0. The second equation is a reduced form for:

$$\varphi = \varphi_i \text{ on } C_i, \quad i = 1...N, \quad \varphi = 0 \text{ on } C_0. \tag{1.5}$$

1.2.2 The ^{G}fem program

```
/* a circle with centre at (0 ,0) and radius 5*/
border(1,0,2*pi,60) begin
  x := 5 * cos(t); y := 5 * sin(t);
end ;
/* Right rectangle */
border(2,0,1,4)     begin x:=1+t; y:=3        end ;
border(2,0,1,24)    begin x:=2; y:=3-6*t      end ;
border(2,0,1,4)     begin x:=2-t; y:=-3    end ;
border(2,0,1,24)    begin x:=1; y:=-3+6*t end ;
/* Left rectangle */
border(3,0,1,4)     begin x:=-2+t; y:=3    end ;
border(3,0,1,24)    begin x:=-1; y:=3-6*t end ;
border(3,0,1,4)     begin x:=-1-t; y:=-3      end ;
border(3,0,1,24)    begin x:=-2;y:=-3+6*t   end ;

buildmesh(800); /* call the mesh generator */

solve(v) begin
  onbdy(1) v = 0;/* input Dirichlet boundary conditions */
  onbdy(2) v = 1;
  onbdy(3) v = -1;
  pde(v) -laplace(v) =0     /*input PDE */
end;                           /*solution */
plot(v);                       /*visualisation */
```

The beginning of the program contains geometric information required by the mesh, and additional information such as logical boundary numbers, and the

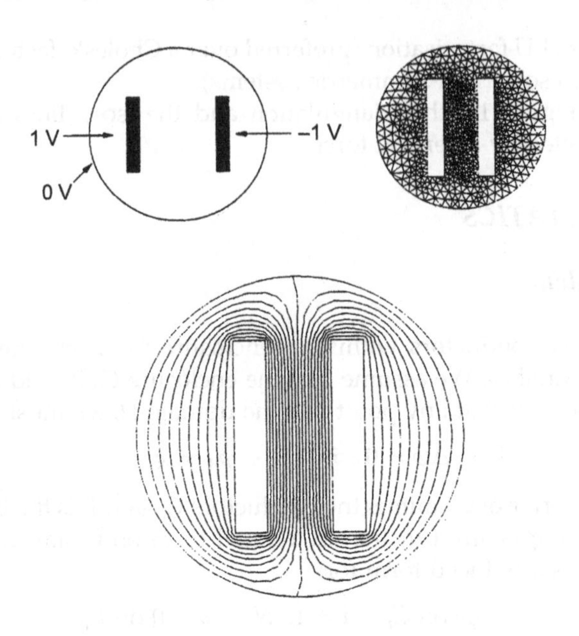

Figure 1.2 A model capacitor

number of vertices on each boundary part. Note that the circle is described counterclockwise, whereas the rectangles are described clockwise because, by convention, the boundary must be oriented so that the computational domain is to its left.

Last, in the above program, Dirichlet boundary conditions equal to +1 volt, and -1 volt, were imposed on the electrodes, whereas the enclosure is at a 0 potential. Figure 1.2 shows the triangulation and the equipotential lines for a model capacitor made of two rectangles.

EXERCISE Use the symmetry of the problem with respect to the axes; triangulate only one quarter of the domain, and set Dirichlet conditions on the vertical axis, and Neumann conditions on the horizontal axis. □

1.3 THERMAL CONDUCTION

1.3.1 Rod with Rectangular Section

We seek the temperature distribution in a rectangular rod, both of whose ends are held at constant temperatures t_0 and $t_1 > t_0$; there is loss of heat through convection and radiation into the surrounding medium, held at a temperature $t_e < t_1$.

In the vertical plane of symmetry passing through the centre of the plate, the problem is bi-dimensional; loss through convection is modelled by a dissipation α proportional to $t_0 + t_1 - 2t_e$. We must solve

$$-\Delta u + \alpha u = 0 \text{ in } \Omega,$$
$$u|_{\Gamma_1} = t_0, \quad u|_{\Gamma_2} = t_1, \quad \frac{\partial u}{\partial n}|_{\Gamma_3} = 0, \tag{1.6}$$

where $\Gamma = \partial\Omega$, Γ_1 and Γ_2 being the extremities and $\Gamma_3 = \partial\Omega - \Gamma_1 - \Gamma_2$ the side edges.

For a rod of length 10, the ^{G}fem program is the following*:

```
border(2,0,1,4)    begin x:= 10;      y := t     end;
border(4,0,1,4)    begin x:= 0;       y := 1-t   end;
border(1,0,10,40)  begin x:= t;       y := 0     end;
border(3,0,10,40)  begin x:= 10 - t;  y := 1     end;
buildmesh(800);

t0 := 10;  t1 := 100;  te := 25;
alpha =0.05 * ( t0 + t1 - 2*te);

solve(v) {
  onbdy(4) v=t0;
  onbdy(2) v = t1;
  onbdy(1,3) dnu(v)=0; /* the 2 horizontal edges */
  pde(v) -laplace(v) +id( v ) * alpha =0;
};
plot(v);
```

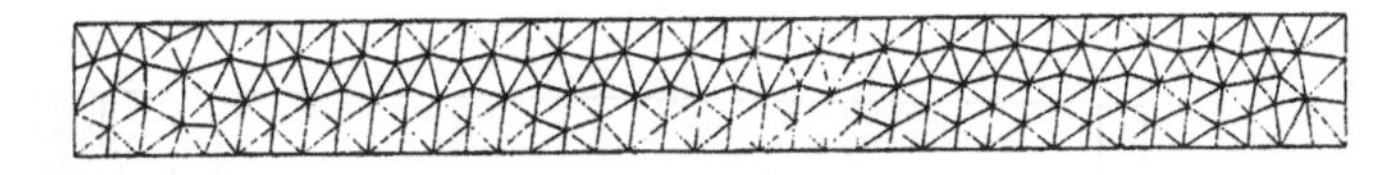

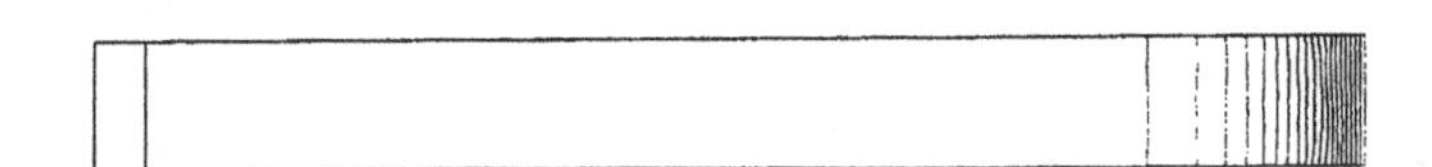

Figure 1.3 Thermal equilibrium in a plate

EXERCISE Check that the solution is independent of x_2, and compare it with the analytical solution. □

1.3.2 Rod with Circular Section

The symmetry is now cylindrical. In cylindrical coordinates, the Laplace operator becomes (r is distance to the axis, z distance along the axis, θ polar angle in a fixed plane perpendicular to the axis):

* Note the difference between assigning with '=' (for a function), and assigning with ':=' (for a scalar), which avoids declaring an array containing nv times the same value (nv= number of vertices). Indeed, in ^{G}fem all functions are continuous and piecewise linear , and thus (cf. Chapter 2) are determined by their values at the triangulation vertices. Constants are a particular type of such functions, for which only one memory location is needed.

$$\Delta u = \frac{1}{r}\partial_r(r\partial_r u) + \frac{1}{r^2}\partial^2_{\theta\theta}u + \partial^2_{zz}. \tag{1.7}$$

Symmetry implies that we lose the dependence with respect to θ; the problem is thus:

$$\begin{aligned} &-\partial_r(r\partial_r u) - \partial_z(r\partial_z u) + r\alpha u = 0 \text{ in } \Omega, \\ &u|_{\Gamma_1} = t_0, \quad u|_{\Gamma_2} = t_1, \quad \frac{\partial u}{\partial n}|_{\Gamma_3} = 0. \end{aligned} \tag{1.8}$$

Note that the equation was multiplied by r in order to put it in 'divergence' form $-\nabla \cdot (r\nabla u)$.

We obtain the following ^{G}fem* program:

```
t0 := 10;  t1 := 100;  te := 25;
alpha =0.1 * ( (t0 + t1)/2 - te);

solve(v){
  onbdy(4) v=t0;
  onbdy(2) v = t1;
  onbdy(1,3) dnu(v)=0; /* see footnote */
pde(v) -laplace(v)*y +id( v ) * alpha * y =0;
};
plot(v);
```

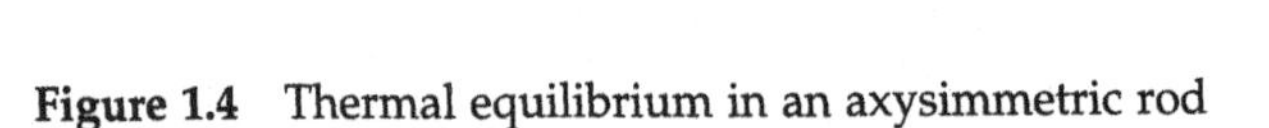

Figure 1.4 Thermal equilibrium in an axysimmetric rod

1.3.3 Convection and radiation

Losses through convection and radiation have, up to now, been modelled rather crudely by a dissipation α. A better way is to use, for convection:

$$\frac{\partial u}{\partial n} = -b(u - t_e), \tag{1.9}$$

and for radiation, the law of black bodies, that is a loss proportional to temperature to the fourth power:

$$\frac{\partial u}{\partial n} - c[(u + 273)^4 - (t_e + 273)^4] = 0. \tag{1.10}$$

The combined condition is thus:

* In the ^{G}fem language, laplace(u)*f actually means $\nabla \cdot (f\nabla u)$

$$\frac{\partial u}{\partial n} + b(u - t_e) - c[(u + 273)^4 - (t_e + 273)^4] = 0. \quad (1.11)$$

The problem is non-linear, and must be solved iteratively. If m denotes the iteration index, a semi-linearisation of condition (1.11) gives

$$\frac{\partial u^{m+1}}{\partial n} + b(u^{m+1} - t_e) - c[u^{m+1} - t_e][u^m + t_e + 546][(u^m + 273)^2 + (t_e + 273)^2] = 0,$$

because we have the identity $a^4 - b^4 = (a - b)(a + b)(a^2 + b^2)$. The iterative process will work with $v = u - t_e$, initialised to 0.

```
t0 := 10;  t1 := 100;  te := 25; tk := 273;
b=0.1; c = -5.0e-9;
u = (t0*(10-x)+t1*x)/10;
iter(10) begin
  w= b - c * (u + te - 2*tk) * ((u-tk)^2 + (te-tk)^2);
  solve(v){
    onbdy(4) v=t0-te;
    onbdy(2) v = t1-te;
    onbdy(1,3) id(v) * w + dnu(v) = 0;
    pde(v) -laplace(v) = 0;
  };
  u = v + te;
  plot(u);
end;
```

Figure 1.5 Thermal equilibrium with radiation

1.3.4 *Non-homogeneous plate*

Let us go back to the rectangular rod, but let us assume this time that it is made up of two plates of different materials, welded onto one another (as in a thermostat). Let Ω_i be the domain occupied by material i with thermal diffusion κ_i. The computational domain Ω is the interior of $\bar{\Omega}_1 \cup \bar{\Omega}_2$.

With dissipation, and at the thermal equilibrium, the temperature equation is:

$$\alpha u - \nabla \cdot (\kappa \nabla u) = 0 \text{ in } \Omega. \quad (1.12)$$

We recall the definition of the div and grad operators:

$$\text{div}v = \nabla . v = \frac{\partial v_1}{\partial x_1} + \frac{\partial v_2}{\partial x_2}, \quad \text{grad}u = \nabla u = \begin{pmatrix} \dfrac{\partial u}{\partial x_1} \\ \dfrac{\partial u}{\partial x_2} \end{pmatrix}. \quad (1.13)$$

Equation (1.12), taken in the distribution sense, contains a jump condition at the interface Σ between Ω_1 and Ω_2:

$$\kappa_1 \frac{\partial u}{\partial n}|_{\Sigma\cap\bar{\Omega}_1} = \kappa_2 \frac{\partial u}{\partial n}|_{\Sigma\cap\bar{\Omega}_2}. \tag{1.14}$$

This condition is automatically taken into account by the finite element method, because the method works with the variational formulation of the problem.

Let h_0 and h_1 be the thicknesses of the two plates.

In order to triangulate such a domain, we need an internal boundary to describe the interface Σ. We shall assign it number 0. We also have to check that its extremities are vertices of the outer boundaries. The best way to make sure of this is to split these outer boundaries in two by the intersection points with the internal boundary.

```
h0 := 1.0; h1 := 1.0;
border(2,0,1,4)    begin x:= 10; y := h0 * t           end;
border(2,0,1,4)    begin x:= 10; y := h0 + h1 * t      end;
border(4,0,1,4)    begin x:= 0;  y := h0 + h1 * (1-t) end;
border(4,0,1,4)    begin x:= 0;  y := h0 * (1-t)      end;
border(1,0,10,40)  begin x:= t;  y := 0                end;
border(3,0,10,40)  begin x:= 10 - t; y := h0 + h1      end;
border(0,0,10,40)  begin x:= t;  y := h0              end;
buildmesh(800);

t0 = 10;  t1 = 100;  alpha:=1;
kappa =0.01 + one(y > h0);
solve(v){
  onbdy(4) v=t0;
  onbdy(2) v = t1;
  onbdy(1) dnu(v)=0;
  onbdy(3) dnu(v)=0;
  pde(v) -laplace(v)*kappa + id(v)*alpha = 0;
};
plot(v);
```

In Figure 1.6 are shown results obtained for a thermal diffusion $\kappa = 0.01$ in Ω_1 and $\kappa = 1.01$ in Ω_2, and for a dissipation $\alpha = 1$.

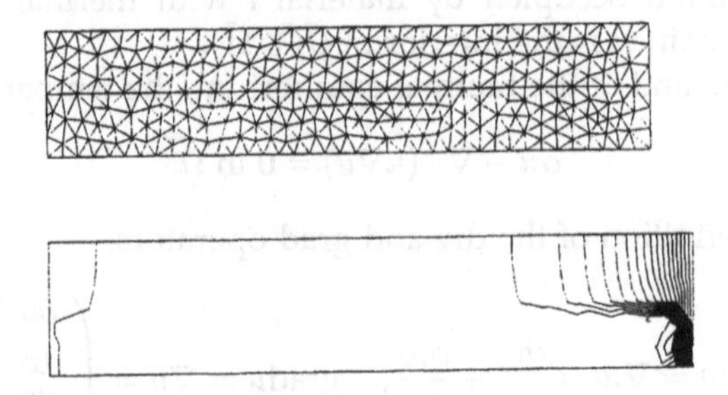

Figure 1.6 Thermics in a non-homogeneous plate

1.4 ACOUSTICS

Pressure variations in air at rest are governed by the wave equation:

$$\frac{\partial^2 u}{\partial t^2} - c^2 \Delta u = f. \tag{1.15}$$

When the solution wave is monochromatic, u is of the form $u(x,t) = v(x)e^{ikt}$ where v is a solution of Helmholtz's equation:

$$\begin{aligned} k^2 v + c^2 \Delta v &= f \text{ in } \Omega, \\ \frac{\partial v}{\partial n}|_\Gamma &= 0. \end{aligned} \tag{1.16}$$

Note the '+' sign in front of the Laplacian. This sign may make the problem ill posed for some values of the coefficients, a phenomenon called 'resonance'.

Any solution of (1.16) is the amplitude of a solution of (1.15) but the converse is not true: one needs specially chosen initial and boundary conditions in order for the solution to be monochromatic.

The speed of sound is c, k is the frequency and f contains the sources.

The following small example simulates sound propagation in the vertical section of a polygonal concert hall with an exponentially decreasing source centred on the stage.

```
border(1,0,1,20) begin x:= 5;      y:= 1+2*t  end;
border(1,0,1,20) begin x:= 5-2*t;  y:= 3      end;
border(1,0,1,20) begin x:= 3-2*t;  y:=3-2*tend;
border(1,0,1,10) begin x:= 1-t;    y:= 1      end;
border(1,0,1,10) begin x:= 0;      y:= 1-t    end;
border(1,0,1,10) begin x:= t;      y:= 0      end;
border(1,0,1,20) begin x:= 1+4*t;  y:= t      end;

buildmesh(1000);

f=exp(-10*((x-0.5)*(x-0.5)+(y-0.5)*(y-0.5)));
solve(v){
  onbdy(1) dnu(v)=0;
  pde(v) laplace(v) + id(v)*0.8 =f;
};
plot(v);
```

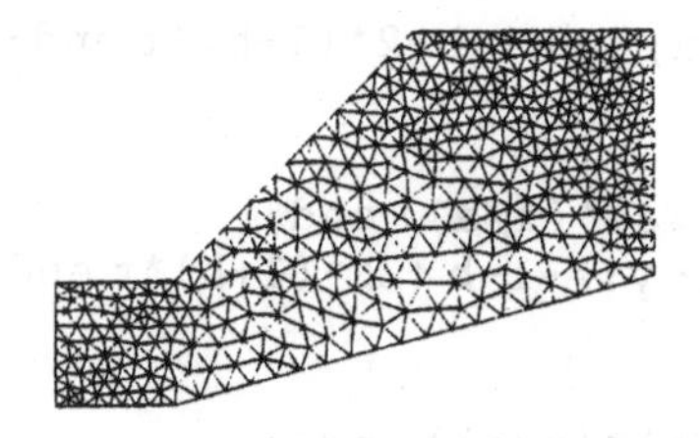
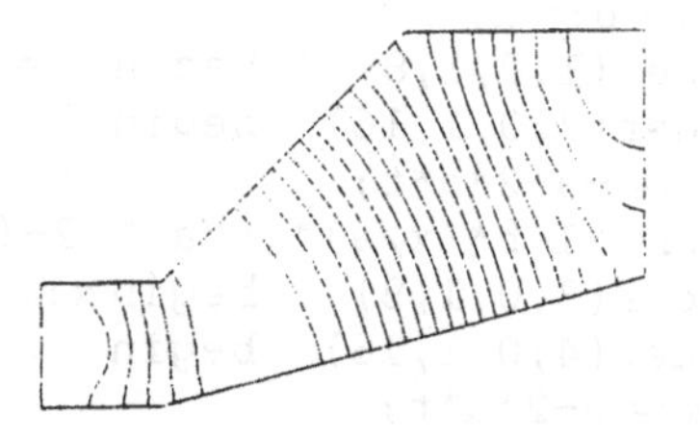

Figure 1.7 Acoustics in a concert hall (vertical section)

EXERCISE Simulate absorbing walls by boundary conditions of the type $va + \partial v/\partial n = g$, where a may be complex. □

1.5 IRROTATIONAL FLOW

1.5.1 Incompressible Flow

Without viscosity and vorticity, incompressible flows have a velocity given by:

$$u = \left(\frac{\partial\psi}{\partial x_2}, -\frac{\partial\psi}{\partial x_1}\right), \text{ where } \psi \text{ is a solution of } \Delta\psi = 0. \tag{1.17}$$

This equation expresses both incompressibility ($\nabla \cdot u = 0$) and absence of vortex ($\nabla \times u = 0$).

As the fluid slips along the walls, normal velocity is zero, which means that ψ satisfies:

$$\psi \text{ constant on the walls.} \tag{1.18}$$

One can also give the flux (the integral of the normal velocity), and this translates into Dirichlet data for ψ or the normal velocity which, if the normal velocity is constant, leads to a linear Dirichlet condition for ψ. Flow in a duct with given inflow and outflow is expressed by

$$-\Delta\psi = 0 \text{ in } \Omega, \quad \psi|_\Gamma = g, \tag{1.19}$$

with linear ψ at the inflow, free outflow and constant ψ on the walls.

Let us consider a nozzle symmetric with respect to both axes, and which passes through points $(0,1)$ and $(a, 1+b)$ with a right-angled junction with the vertical boundaries. The equation of the boundary is:

$$y = \pm b\left(\frac{x}{a}\right)^2\left(2 - \left(\frac{x}{a}\right)^2\right) + 1, \quad -a < x < a \tag{1.20}$$

If the inflow velocity is unitary, we must solve

```
a:= 6; b:= 1;
border(1,0,1,8)    begin x:= -a; y:= 1+b-2*(1+b)*t end;
border(2,0,1,16)   begin
   x:= -a+2*a*t;
   y:= -1-b*(x/a)*(x/a)*(2-(x/a)*(x/a)) end;
border(3,0,1,8)    begin x:= a; y:=-1-b + 2 *(1+b)*t end;
border(4,0,1,16)   begin
   x:= a-2*a*t;
   y:= 1+b*(x/a)*(x/a)*(2-(x/a)*(x/a)) end;
buildmesh(800);
```

```
solve(psi){
  onbdy(1) psi=y;
  onbdy(2) psi=-1-b;
  onbdy(3) dnu(psi)=0; /* free outflow */
  onbdy(4) psi=1+b;
  pde(psi) -laplace(psi)=0;
};
plot(psi);
```

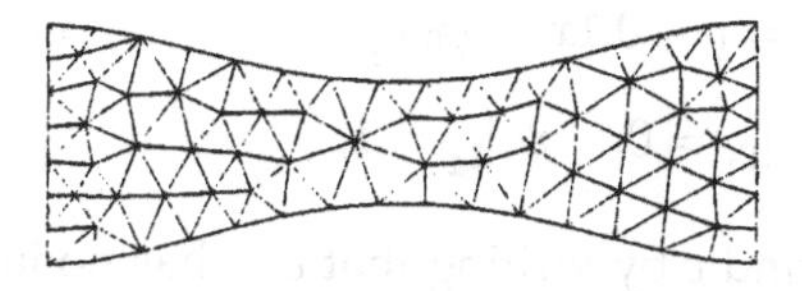
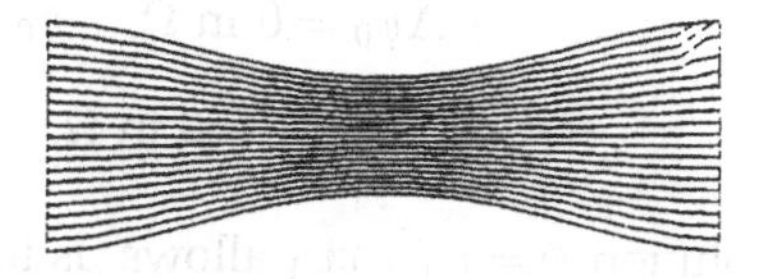

Figure 1.8 Potential flow: stream lines

EXERCISE Solve the same problem on only a quarter of the domain by making use of symmetries. □

1.5.2 *Airfoil with lift*

In more complex situations the value of the constant ψ on the walls may not be known. This is the case of an airfoil.

Let us consider a wing profile S in a uniform flow. Infinity will be represented by a large circle Γ_∞. As previously, we must solve

$$\begin{aligned} \Delta\psi &= 0 \text{ in } \Omega, \\ \psi|_S &= c, \\ \psi|_{\Gamma_\infty} &= u_{\infty 1} y - u_{\infty 2} x, \end{aligned} \tag{1.21}$$

where Ω is the area occupied by the fluid, u_∞ is the air speed at infinity, c is a constant to be determined so that $\partial_n \psi$ is continuous at the trailing edge P of S (so called Kutta–Joukowski condition). Lift is proportional to c.

To find c we use a superposition method. As all equations in (1.21) are linear, the solution ψ_c is a linear function of c

$$\psi_c = \psi_0 + c\psi_1, \tag{1.22}$$

where ψ_0 is a solution of (1.21) with $c = 0$ and ψ_1 is a solution with $c = 1$ and zero speed at infinity. With these two fields computed, we shall determine c by requiring the continuity of $\partial\psi/\partial n$ at the trailing edge.

EXAMPLE Lift of a NACA0012.

An equation for the upper surface of a NACA0012 (this is a classical wing profile in aerodynamics; the rear of the wing is called the trailing edge) is:

$$y := 0.17735\sqrt{x} - 0.075597x - 0.212836x^2 + 0.17363x^3 - 0.06254x^4. \tag{1.23}$$

Taking an incidence angle α such that $\tan\alpha = 0.1$, we must solve

$$-\Delta\psi = 0 \text{ in } \Omega, \quad \psi|_{\Gamma_1} = y - 0.1x, \quad \psi|_{\Gamma_2} = c, \tag{1.24}$$

where Γ_2 is the wing profile and Γ_1 is an approximation of infinity. One finds c by solving:

$$-\Delta\psi_0 = 0 \text{ in } \Omega, \quad \psi_0|_{\Gamma_1} = y - 0.1x, \quad \psi_0|_{\Gamma_2} = 0, \tag{1.25}$$

$$-\Delta\psi_1 = 0 \text{ in } \Omega, \quad \psi_1|_{\Gamma_1} = 0, \quad \psi_1|_{\Gamma_2} = 1. \tag{1.26}$$

The solution $\psi = \psi_0 + c\psi_1$ allows us to find c by writing that $\partial_n\psi$ has no jump at the trailing edge $P = (1,0)$. We have $\partial_n\psi \approx -(\psi(P^+) - \psi(P))/\delta$ where P^+ is the point just above P in the direction normal to the profile at a distance δ. Thus the jump of $\partial_n\psi$ is $(\psi_0|_{P^+} + c(\psi_1|_{P^+} - 1)) + (\psi_0|_{P^-} + c(\psi_1|_{P^-} - 1))$ divided by δ because the normal changes sign between the lower and upper surfaces. Thus

$$c = -\frac{\psi_0|_{P^+} + \psi_0|_{P^-}}{(\psi_1|_{P^+} + \psi_1|_{P^-} - 2)}, \tag{1.27}$$

which can be programmed as:

$$c = -\frac{\psi_0(0.99, 0.01) + \psi_0(0.99, -0.01)}{(\psi_1(0.99, 0.01) + \psi_1(0.99, -0.01) - 2)}. \tag{1.28}$$

```
border(1,0, 2 * pi,40)      begin
              x:=5*cos(t); y:=5*sin(t) end ;

border(2,0,2,71) begin
  if(t<=1) then /* upper surface */
  begin x := t; y := 0.17735*sqrt(t)-0.075597*t
       - 0.212836*(t^2)+0.17363*(t^3)-0.06254*(t^4);
  end
  else begin /* lower surface */
       x := 2-t; y:= -(0.17735*sqrt(2-t)-0.075597
       *(2-t)-0.212836*((2-t)^2)+0.17363*((2-t)^3)
       -0.06254*(2-t)^4)
     end ;
  end ;
 buildmesh(1800);

solve(psi0){
  onbdy(1) psi0 = y-0.1*x;
  onbdy(2) psi0 = 0;
  pde(psi0) -laplace(psi0) = 0;
};

solve(psi1,-1) /* -1 => matrix already factored */
onbdy(1) psi1 = 0;
```

```
    onbdy(2) psi1 = 1;
    pde(psi1) -laplace(psi1) = 0;
  };
  beta := - (psi0(0.99,0.01)+psi0(0.99,-0.01))
            / (psi1(0.99,0.01)+ psi1(0.99,-0.01)-2);
  psi = beta*psi1 +psi0;
  cp = - dx(psi)^2 - dy(psi)^2 ; /* Pressure */
  plot(cp);
```

In Figure 1.9, we have chosen to represent pressure, which is equal, up to a constant, to $-|\nabla\psi|^2$.

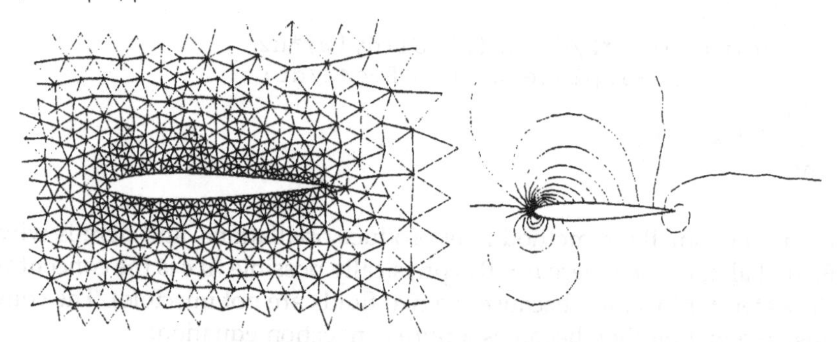

Figure 1.9 Triangulation and pressure lines around a NACA0012

1.6. CONVECTION

1.6.1 Numerical approximation

Let $u(x)$ be a velocity field and $c(x,t)$ be the concentration of a pollutant which is present at the initial time with a concentration $c^0(x)$, or is generated by a source f, or else by a boundary source g. The convection–diffusion equation for c is written

$$\frac{\partial c}{\partial t} + u.\nabla c - \nu\Delta c = f \text{ in } \Omega\times]0,T[, \tag{1.29}$$

where ν is the diffusion coefficient. Here the fluid is incompressible, otherwise one must replace $u\cdot\nabla c$ by $\nabla\cdot(uc)$.

The following initial and boundary conditions define a unique c:

$$c(x,0) = c^0(x) \quad \forall x\in\Omega, \tag{1.30}$$

$$c|_{\Gamma_1} = c_\Gamma, \quad \frac{\partial c}{\partial n}|_{\Gamma_2} = g, \quad \text{with} \quad \Gamma_1\cup\Gamma_2 = \Gamma, \overline{\Gamma_1}\cap\overline{\Gamma_2} = \emptyset. \tag{1.31}$$

There are several ways for discretising (1.29) in time. The Crank–Nicolson scheme combines accuracy and stability (cf. Chapter 7):

$$\frac{1}{\delta t}[c^{m+1} - c^m] + \frac{1}{2}u\cdot\nabla(c^{m+1} + c^m) - \frac{\nu}{2}\Delta(c^{m+1} + c^m) = f. \tag{1.32}$$

By introducing the auxiliary variable $w = (c^{m+1} + c^m)/2$ we can rewrite the scheme as:

$$\frac{2}{\delta t}[w - c^m] + u \cdot \nabla w - \nu \Delta w = f, \quad c^{m+1} = 2w - c^m. \tag{1.33}$$

We can program this as:

```
iter(...)
begin...
  solve(w){
    ...
    pde(w) id(w)*2/dt+dx(w)*u1+dy(w)*u2
               -laplace(w)*nu = f+c*2/dt;
  };
  c = 2 * w - c;
end
```

If ν is very small the method may produce oscillations, and here is why:

First of all the case where $\nu = 0$ is physically possible (if c is the concentration of oil in water, for instance, because the two fluids are not miscible). The convection diffusion equation then becomes a pure convection equation:

$$\frac{\partial c}{\partial t} + u \cdot \nabla c = f \text{ in } \Omega \times]0, T[. \tag{1.34}$$

This is not well-posed in conjunction with (1.31). One needs instead

$$c(x, 0) = c^0(x), \quad \forall x \in \Omega \tag{1.35}$$

$$c|_{\Gamma_1} = c_\Gamma, \frac{\partial c}{\partial n}|_{\Gamma_2} = g \quad \text{with} \quad \Gamma_1 \cup \Gamma_2 = \Gamma^-, \overline{\Gamma_1 \cap \Gamma_2} = \emptyset, \tag{1.36}$$

where Γ^- is the part of the boundary where the fluid enters Ω,

$$\Gamma^- = \{x \in \Gamma : u(x).n(x) < 0\}, \tag{1.37}$$

$n(x)$ being the outer normal to Γ at point x.

This suggests that we must use downstream information, and thus a non-centred scheme. For instance, with the method of characteristics:

$$\partial_t c + u \cdot \nabla c = \frac{Dc}{Dt}(x, t) \approx \frac{c^{m+1}(x) - c^m(x - u(x)\delta t)}{\delta t} \quad \text{at } t = t^{m+1}. \tag{1.38}$$

To program this idea Gfem provides a function *convect*, that builds from c the function $c(x - u(x)dt)$; call *convect* as

```
convect(c,u1,u2,dt)
```

where $u1, u2$ are the two components of the velocity and dt is the time step. Another scheme for the problem is thus:

```
iter(...)
begin...
  solve(w){...
    pde(w) id(w) * 2/dt -laplace(w) * nu
        = f + c/dt + convect(c,u1,u2,dt) */dt;
  };
  c = 2 * w - c;
end
```

To illustrate the above, let us consider a nozzle where the flow is irrotational and inviscid. The nozzle divides into two branches and we want to know where the pollutant, initially concentrated around point $(0.8a, 0)$, goes.

```
nowait; a:= 6; b:= 1; c:=0.5;
border(1,0,1,8)  { x:= -a; y:= 1+b - 2*(1+b)*t };
border(2,0,1,46){ x:= -a+2*a*t;
      y:= -1-b*(x/a)*(x/a)*(3-2*abs(x)/a ) end;
border(3,0,1,8)  { x:= a; y:=-1-b + (1+ b )* t };
border(4,0,1,20){ x:= a - a*t; y:=0 end};
border(4,0,pi,8){ x:=-c*sin(t)/2;y:=c/2-c*cos(t)/2};
border(4,0,1,30){ x:= a*t; y:=c };
border(3,0,1,8)  { x:= a; y:=c + (1+ b-c )* t };
border(5,0,1,55){ x:= a-2*a*t;
      y:= 1+b*(x/a)*(x/a)*(3-2*abs(x)/a) };
buildmesh(800);
solve(psi){
onbdy(1) psi = y;
onbdy(2) psi =-1 -b;
onbdy(3) dnu(psi) =0;
onbdy(4) psi = 0;
onbdy(5) psi =1+b;
pde(psi) -laplace(psi) =0;
};

u1 = dy(psi); u2 = -dx(psi);
dt =0.4; plot(psi); nu=0.01; j:=1;
yy=(x+a*0.8)^2 + y^2;
phi = exp(max((-4.0*yy),-10));

iter(10) begin
  f = convect(phi,u1,u2,dt)/dt;
  solve(w,j) {
      onbdy(1,2,4,5)w=0;
      onbdy(3)dnu(w)=0;
      pde(w) id(w) * 2/dt -laplace(w) * nu
      = phi/dt +f;
  };
  phi = 2 * w - phi; plot(phi); j:=-1;
end;
```

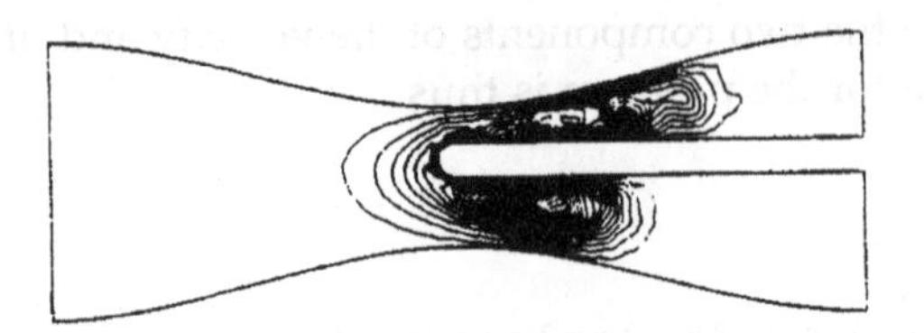

Figure 1.10 Convection in a nozzle

1.6.2 *The rotating bell*

A classical convection problem is that of the 'rotating bell'.

Let Ω be the unit disk centred at 0, with its centre rotating with speed

$$u_1 = y, \quad u_2 = -x. \tag{1.39}$$

We consider the problem:

$$\partial_t c + u.\nabla c = 0 \text{ in } \Omega, \tag{1.40}$$

$$c(t = 0) = c^0. \tag{1.41}$$

The exact solution is $c(x, y, t) = c^0(X(t), Y(t))$ where (X, Y) equals (x, y) rotated around the origin by an angle $\theta = t$. So, if c^0 in a 3D perspective looks like a bell, then c will have exactly the same shape, but rotated by the same amount.

The game consists in solving the equation until $T = 2\pi$, that is for a full revolution and to compare the final solution with the initial one; they should be equal.

```
nowait; twopi := 2 * pi;
border(1,0,twopi,100)
  {      x := cos(t); y := sin(t) };
buildmesh(800);

r2 = (x-0.3)^2 +(y-0.3)^2;
v = exp(-10*r2); /* initial condition */
plot(v); dt := 0.3; u1 = y; u2 = -x; i:=1;

iter(20) begin
solve(v,i){
        pde(v) id(v) = convect(v,u1,u2,dt);
};
i:=-1; plot(v);
end;
```

Remark 1.1 For this example, results will be much improved if one adds at the beginning of the program the instruction 'precise' which results in increasing the degree of the interpolation functions. □

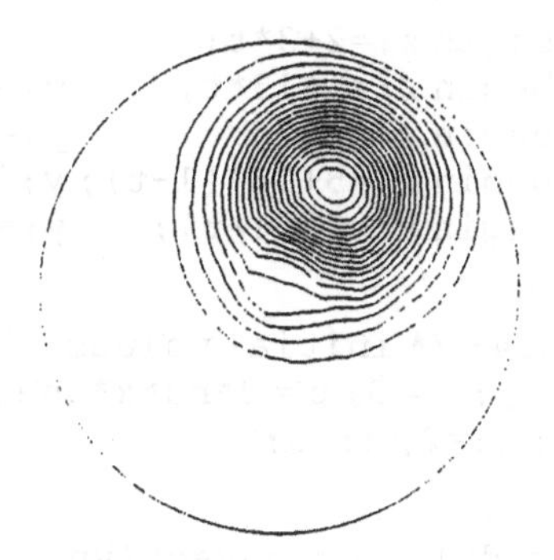

Figure 1.11 Rotating bell after one circuit; its height went from 1. to 0.9

1.7 NAVIER–STOKES EQUATIONS

1.7.1 Generalities

An incompressible viscous fluid satisfies:

$$\partial_t u + u \cdot \nabla u + \nabla p - \nu \Delta u = 0, \quad \nabla \cdot u = 0 \quad \text{in } \Omega \times]0, T[, \tag{1.42}$$

$$u|_{t=0} = u^0, \quad u|_\Gamma = u_\Gamma. \tag{1.43}$$

A possible algorithm is

$$\frac{1}{\delta t}[u^{m+1} - u^m o X^m] + \nabla p^m - \nu \Delta u^m = 0, \quad u|_\Gamma = u_\Gamma, \tag{1.44}$$

$$-\Delta p^{m+1} = -\nabla \cdot u^m o X^m, \quad \partial_n p^{m+1} = 0, \tag{1.45}$$

where $uoX(x) = u(x - u(x)\delta t)$ since $\partial_t u + u \cdot \nabla u$ is approximated by the method of characteristics, as in (1.38).

Remark 1.2 For exterior problems, where a circle approximates infinity, one should distinguish between incoming boundaries where one will stipulate $u, \partial_n p$, and outgoing boundaries where one will stipulate $\partial_n u, p$. (Projection algorithms allow boundary conditions for p.) □

1.7.2 Example of the Backward Facing Step

The geometry is that of a channel with a backward facing step so that the inflow section is smaller than the outflow section. This geometry produces a fluid recirculation zone that must be captured correctly.

This can only be done if the triangulation is sufficiently fine, or well adapted to the flow.

```
border(1,0,1,6) begin x:=0;      y:=1-t  end;
border(2,0,1,15)begin x:=2*t;    y:=0    end;
border(2,0,1,10)begin x:=2;      y:=-t   end;
```

```
border(2,0,1,20) begin x:=2+3*t;       y:=-1        end;
border(2,0,1,35) begin x:=5+15*t;      y:=-1        end;
border(3,0,1,10) begin x:=20;          y:=-1+2*t    end;
border(4,0,1,35) begin x:=5+15*(1-t); y:=1         end;
border(4,0,1,40) begin x:=5*(1-t);     y:=1         end;
buildmesh(1500);

nu := 0.001; dt := 0.4; /* initial values */
u = y*(1-y)*one(y>0); v = 0; p = 2*nu*x*one(y>0);
un = u; vn = v; i:=1; j:=2; k:=3;

iter(180) begin
 f=convect(un,u,v,dt);  g=convect(vn,u,v,dt);
 solve(u,i)           /*Horizontal velocity */
             onbdy(1) u = y*(1-y);
             onbdy(2,4) u = 0;
             onbdy(3)dnu(u)=0;
  pde(u) id(u)/dt-laplace(u)*nu = f/dt -dx(p);
 }; plot(u);

 solve(v,j)           /*Vertical velocity */
             onbdy(1,2,3,4) v = 0;
  pde(v) id(v)/dt-laplace(v)*nu = g/dt -dy(p);
 };

    solve(p,k)   {        /*Pressure */
             onbdy(1,2,4) dnu(p) = 0; onbdy(3) p=0;
  pde(p) -laplace(p)= -(dx(f) + dy(g))/dt;
 };
 un = u; vn = v; i:=-1; j:=-2; k:=-3;
 end ;
```

We show in Figure 1.12 the numerical results obtained for a Reynolds number of 250.

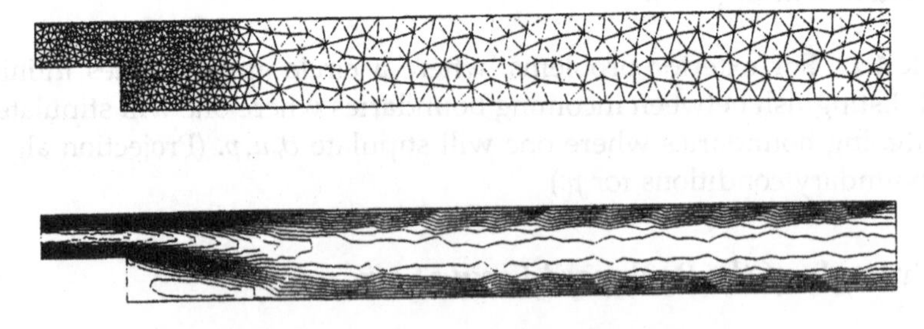

Figure 1.12 Level lines of horizontal velocity

1.8 EXAMPLES WITH SYSTEMS

1.8.1 Elasticity

Solid objects deform under the action of applied forces: a point in the solid, originally at (x, y, z) will come to (X, Y, Z) after some time; the vector

$U = (u, v, w) = (X - x, Y - y, Z - z)$ is called the displacement. When the displacement is small and the solid is elastic, Hooke's law gives a relationship between the stress tensor σ and the strain tensor ϵ

$$\sigma_{ij} = \lambda \delta_{ij} \nabla . U + \mu \epsilon_{ij}, \tag{1.46}$$

where $\delta_{ij} = 1$ if $i = j$, 0 otherwise, with

$$\epsilon_{ij} = \frac{1}{2}\left(\frac{\partial U_i}{\partial x_j} + \frac{\partial U_j}{\partial x_i}\right), \tag{1.47}$$

and where λ, μ are two constants that describe the mechanical properties of the solid, and are themselves related to the better known constants: E, Young's modulus; and ν, Poisson's ratio:

$$\mu = \frac{E}{1+\nu}, \quad \lambda = \frac{E\nu}{(1+\nu)(1-2\nu)}. \tag{1.48}$$

1.8.2 Lame's system

Let us consider a beam with axis Oz and with perpendicular section Ω. The components along x and y of the strain $\mathbf{u}(x)$ in a section Ω subject to forces f perpendicular to the axis are governed by

$$-\lambda \Delta \mathbf{u} - \mu \nabla(\nabla . \mathbf{u}) = \mathbf{f} \text{ in } \Omega, \tag{1.49}$$

where λ, μ are the Lamé coefficients introduced above.

EXAMPLE 1.1

Let us illustrate this by a beam with rectangular section that has a fracture at its centre and subjected to a vertical traction along its horizontal edges. The vertical edges are free. We only solve the equations in one quarter of the domain thanks to symmetries. We denote respectively by $\Gamma_1, \Gamma_2, \Gamma_3, \Gamma_4, \Gamma_5$ the fracture, the remainder of the lower edge, the right vertical edge, the upper vertical edge and the left vertical edge. The boundary conditions are

$$\begin{aligned}
&\lambda \frac{\partial \mathbf{u}}{\partial n} + \mu(\nabla \cdot \mathbf{u})\mathbf{n} = 0 \text{ on } \Gamma_1, \\
&v = 0, \lambda \frac{\partial u}{\partial n} + \mu \nabla \cdot \mathbf{u} = 0 \text{ on } \Gamma_2 \\
&u = 0, \lambda \frac{\partial v}{\partial n} + \mu \nabla \cdot \mathbf{u} = 0 \text{ on } \Gamma_3, \Gamma_5 \\
&u = 0 \text{ on the left vertical edge} \\
&\lambda \frac{\partial \mathbf{u}}{\partial n} + \mu(\nabla \cdot \mathbf{u})\mathbf{n} = (0,1)^T \text{ on } \Gamma_4
\end{aligned} \tag{1.50}$$

Here $\mathbf{u} = (u, v)$ has two components. Lamé's equations are:

$$
\begin{aligned}
-\lambda\Delta u - \mu\frac{\partial^2 u}{\partial x^2} - \mu\frac{\partial^2 v}{\partial x \partial y} &= f_1, \\
-\lambda\Delta v - \mu\frac{\partial^2 v}{\partial y^2} - \mu\frac{\partial^2 u}{\partial x \partial y} &= f_2.
\end{aligned}
\tag{1.51}
$$

The above two equations are strongly coupled by their mixed derivatives, and thus any iterative solution on each of the components is risky. One should rather use ^{G}fem's system approach and write:

```
border(1,0,1,20)  begin x:=1-(t-1)^2;  y:=0;     end;
border(2,1,2,20)  begin x:=(t-1)^2+1;  y:=0;     end;
border(3,0,1,5)   begin x:=2;          y:=t;     end;
border(4,0,2,10)  begin x:=2-t;        y:=1;     end;
border(5,0,1,5)   begin x:=0;          y:=1-t;   end;
buildmesh(500);

E:= 2.15; sigma := 0.29; lambda :=E/(2*(1+sigma));
nu := E*sigma/((1+sigma)*(1-2*sigma));
mu := nu+sigma;

solve(u,v)
begin
      onbdy(3,5) u=0;
      onbdy(2) v=0;
      onbdy(4) dnu(v) = 1;
      pde(u) -laplace(u)*mu - dxx(u)*nu-dxy(v)*nu =0;
      pde(v) -laplace(v)*mu - dyx(u)*nu-dyy(v)*nu =0;
end;
plotmesh;
x=x+0.1*u;
y=y+0.1*v;
plotmesh;
```

The numerical results are shown in figure 1.13.

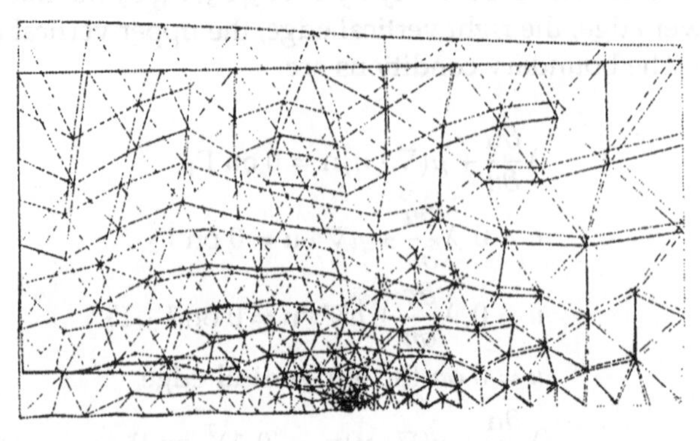

Figure 1.13 Displacements in a quarter of a beam with a fracture

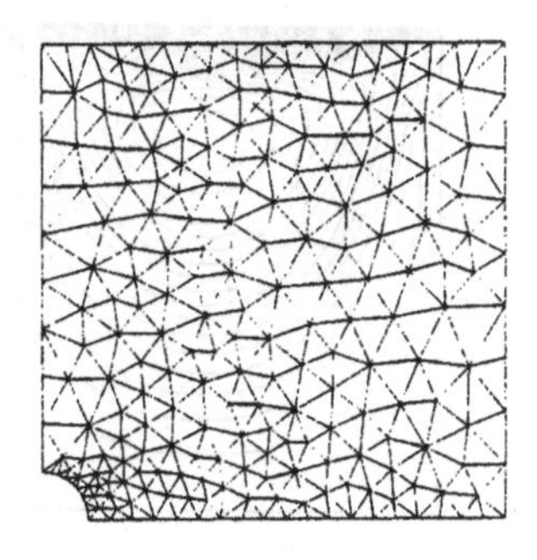
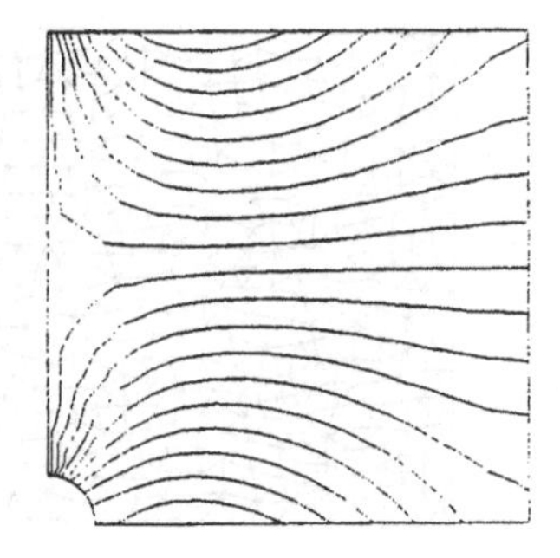

Figure 1.14 Triangulation and vertical iso-displacements of the pierced plate

EXAMPLE 1.2
A classical example [Zienkiewicz (1971)] is to consider an infinite plate subject to a horizontal constant compression $\sigma_x = -1$. The plate has a hole.

For the computation we take a large rectangular plate and symmetry allows us to consider only one quarter of the plate. We apply symmetry conditions on the axes, and nothing on the hole. Thus we have the following program:

```
border(1,0,1.8,10)begin x:=0; y:=2-t end;
border(2,pi/2,0,10) begin
              x:= 0.2*cos(t); y:= 0.2*sin(t) end;
border(3,0,1.8,20)begin x:=0.2+t; y:=0 end;
border(4,0,2,10)     begin x:=2; y:=t end;
border(5,0,2,20)     begin x:=2 - t; y:= 2 end;
buildmesh(800);

E:= 1;
nu := 0.29;
mu :=E/(1+nu);
lambda := E*nu/((1+nu)*(1-2*nu));

solve(u,v)
begin
    onbdy(1) u=0;
    onbdy(1) v=0;
    pde(u) -laplace(u)*lambda-dxx(u)*mu-dxy(v)*mu=-1;
    pde(v) -laplace(v)*lambda-dyx(u)*mu-dyy(v)*mu =0;
end;
plot(u); plot(v);
```

1.8.3 Stokes Problem

In the case of a flow invariant with respect to the third coordinate (two-dimensional flow), flows with a low Reynolds number (for instance micro-organisms) satisfy, for a stream function ψ, a velocity $u = (u_1, u_2)^T$ and a vorticity ω:

$$u_1 = \partial_y \psi, \quad u_2 = -\partial_x \psi, \tag{1.52}$$

$$-\Delta\psi = \omega, \quad -\Delta\omega = 0. \tag{1.53}$$

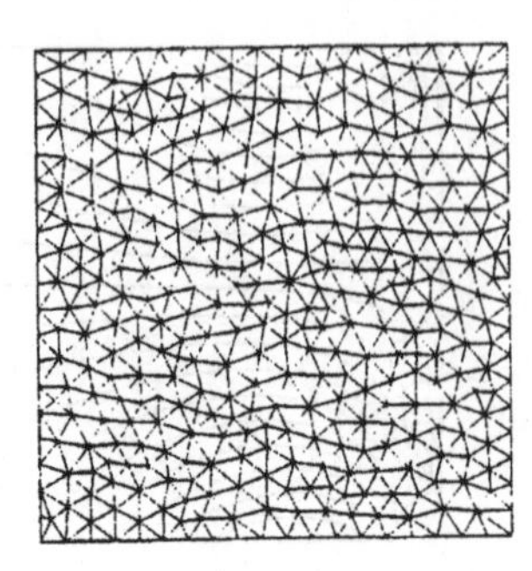
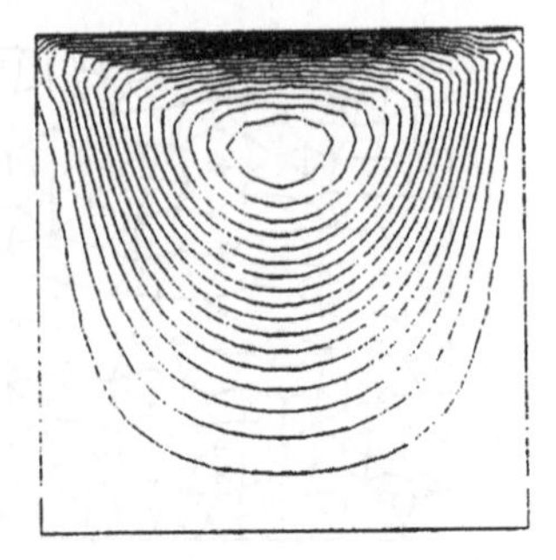

Figure 1.15 The Stokes cavity problem

Boundary conditions are:

$$\psi = \psi_\Gamma, \quad \frac{\partial \psi}{\partial n} = g \text{ on } \Gamma, \tag{1.54}$$

where ψ_Γ and g are determined from the velocity on the edges. Let us note that there are two conditions on ψ and none on ω. This system is actually of fourth order. We can reduce it to a second order system by penalization if we replace (1.54) by:

$$\psi + \epsilon \frac{\partial \omega}{\partial n} = \psi_\Gamma, \quad \frac{\partial \psi}{\partial n} = g \text{ on } \Gamma, \tag{1.55}$$

with $\epsilon << 1$.

In the example below, the domain is a cavity whose upper face is in translation with speed $(1,0)^T$.

```
border(1,0,1,20){ x:=t; y:=0};
border(1,0,1,20){ x:=1; y:=t};
border(2,0,1,20){ x:=1-t; y:=1};
border(1,0,1,20){ x:=0; y:=1-t};
buildmesh(1000);

eps =0.00001;

solve(p,o) begin
  onbdy(1,2) id(p)/eps + dnu(o) = 0;
  onbdy(1) dnu(p) = 0;
  onbdy(2) dnu(p) =1;
  pde(o) -laplace(o) =0;
  pde(p) id(o) + laplace(p) = 0;
end;

plot(o); plot(p);
```

1.9 AN EXAMPLE WITH COMPLEX NUMBERS

In a microwave oven, heat comes from molecular excitation by an electromagnetic field. For a plane monochromatic wave, the amplitude is given by Helmholtz's equation:

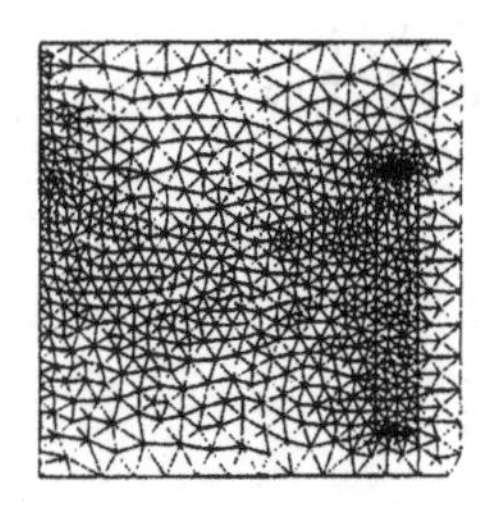
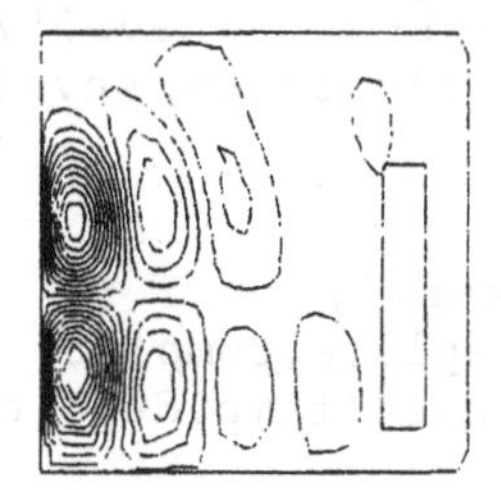
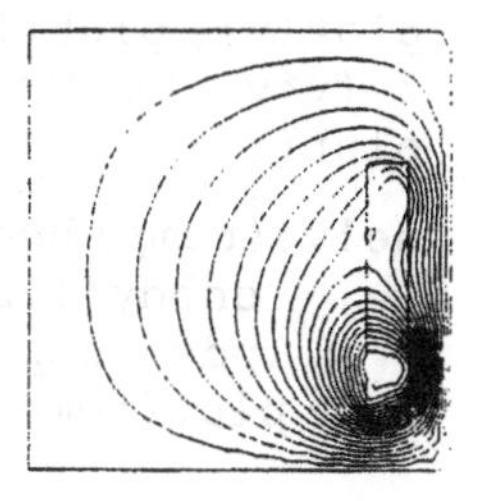

Figure 1.16 A microwave oven: wave (left) and temperature (right)

$$\beta v + \Delta v = 0. \tag{1.56}$$

We consider a rectangular oven where the wave is emitted by part of the upper wall. So the boundary of the domain is made up of a part Γ_1 where $v = 0$ and of another part $\Gamma_2 = [c, d]$ where for instance $v = \sin\left(\pi \frac{y - c}{c - d}\right)$.

Within an object to be cooked, denoted by B, the heat source is proportional to v^2. At equilibrium, one has

$$-\Delta\theta = v^2 I_B, \quad \theta_\Gamma = 0, \tag{1.57}$$

where I_B is 1 in the object and 0 elsewhere.

In the program below $\beta = 1/(1 - I/2)$ in the air and $2/(1 - I/2)$ in the object ($I = \sqrt{-1}$):

```
complex; a:=20; b:=20; c:=15; d:=8;
border(1,0,6,81){
if(t<=1)then x:=a*t; y:=0};
if((t>1)and(t<=2)) then {x:=a; y:= b*(t-1)};
if((t>2)and(t<=3)) then {x:=a*(3-t);y:=b};
if((t>3)and(t<=4)) then {x:=0;y:=b-(b-c)*(t-3)};
if((t>4)and(t<=5)) then {x:=0; y:=c-(c-d)*(t-4); ib:=2;
if(t>5) then {x:=0; y:= d*(6-t)};
};
e:=2; l:=12; f:=2; g:=2;
border(3,0,8,63){
if(t<=1) then {x:=a-f+e*(t-1); y:=g};
if((t>1)and(t<=4)) then {x:=a-f; y:=g+l*(t-1)/3};
if((t>4)and(t<=5)) then {x:=a-f-e*(t-4); y:=l+g};
if(t>5) then {x:=a-e-f; y:= l+g-l*(t-5)/3};
};
buildmesh(2000);

solve(v) begin
   onbdy(1) v=0;
   onbdy(2) v=sin(pi*(y-c)/(c-d));
pde(v) id(v)*(1+region)+laplace(v)*(1-0.5*I)=0;
end;
```

```
plot(v); plot(Im(v));        save('v.dta',v);
f=(v*v + Im(v)*Im(v))*region; save('f.dta',f);
plot(f);

solve(temp) begin
     onbdy(1,2)temp=0;
     pde(temp) -laplace(temp)=f;
end; plot(temp); save('temp.dta',temp);
```

1.10 CLASSIFICATION OF THE EQUATIONS

1.10.1 Elliptic, Parabolic and Hyperbolic Equations

A partial differential equation (PDE) is a relation between a function of several variables and its derivatives.

$$F\left(\varphi(x), \frac{\partial\varphi}{\partial x_1}(x), \cdots, \frac{\partial\varphi}{\partial x_d}(x), \frac{\partial^2\varphi}{\partial x_1^2}(x), \cdots, \frac{\partial^m\varphi}{\partial x_d^m}(x)\right) = 0 \quad \forall x \in \Omega \subset \mathbb{R}^d. \tag{1.58}$$

The domain over which the equation is taken, here Ω, is called the *domain* of the PDE. The highest derivation index in (1.58), here m, is called the *order*. If F and φ are vector valued functions, then the PDE is actually a *system* of PDEs.

Unless indicated otherwise, here by convention *one* PDE corresponds to one scalar valued F and φ. If F is linear with respect to its arguments, then the PDE is said to be *linear*.

The general form of a second order, linear scalar PDE is*

$$\alpha\varphi + a \cdot \nabla\varphi + B : \nabla(\nabla\varphi) = f \quad \text{in} \quad \Omega \subset \mathbb{R}^d, \tag{1.59}$$

where $f(x), \alpha(x) \in \mathbb{R}, a(x) \in \mathbb{R}, B(x) \in \mathbb{R}^{d\times d}$ are the PDE *coefficients*. If the coefficients are independent of x, the PDE is said to have *constant coefficients*.

To a PDE such as (1.59), we associate a quadratic form, by replacing φ by 1, $\partial\varphi/\partial x_i$ by z_i and $\partial^2\varphi/\partial x_i\partial x_j$ by $z_i z_j$, where z is a vector in $\mathbb{R}^d$:

$$\alpha + a \cdot z + z^T B z = f. \tag{1.60}$$

If (1.60) is the equation of an ellipse (ellipsoid if $d \geq 2$), the PDE is said to be *elliptic*; if it is the equation of a parabola or a hyperbola, the PDE is said to be *parabolic* or *hyperbolic*. If $B \equiv 0$, the degree is no longer 2 but 1, and for reasons that will appear more clearly later, the PDE is still said to be hyperbolic.

These concepts can be generalised to systems, by studying whether or not the polynomial system $P(z)$ associated with the PDE system has branches at infinity (ellipsoids have no branches at infinity, paraboloids have one, and hyperboloids have several).

* $\nabla(\nabla\varphi)$ is a second order tensor (*i.e.* represented by a matrix) with elements $\dfrac{\partial^2\varphi}{\partial x_i \partial x_j}$ and A:B means $\sum_{i,j=1}^d a_{ij}b_{ij}$

If the PDE is not linear, it is said to be *non-linear*. Those are said to be locally elliptic, parabolic, or hyperbolic according to the type of the linearized equation. For example, for the non-linear equation

$$\frac{\partial^2\varphi}{\partial t^2} - \frac{\partial\varphi}{\partial x}\frac{\partial^2\varphi}{\partial x^2} = 1, \tag{1.61}$$

we have $d = 2, x_1 = t, x_2 = x$ and its linearised form is:

$$\frac{\partial^2 u}{\partial t^2} - \frac{\partial u}{\partial x}\frac{\partial^2\varphi}{\partial x^2} - \frac{\partial\varphi}{\partial x}\frac{\partial^2 u}{\partial x^2} = 0, \tag{1.62}$$

which, for the unknown u, is locally elliptic if $\frac{\partial\varphi}{\partial x} < 0$ and locally hyperbolic if $\frac{\partial\varphi}{\partial x} > 0$.

1.10.2 Examples

Laplace's equation is elliptic:

$$\Delta\varphi \equiv \frac{\partial^2\varphi}{\partial x_1^2} + \frac{\partial^2\varphi}{\partial x_2^2} + \cdots + \frac{\partial^2\varphi}{\partial x_d^2} = f, \quad \forall x \in \Omega \subset \mathbb{R}^d. \tag{1.63}$$

The *heat* equation is parabolic in $Q = \Omega \times]0, T[\subset \mathbb{R}^{d+1}$:

$$\frac{\partial\varphi}{\partial t} - \mu\Delta\varphi = f \quad \forall x \in \Omega \subset \mathbb{R}^d, \quad \forall t \in]0, T[. \tag{1.64}$$

If $\mu > 0$, the *wave* equation is hyperbolic:

$$\frac{\partial^2\varphi}{\partial t^2} - \mu\Delta\varphi = f \quad \text{in} \quad Q. \tag{1.65}$$

The *convection diffusion* equation is parabolic if $\mu \neq 0$ and hyperbolic otherwise:

$$\frac{\partial\varphi}{\partial t} + a\nabla\varphi - \mu\Delta\varphi = f. \tag{1.66}$$

The *bi-harmonic* equation is elliptic:

$$\Delta(\Delta\varphi) = f \quad \text{in} \quad \Omega. \tag{1.67}$$

1.10.3 Boundary Conditions

A relation such as (1.58) is not sufficient to define φ. Additional information on the boundary $\Gamma = \partial\Omega$ of Ω, or on part of Γ is necessary.

Such information is called a *boundary condition*. For example,

$$\varphi(x) \text{ given}, \forall x \in \Gamma, \tag{1.68}$$

is called a *Dirichlet boundary condition*. The *Neumann* condition is

$$\frac{\partial \varphi}{\partial n}(x) \text{ given on } \Gamma \text{ for (1.63) and } n \cdot B\nabla\varphi \text{ given for (1.59)}, \tag{1.69}$$

where n is the normal at $x \in \Gamma$ directed towards the exterior of Ω (by definition $\frac{\partial \varphi}{\partial n} = \nabla\varphi \cdot n$).

Another classical condition, called a Robin condition (or *Fourier* boundary condition) is written as:

$$\varphi(x) + \beta(x)\frac{\partial \varphi}{\partial n}(x) \text{ given on } \Gamma. \tag{1.70}$$

Finding a set of boundary conditions for (1.58) that defines a unique φ is a difficult art for which we can only give here general directions.

In general, an elliptic equation is well posed (i.e. φ is unique) with (1.68), (1.69) or (1.70) on the whole boundary of Ω, or on part of it, as long as only one of those conditions is given at each boundary point.

Thus, Laplace's equation (1.63) is well posed with (1.68) or (1.69) but also with

$$\varphi \text{ given on } \Gamma_1, \quad \frac{\partial \varphi}{\partial n} \text{ given on } \Gamma_2, \quad \Gamma_1 \cup \Gamma_2 = \Gamma, \quad \dot{\Gamma}_1 \cap \dot{\Gamma}_2 = \emptyset. \tag{1.71}$$

Parabolic and hyperbolic equations never require boundary conditions on all of $\Gamma \times]0, T[$. For instance, the heat equation (1.64) is well posed with

$$\varphi \text{ given at } t = 0 \text{ and (1.68) or (1.69) or (1.70) or (1.71) on } \partial\Omega. \tag{1.72}$$

The wave equation (1.65) is well posed with

$$\varphi \text{ and } \frac{\partial \varphi}{\partial t} \text{ given at } t = 0 \text{ and (1.68) or (1.69) or (1.70) or (1.71) on } \partial\Omega. \tag{1.73}$$

Part A

PROGRAMMING THE MODEL PROBLEM BY A FINITE ELEMENT METHOD

2 Introduction to the Finite Element Method: Energy Minimisation

Summary

We study a first order finite element method for the solution of the Laplacian with Dirichlet boundary conditions. The problem is expressed in terms of energy minimisation, then solved numerically by conjugate gradient. No matrix storage is required in this approach.

2.1 *STATEMENT OF THE PROBLEM*

We describe what we have called the 'model problem', i.e. the Laplacian with Dirichlet boundary conditions, by first specifying the data of the problem. For simplicity's sake, we work in two space dimensions, even though the method can easily be generalised to three dimensions.

DATA OF THE PROBLEM The data of the problem are the following:

Ω: bounded domain in $\mathbb{R}^2$ with sufficiently smooth boundary Γ (cf. Figure 2.1),
f: a function defined on Ω, real valued and square integrable,
g: a function defined on Γ, real valued and at least square integrable.

OBJECTIVES Let φ be the solution of the model problem

$$-\Delta\varphi(\vec{x}) = f(\vec{x}), \quad \forall \vec{x} \in \Omega, \tag{2.1}$$

$$\varphi(\vec{x}) = g(\vec{x}), \quad \forall \vec{x} \in \Gamma, \tag{2.2}$$

where the Laplacian is defined by:

$$(\Delta\varphi)(\vec{x}) = \left(\frac{\partial^2\varphi}{\partial x^2} + \frac{\partial^2\varphi}{\partial y^2}\right)(\vec{x}), \quad \vec{x} = \begin{pmatrix} x \\ y \end{pmatrix}.$$

We intend to compute an approximation φ_h to φ.

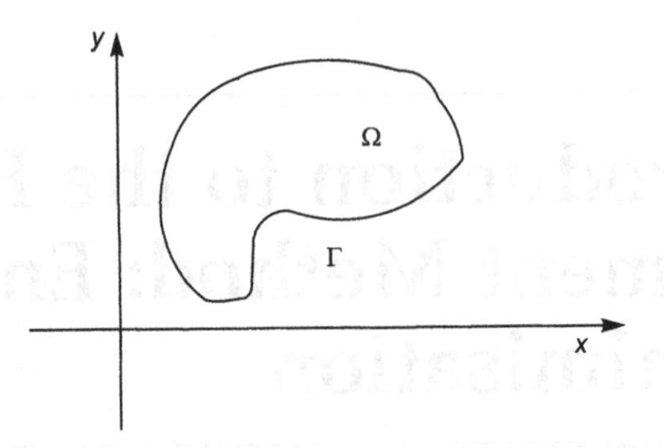

Figure 2.1 A bi-dimensional domain Ω and its boundary Γ

IDEA OF THE METHOD: Let us briefly describe the method we use, without going into details, as it forms the object of the present chapter. We first show that problem (2.1)–(2.2) is equivalent to a minimisation problem for the quadratic functional

$$E(\varphi) = \frac{1}{2}\int_{\Omega} |\nabla\varphi(\vec{x})|^2 \mathrm{d}\vec{x} - \int_{\Omega} f(\vec{x})\varphi(\vec{x})\mathrm{d}\vec{x}, \tag{2.3}$$

defined on a function space H comprised of 'sufficiently smooth' functions, that 'coincide' with g on the boundary Γ. In this formulation the problem becomes

$$\min_{\varphi\in H} E(\varphi),$$

Then, after having covered the domain Ω by small triangles, of maximum size h (h is a small parameter, destined to go to zero), we define an approximate function space H_h, composed of functions continuous on the whole computational domain, linear on each triangle, and that coincide with g on Γ. The functions in this space H_h are fully determined by their values ϕ_i at each of the internal (i.e. those not located on the boundary) vertices of the so formed triangles, so that the approximate problem is reduced to an optimisation problem of finite dimension, of the type:

$$\min_{\Phi\in R^N} E_h(\Phi),$$

where Φ is the vector with components ϕ_i. We show how to compute this functional E_h, and we propose an iterative method to solve this problem numerically. The method we propose is not optimal, mostly because of computer time, but it is simple to implement, since it requires no matrix storage. We shall consider this matrix storage problem in the next chapter, where we treat problem (2.1)–(2.2) in variational form. Before we do that, we shall state precisely the mathematical aspects of the problem, that is the function spaces we use, as well as the quadratic functional we must minimise. The following two sections may present technical difficulties (for instance Remark 2.3) that can be omitted on first reading.

2.2 TRANSFORMATION OF THE PROBLEM

We consider the following function space

$$H_g^1(\Omega) = \{w : \Omega \to \mathbb{R}, \quad w \in H^1(\Omega), \quad \text{and} \quad \forall \vec{x} \in \Gamma, w(\vec{x}) = g(\vec{x})\}, \tag{2.4}$$

where $H^1(\Omega)$ is the space of real functions defined on Ω that are square integrable, and whose first derivatives with respect to each variable are also square integrable, in other words:

$$H^1(\Omega) = \left\{ w \in L^2(\Omega), \quad \frac{\partial w}{\partial x}, \frac{\partial w}{\partial y} \in L^2(\Omega) \right\}. \tag{2.5}$$

It is indeed necessary to consider functions φ in this space, so as to give a meaning to the integrals that appear in (2.3). In practice, $H^1(\Omega)$ will denote a set of 'sufficiently smooth' functions. Let us note that the space $H_g^1(\Omega)$ is an affine space in the general case.

We shall also use the space $H_0^1(\Omega)$, defined by (2.4) with $g = 0$, which is a vector space; it is defined by:

$$H_0^1(\Omega) = \{w \in H^1(\Omega) \quad \text{and} \quad w(\vec{x}) = 0, \quad \forall \vec{x} \in \Gamma\}. \tag{2.6}$$

We shall now write two formulations of problem (2.1)–(2.2).

Let us denote by E the *energy functional* defined on the space $H_g^1(\Omega)$ by

$$E(\varphi) = \frac{1}{2} \int_\Omega |\nabla \varphi(\vec{x})|^2 \mathrm{d}\vec{x} \; - \int_\Omega f(\vec{x}) \varphi(\vec{x}) \, \mathrm{d}\vec{x}, \tag{2.7}$$

$\nabla \varphi(\vec{x})$ being the gradient of φ taken at point $\vec{x}$, i.e. the vector with components $(\partial\varphi/\partial x)(\vec{x})$ and $(\partial\varphi/\partial y)(\vec{x})$; naturally, $|\nabla\varphi(\vec{x})|^2 = \left(\frac{\partial \varphi}{\partial x}(\vec{x})\right)^2 + \left(\frac{\partial \varphi}{\partial y}(\vec{x})\right)^2$.

We are interested in the minimisation problem

$$\text{find } \varphi \in H_g^1(\Omega) \quad \text{such that} \quad \forall \psi \in H_g^1(\Omega), \quad E(\varphi) \leq E(\psi),$$

which is usually written symbolically in the following way:

$$\min_{\varphi \in H_g^1(\Omega)} E(\varphi). \tag{2.8}$$

We show that problem (2.1)–(2.2) is equivalent to problem (2.8) (under some additional smoothness assumptions), by using an intermediate problem, called *variational formulation* of problem (2.1)–(2.2), and that we write

$$\text{find } \varphi \in H^1_g(\Omega) \quad \text{such that}$$
$$\forall w \in H^1_0(\Omega), \qquad \int_\Omega \nabla\varphi(\vec{x}) \cdot \nabla w(\vec{x}) \mathrm{d}\vec{x} = \int_\Omega f(\vec{x}) w(\vec{x}) \mathrm{d}\vec{x}, \tag{2.9}$$

where the notation $\nabla\varphi(\vec{x}) \cdot \nabla w(\vec{x})$ denotes the inner product in $\mathbb{R}^2$ of vectors $\nabla\varphi(\vec{x})$ and $\nabla w(\vec{x})$, i.e.

$$\nabla\varphi(\vec{x}) \cdot \nabla w(\vec{x}) = \frac{\partial\varphi}{\partial x}(\vec{x}) \frac{\partial w}{\partial x}(\vec{x}) + \frac{\partial\varphi}{\partial y}(\vec{x}) \frac{\partial w}{\partial y}(\vec{x}).$$

A first result allows us to link problem (2.1)–(2.2) with variational problem (2.9), modulo some additional smoothness assumptions. Towards this aim let us define the space $H^2(\Omega)$ as the set of functions in $H^1(\Omega)$ all of whose second order partial derivatives are square integrable on Ω, i.e.

$$H^2(\Omega) = \left\{ w \in H^1(\Omega), \ \frac{\partial^2 w}{\partial x^2}, \ \frac{\partial^2 w}{\partial x \partial y}, \ \frac{\partial^2 w}{\partial y^2} \in L^2(\Omega) \right\}. \tag{2.10}$$

Theorem 2.1 If $\varphi \in H^2(\Omega)$, φ solves (2.1)–(2.2) if and only if φ solves (2.9).

PROOF Let w be any function in the space $H^1_0(\Omega)$. If φ solves (2.1) and if $\Delta\varphi \in L^2(\Omega)$, we can multiply equation (2.1) by w to obtain

$$\int_\Omega -(\Delta\varphi w)(\vec{x}) \mathrm{d}\vec{x} = \int_\Omega (fw)(\vec{x}) \mathrm{d}\vec{x}. \tag{2.11}$$

Let us recall Green's formula

$$\int_\Omega -(\Delta\varphi w)(\vec{x}) \mathrm{d}\vec{x} = \int_\Omega (\nabla\varphi \cdot \nabla w)(\vec{x}) \mathrm{d}\vec{x} - \int_\Gamma \left(\frac{\partial\varphi}{\partial n} w \right)(\vec{x}) \mathrm{d}\gamma(\vec{x}), \tag{2.12}$$

which is nothing but the generalisation of the integration by parts formula in one dimension. In this formula, $\frac{\partial\varphi}{\partial n}(\vec{x}) = \nabla\varphi(\vec{x}) \cdot n(\vec{x})$, where $n(\vec{x})$ is the normal at point $\vec{x}$ of Γ oriented towards the exterior of Ω. Using (2.12) allows us to transform equation (2.11) in the following way:

$$\int_\Omega (\nabla\varphi \cdot \nabla w)(\vec{x}) \mathrm{d}\vec{x} - \int_\Gamma \left(\frac{\partial\varphi}{\partial n} w \right)(\vec{x}) \mathrm{d}\gamma(\vec{x}) = \int_\Omega (fw)(\vec{x}) \mathrm{d}\vec{x}. \tag{2.13}$$

As w belongs to $H^1_0(\Omega)$, the second integral on the left hand side of this equation is zero, since $w|_\Gamma = 0$, which gives (2.9).

The converse requires some knowledge of functional analysis. If (2.9) holds for any function w in $H^1_0(\Omega)$, it holds in particular for any function in the space $D(\Omega)$

of functions of class C^∞ with compact support in Ω, whose dual is the space $D'(\Omega)$ of distributions on Ω. If $\Delta\varphi \in L^2(\Omega)$, and if we denote by $< .,. >$ the duality bracket between these two spaces, we can write equation (2.9) as (cf. (2.12)),

$$< -\Delta\varphi - f, w >= 0;$$

then $-\Delta\varphi - f$ being a function in $L^2(\Omega)$, we deduce from this, according to a classical functional analysis result [Brezis (1987)], that it is zero almost everywhere. □

The following theorem enables us to link both formulations (2.8) and (2.9).

Theorem 2.2 Problem (2.8) is equivalent to problem (2.9).

PROOF By definition φ is a solution of (2.8) if and only if, for any function $w \in H_0^1(\Omega)$ and any real number λ, the following inequality

$$E(\varphi) \leq E(\varphi + \lambda w) \tag{2.14}$$

is verified; this is due to the fact that:

$$\varphi \in H_g^1(\Omega), \quad w \in H_0^1(\Omega) \Rightarrow \quad \forall\lambda \in \mathbb{R}, \quad \varphi + \lambda w \in H_g^1(\Omega). \tag{2.15}$$

Now, by definition of E, we have, suppressing the integration variables to simplify the notation,

$$\begin{cases} E(\varphi + \lambda w) = \dfrac{1}{2}\displaystyle\int_\Omega |\nabla(\varphi + \lambda w)|^2 - \int_\Omega f(\varphi + \lambda w) \\ \qquad = \dfrac{1}{2}\displaystyle\int_\Omega |\nabla\varphi|^2 + \lambda\int_\Omega \nabla\varphi\cdot\nabla w + \frac{\lambda^2}{2}\int_\Omega |\nabla w|^2 - \int_\Omega f\varphi - \lambda\int_\Omega fw \\ \qquad = E(\varphi) + \lambda\Big(\displaystyle\int_\Omega \nabla\varphi\cdot\nabla w - \int_\Omega fw\Big) + \frac{\lambda^2}{2}\int_\Omega |\nabla w|^2, \end{cases} \tag{2.16}$$

so that (2.14) becomes:

$$\lambda\Big(\int_\Omega \nabla\varphi\cdot\nabla w - \int_\Omega fw\Big) + \frac{\lambda^2}{2}\int_\Omega |\nabla w|^2 \geq 0. \tag{2.17}$$

Let us take $\lambda = \lambda_n$, where $(\lambda_n)_n$ is a sequence of *positive* real numbers going to zero as $n \to \infty$. After dividing by λ_n, we rewrite inequality (2.17), in the limit $n \to \infty$:

$$\int_\Omega \nabla\varphi\cdot\nabla w - \int_\Omega fw \geq 0. \tag{2.18}$$

In a similar way, by first taking $\lambda = -\lambda_n$, we obtain the converse inequality

$$\int_{\Omega} \nabla\varphi \cdot \nabla w - \int_{\Omega} fw \leq 0, \tag{2.19}$$

so that equality (2.9) is satisfied. The converse is immediate because of (2.16). □

Remark 2.3 A few comments on this theoretical part.
(i) *Remarks about Theorem (2.1)*
The conclusions of Theorem 2.1 remain valid if, instead of assuming $\varphi \in H^2(\Omega)$, we only have $\varphi \in H^1(\Omega)$ and $\Delta\varphi \in L^2(\Omega)$. To prove this, it suffices to replace in Green's formula (2.12) the integral over Γ by a duality bracket between the space $H^{1/2}(\Gamma)$ of traces on Γ of $H^1(\Omega)$ functions

$$H^{1/2}(\Gamma) = \{w|_{\Gamma}, \quad w \in H^1(\Omega)\}, \tag{2.20}$$

and its dual space $H^{-1/2}(\Gamma)$. One can indeed show that if $\varphi \in H^1(\Omega)$, one only has $\frac{\partial\varphi}{\partial n} \in H^{-1/2}(\Gamma)$.
(ii) *Existence of a solution to the variational problem, traces and extensions*
Existence is proved after extending the boundary condition g to a function in the space $H^1(\Omega)$, which supposes g is a little smoother than $L^2(\Gamma)$; actually, g must be in the space $H^{1/2}(\Gamma)$.

These notions, mathematically quite technical, will not be further developed here; we refer to the abundant literature on this topic [Raviart and Thomas (1987), Brenner and Scott (1994), Brezis (1987), Dautray and Lions (1990)]. We will content ourselves with recalling that there exists a continuous linear mapping, usually called 'trace mapping' and denoted by γ_0, from $H^1(\Omega)$ into $L^2(\Gamma)$ whose kernel is the space $H_0^1(\Omega)$ and whose range is precisely $H^{1/2}(\Gamma)$. Conversely, any function g in $H^{1/2}(\Gamma)$ can be extended to a function u in $H^1(\Omega)$.
(iii) *Existence of a solution to the energy minimisation problem*
We can also show directly the existence of a solution to problem (2.8) by considering a minimising sequence and using the convexity of the functional E and of the space $H_g^1(\Omega)$, as well as the fact that this space is closed for the weak topology. □

2.3 APPROXIMATION BY THE GALERKIN METHOD

To discretise (2.8), we use a method called *Galerkin's method*, which consists in introducing a sequence of spaces $(H_{gh}^1)_h$ of *finite dimension* N_h 'approximating' (in a sense we shall define) the space $H_g^1(\Omega)$ when $N_h \to \infty$. (It is more convenient to index the space with the 'size' h of the mesh, as this goes to 0, rather than by its dimension).

More generally, let H be a subspace of $H^1(\Omega)$, provided with the norm

$$\|\varphi\|_1 = \left(\int_{\Omega} (\varphi(\vec{x}))^2 \, d\vec{x} + \int_{\Omega} |\nabla\varphi(\vec{x})|^2 \, d\vec{x} \right)^{\frac{1}{2}}, \tag{2.21}$$

and let $(H_h)_h$ be a sequence of finite dimensional spaces, we shall use the following definition:

Definition 2.4 We say the sequence $(H_h)_h$ is an *internal approximation* of H if

(i) $\forall \phi \in H, \quad \exists \phi_h \in H_h$ such that $\|\phi - \phi_h\|_1 \to 0$ when $h \to 0$,

(ii) $\forall h, \quad H_h \subset H$.

To make proofs simpler, we shall assume in this section that the function g is defined on all of Ω and that $g \in H^1(\Omega)$ (otherwise, the following proofs remain valid if we replace g by any of its extensions). Let us give ourselves a sequence $(H_h^1)_h$ of internal approximation of $H^1(\Omega)$; according to (i), there exists $g_h \in H_h^1$ such that $\|g_h - g\|_1 \to 0$ when $h \to 0$. We now consider the following spaces:

$$H_{gh}^1 = \{w_h \in H_h^1, w_h(\vec{x}) = g_h(\vec{x}), \quad \forall \vec{x} \in \Gamma\}, \tag{2.22}$$

$$H_{0h}^1 = \{w_h \in H_h^1, w_h(\vec{x}) = 0, \quad \forall \vec{x} \in \Gamma\}. \tag{2.23}$$

Let us note that we have:

$$H_{0h}^1 \subset H_0^1(\Omega); \tag{2.24}$$

on the other hand, this inclusion is false in the general case, i.e. for any arbitrary g.

We first show that the previous minimisation problem, restricted to the space H_{gh}^1, i.e.

$$\min_{\varphi_h \in H_{gh}^1} E(\varphi_h), \tag{2.25}$$

admits one and only one solution. To do this, we need a lemma, called *Poincaré's inequality* which is easy to prove in dimension 1, and whose proof in the general case can be found in Necas (1967), Dautray and Lions (1990):

Lemma 2.5 (Poincaré's inequality) If Ω is a bounded domain, then there exists a strictly positive constant $C(\Omega)$, such that:

$$\forall w \in H_0^1(\Omega), \quad \int_\Omega w^2(\vec{x}) d\vec{x} \le C(\Omega) \int_\Omega |\nabla w(\vec{x})|^2 d\vec{x}. \tag{2.26}$$

EXERCISE Prove this inequality in one dimension; the proof is simple, but instructive, as it helps to exhibit the fundamental nature of the two following assumptions: Ω bounded and $w|_\Gamma = 0$.

Theorem 2.6 Problem (2.25) has one and only one solution.

PROOF Let $(w^i)_1^{N_h}$ be a basis of the space H_{0h}^1 (any finite dimensional vector space has a basis). Then

$$H_h^1 = \left\{ \varphi_h : \exists(\varphi_1, ..., \varphi_{N_h}) \in \mathbb{R}^{N_h} \; \varphi_h(\vec{x}) = g_h(\vec{x}) + \sum_{i=1}^{N_h} \varphi_i w^i(\vec{x}) \right\}, \tag{2.27}$$

and we rewrite $E(\varphi_h)$ as

$$E(\varphi_h) = \frac{1}{2}\int_\Omega |\nabla\left(g_h + \sum_{i=1}^{N_h} \varphi_i w^i\right)|^2 - \int_\Omega f\left(g_h + \sum_{i=1}^{N_h} \varphi_i w^i\right) = E_h(\Phi),$$

where E_h is the quadratic functional quadratic functional of the variable $\Phi = (\varphi_1, ..., \varphi_{N_h}) \in \mathbb{R}^{N_h}$, defined by:

$$\begin{aligned} E_h(\Phi) = \frac{1}{2}\sum_{i,j=1}^{N_h} \varphi_i\varphi_j \int_\Omega \nabla w^i \nabla w^j + \sum_{i=1}^{N_h} \varphi_i\left(\int_\Omega \nabla g_h \nabla w^i - \int_\Omega f w^i\right) \\ + \frac{1}{2}\int_\Omega |\nabla g_h|^2 - \int_\Omega f g_h, \end{aligned} \tag{2.28}$$

where to simplify notation, we have suppressed the integration variables. Now that the minimisation problem (2.25) has been reduced to a simple finite dimensional minimisation problem, it will have a solution if and only if the matrix of second partial derivatives of E_h, $A = \partial^2 E_h/\partial\varphi_i\partial\varphi_j$, is positive definite, which we shall prove. According to (2.28), we have:

$$A_{ij} = \int_\Omega \nabla w^i \cdot \nabla w^j. \tag{2.29}$$

Let $z = (z_1, ..., z_{N_h})$ be any vector in $\mathbb{R}^{N_h}$; we have

$$z^T A z = \sum_{i,j=1}^{N_h} z_i A_{ij} z_j = \sum_{i,j=1}^{N_h} \int_\Omega \nabla w^i \cdot \nabla w^j z_i z_j = \int_\Omega |\nabla z_h|^2, \tag{2.30}$$

where we have set $z_h(\vec{x}) = \sum_{i=1}^{N_h} z_i w^i(\vec{x})$. By construction, z_h is in the space H_{0h}^1, and thus also, by (2.24), in $H_0^1(\Omega)$.

According to (2.30), we see that for any z, $z^T A z$ is a positive quantity, that can only become zero if ∇z_h is zero almost everywhere; but since z_h is in $H_0^1(\Omega)$, this can only happen, because of Poincaré's inequality, if z_h is zero almost everywhere, i.e. if z is zero, which proves that A is positive definite. □

As in the continuous case, we link the previous minimisation problem with a *discrete variational formulation*; more precisely we have the following result, which is the discrete analogue of theorem 2.2:

Theorem 2.7 (discrete variational formulation) Problem (2.25) is equivalent to:

$$\text{find} \quad \varphi_h \in H^1_{gh} \quad \text{such that}$$

$$\forall w_h \in H^1_{0h}, \qquad \int_\Omega \nabla\varphi_h(\vec{x}) \cdot \nabla w_h(\vec{x})\, d\vec{x} = \int_\Omega f(\vec{x}) w_h(\vec{x})\, d\vec{x}. \tag{2.31}$$

PROOF We summarise the proof, which is identical to that of Theorem 2.2. Let λ be any real number and let w_h be an arbitrary function from the space $H^1_0(\Omega)$; relations (2.14) and (2.16) become here:

$$0 \leq E(\varphi_h + \lambda w_h) - E(\varphi_h) = \lambda\Big(\int_\Omega \nabla\varphi_h \nabla w_h - \int_\Omega f w_h\Big) + \frac{\lambda^2}{2}\int_\Omega |\nabla w_h|^2. \tag{2.32}$$

Taking $\lambda = \lambda_n$, $\lambda_n > 0 \to 0$, and dividing by λ_n, we find in the limit ($n \to \infty$) that:

$$\int_\Omega \nabla\varphi_h \cdot \nabla w_h = \int_\Omega f w_h \geq 0. \tag{2.33}$$

Then, taking $\lambda = -\lambda_n \to 0$ in (2.32), we find (2.33) but with the opposite sign, which gives (2.31); the converse is immediate. □

Is the solution φ_h of (2.25) a good approximation to the solution φ of the initial problem (2.8)? In other words, if the mesh size becomes very small ($h \to 0$), does $\varphi_h \to \varphi$ hold? The answer lies in Theorem 2.8 below, whose proof we shall only give, for simplicity's sake, in the homogeneous case, i.e. $g = 0$.

Theorem 2.8

$$\|\varphi - \varphi_h\|_1 \to 0, \quad \textit{when} \quad h \to 0.$$

More precisely, in the homogeneous case, (i.e. $g = 0$), there exists a constant $C > 0$, independent of φ and of h such that

$$\|\varphi - \varphi_h\|_1 \quad \leq \quad C \inf_{\psi_h \in H^1_{0h}} \|\varphi - \psi_h\|_1. \tag{2.34}$$

PROOF We give the details of the proof in the case $g = 0$.

By subtracting (2.31) from (2.9) with $w = w_h$, which is permissible by virtue of (2.24), we obtain:

$$\int_\Omega \nabla(\varphi - \varphi_h) \cdot \nabla w_h = 0, \quad \forall w_h \in H^1_{0h}. \tag{2.35}$$

This equality means that φ_h achieves the minimum, over the space H^1_{0h}, of the functional

$$F(\psi_h) = \frac{1}{2}\int_\Omega |\nabla(\varphi - \psi_h)|^2; \tag{2.36}$$

indeed, still by an argument analogous to that used in Theorem 2.7, we have

$$\begin{aligned} F(\varphi_h + \lambda w_h) &= F(\varphi_h) + \lambda \int_\Omega \nabla(\varphi - \varphi_h)\nabla w_h + \frac{\lambda^2}{2}\int_\Omega |\nabla w_h|^2 \\ &\geq F(\varphi_h) \quad \text{if and only if (2.35) holds.} \end{aligned} \tag{2.37}$$

By using the definition of $\|.\|_1$, we thus deduce that

$$\forall \psi_h \in H^1_{0h}, \quad F(\varphi_h) \leq F(\psi_h) \leq \frac{1}{2}\|\varphi - \psi_h\|_1^2. \tag{2.38}$$

Poincaré's inequality (2.26) applied to $\varphi - \varphi_h \in H^1_0(\Omega)$ then provides us with the relation

$$\|\varphi - \varphi_h\|_1^2 \quad = \int_\Omega |\varphi - \varphi_h|^2 + \int_\Omega |\nabla(\varphi - \varphi_h)|^2 \leq 2(1 + C(\Omega))F(\varphi_h), \tag{2.39}$$

which, combined to (2.38), gives

$$\forall \psi_h \in H^1_{0h}, \quad \|\varphi - \varphi_h\|_1^2 \leq (1 + C(\Omega)) \quad \|\varphi - \psi_h\|_1^2, \tag{2.40}$$

which proves (2.34) with $C = (1 + C(\Omega))^{1/2}$. Let us take for granted the following result whose proof is left as an exercise, and is given in Lucquin-Pironneau (1997):

Lemma 2.9 The space H^1_{0h} is an internal approximation of $H^1_0(\Omega)$.

As $\varphi \in H^1_0(\Omega)$, we can associate to it, by Definition 2.4 (i), a function ϕ_h in H^1_{0h} such that

$$\|\varphi - \phi_h\|_1 \to 0, \text{ when } h \to 0.$$

The use of relation (2.40) for $\psi_h = \phi_h$ then ends the proof of the theorem. □

ORIENTATION We can now solve (2.25) on a computer. For this task, we could use a standard optimisation program; the only thing left would be to write a subroutine to compute $E(\varphi_h)$ when the φ_i are known. To do that, we could use the form (2.28) of the functional, by specifying how to compute the various integrals that occur.

This is precisely what we shall do in the next paragraph, but we shall do it for a very particular choice of basis functions $w^i(x)$, for which choice most of the integrals occurring in (2.28) will be zero.

2.4 *FIRST ORDER FINITE ELEMENTS: TRIANGULATION, INTERPOLATION, QUADRATURE FORMULAE*

2.4.1 *Triangulation; Approximation by P^1 Finite Elements*

We first decompose the domain Ω into small elements. We choose them to be of triangular shape, but we could also consider a decomposition of Ω into small

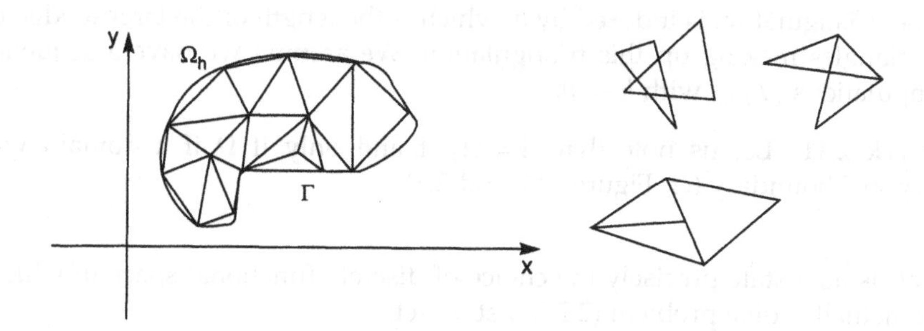

Figure 2.2 A triangulation of Ω and some examples of forbidden triangles

rectangles, for instance. The triangles we consider must satisfy some constraints that we state in the next definition.

Definition 2.10 We call *triangulation* on Ω a family $\mathcal{T}_h$ of triangles T_k, $k = 1, ..., n_t$, having the following properties:

(i) the intersection between two distinct triangles is either empty or reduced to a whole edge or to a point (cf. Figure 2.2);
(ii) all corners of the boundary $\Gamma = \partial\Omega$ are vertices of triangles of $\mathcal{T}_h$;
(iii) conversely, let $\Omega_h = \cup_{k=1}^{k=n_t} T_k$ (note that Ω_h is closed); all corners of $\Gamma_h = \partial\Omega_h$ must be in Γ (cf. Figure 2.2);
(iv) all triangles are non-degenerate, i.e. they all have a non-zero area.

We shall sometimes meet the terminology *admissible triangulation* to describe such a triangulation. The vertices of all triangles of $\mathcal{T}_h$ are the *vertices* of the triangulation. As this is a P^1 finite element approximation, we shall see later (Proposition 2.14) that these vertices are also the *nodes* of the triangulation, i.e. the points where we shall compute the solution of the discrete problem. Thus the notions of vertex and node coincide here, though they would not for a finite element approximation of arbitrary order. The triangulation defined above is composed of n_t triangles; we shall also denote by n_v the total number of vertices in this triangulation, and by n_b the number of vertices located on the numerical boundary Γ_h; the latter vertices are called *boundary vertices*, whereas the others are usually called *internal vertices* of the mesh.

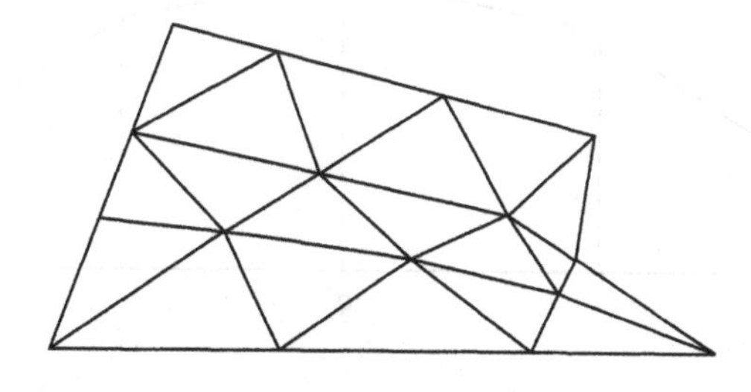

Figure 2.3 Triangulation of a polygonal domain

The triangulation is indexed by h, which is the length of the largest edge of all the triangles making up this triangulation. We assume we have a sequence of triangulations $(\mathcal{T}_h)_h$, with $h \to 0$.

Remark 2.11 Let us note that $\bar{\Omega} = \Omega_h$ if and only if Ω if a domain with a polygonal boundary (cf. Figures 2.2 and 2.3) □

Let us now state precisely the choice of discrete functional space in which we shall actually solve problem (2.25). Let us set:

$$\begin{aligned} H_h^1 &= \{\varphi_h : \Omega_h \to \mathbb{R}, \varphi_h \text{ continuous on } \Omega_h, \text{ and} \\ &\qquad \forall k \in \{1, ..., n_t\}, \varphi_{h|T_k} \text{ linear function}\}, \\ H_{gh}^1 &= \{\varphi_h \in H_h^1 \text{ and } \forall q^i \text{ vertex of } \Gamma_h, \quad \varphi_h(q^i) = g(q^i)\}. \end{aligned} \tag{2.41}$$

We shall seek the approximate solution φ_h of (2.25) in the space H_{gh}^1; on each triangle, φ_h will thus be in the set $P^1(T_k)$ of polynomials of degree less than or equal to 1 in the set of variables (x and y), whence the terminology P^1 *finite element approximation.*

Remark 2.12 Functions in H_{gh}^1 are not globally linear in all of Ω_h but only piecewise linear; however they are globally continuous. Last, the notation is coherent with that used in the previous section, because, since functions in H_{gh}^1 are linear on each triangle, they are infinitely differentiable; moreover, their first derivatives in the distribution sense present no Dirac measure, since the functions are globally continuous. The functions in our new discrete space are thus actually in $H^1(\Omega_h)$. □

Remark 2.13 (analogy in one dimension) In one dimension, Ω is an interval $]a, b[$ and a triangulation of this interval is a partition $[a, b] = \cup_{i=1}^{n_v-1} [q^i, q^{i+1}]$ with $q^1 = a$, $q^{n_v} = b$.

A function in H_{gh}^1 has a graph formed of straight line segments (cf. Figure 2.4). On this figure it is clear that φ_h is completely determined by its values at each of the internal vertices $(q^i)_2^{n_v-1}$ of the triangulation. □

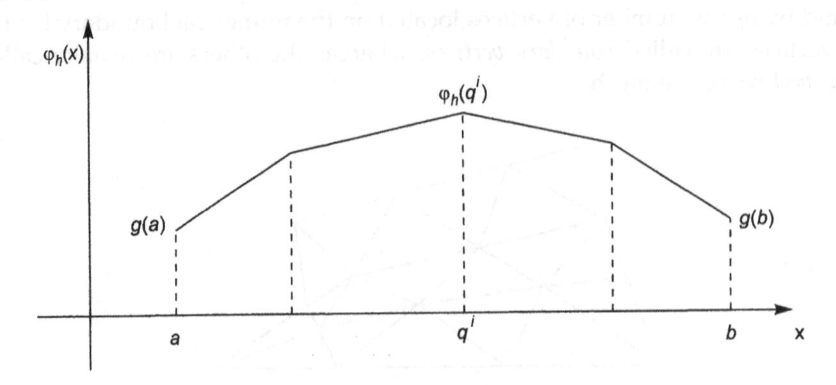

Figure 2.4 The one-dimensional analogue

This property is also true in two dimensions, and this is the object of proposition 2.14.

Proposition 2.14 *Functions in H_h^1 are entirely determined by their values at each of the vertices of the mesh. The dimension of the space H_h^1 is equal to the number of vertices n_v of the triangulation and the dimension of the space H_{0h}^1 is equal to $N_h = n_v - n_b$, where n_b is the number of boundary vertices.*

PROOF Let us begin by proving the first point of the proposition, and let us consider a function φ_h in the space H_h^1. Let T_k be a triangle with vertices denoted by $q^{i_j} = (q_x^{i_j}, q_y^{i_j})$, for $j = 1, 2, 3$ (cf. Figure 2.5). Let us set $\varphi_{i_j} = \varphi_h(q^{i_j})$. By definition, $\varphi_h|_{T_k}$ is a linear function of $\vec{x}$, which means that there are 3 scalars α, β, γ such that:

$$\forall \vec{x} = (x, y) \in T_k, \quad \varphi_h(\vec{x}) = \alpha x + \beta y + \gamma. \tag{2.42}$$

If we write in particular (2.42) for $\vec{x} = q^{i_j}$, with $j = 1, 2, 3$, we obtain a system of three equations in three unknowns α, β, γ:

$$\begin{aligned} \alpha q_x^{i_1} + \beta q_y^{i_1} + \gamma &= \varphi_{i_1}, \\ \alpha q_x^{i_2} + \beta q_y^{i_2} + \gamma &= \varphi_{i_2}, \\ \alpha q_x^{i_3} + \beta q_y^{i_3} + \gamma &= \varphi_{i_3}. \end{aligned} \tag{2.43}$$

The determinant of this system is equal to twice the area $|T_k|$ of the triangle T_k, if its vertices (i_1, i_2, i_3) are oriented in the positive direction, (i.e. counterclockwise); in the general case, we have:

$$\begin{vmatrix} q_x^{i_1} & q_y^{i_1} & 1 \\ q_x^{i_2} & q_y^{i_2} & 1 \\ q_x^{i_3} & q_y^{i_3} & 1 \end{vmatrix} = (q_x^{i_2} - q_x^{i_1})(q_y^{i_3} - q_y^{i_1}) - (q_y^{i_2} - q_y^{i_1})(q_x^{i_3} - q_x^{i_1}) = \pm 2\ |T_k|. \tag{2.44}$$

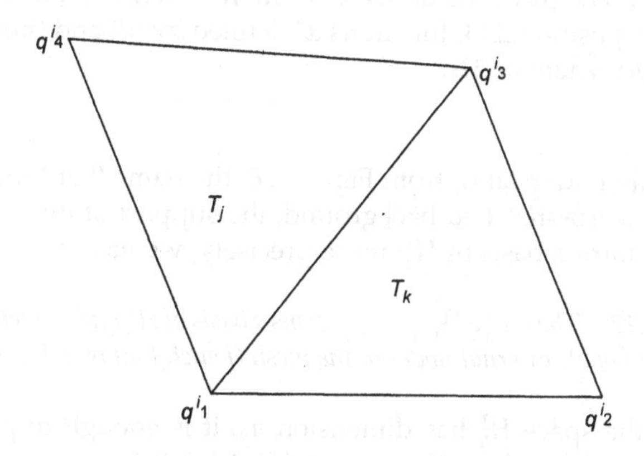

Figure 2.5 Two contiguous triangles

Since we assume the triangulation to be admissible, the triangle T_k is nondegenerate and system (2.43) then has a unique solution. The restriction of φ_h to the triangle T_k is entirely determined by the values taken by this function at each of the three vertices of T_k, which proves the first point in the proposition.

The previous result shows that the dimension of the space H_h^1 is less than or equal to n_v. To show it is actually equal to n_v, it is enough to show that the functions we constructed above, triangle per triangle, via formulae (2.42) and (2.43), are globally continuous on all of Ω_h, or in other words that they are continuous at the interface between two triangles.

Let us denote by T_j the triangle that has in common with T_k the edge $[q^{i_1}, q^{i_3}]$ for instance (cf. Figure 2.5). Then, according to (2.42), (2.43), φ_h has, on the edge $[q^{i_1}, q^{i_3}]$, a 'left' value (computed on T_j) and a 'right' value (computed on T_k), and we must show that these two values are equal.

To do this, it is enough to show that the restriction of φ_h to $[q^{i_1}, q^{i_3}]$ only depends on the values φ_{i_1}, φ_{i_3} taken at q_1^i and q^{i_3}, and not on those taken at q^{i_2} or q^{i_4}. Indeed, this is true, since if $\vec{x} \in [q^{i_1}, q^{i_3}]$, $\vec{x}$ can be written in the form $\vec{x} = \lambda q^{i_1} + (1-\lambda) q^{i_3}$ and since the restriction of φ_h to the edge $[q^{i_1}, q^{i_3}]$ is linear, we necessarily have

$$\varphi_h(\vec{x}) = \lambda \varphi_{i_1} + (1-\lambda)\varphi_{i_3}, \tag{2.45}$$

which simply proves the result.

Now if φ_h is in the space H_{0h}^1, its values at each of the boundary vertices is entirely determined (it is zero), which shows that the dimension of this space is equal to $n_v - n_b$. □

We shall now construct a basis for the space H_h^1.

Definition 2.15 Let us define the ith *hat function* w^i associated with node i in the following way:
(i) w^i is continuous from Ω_h into $\mathbb{R}$;
(ii) w^i is linear on T_k, for each index $k \in \{1, ..., n_t\}$; (2.46)
(iii) $w^i(q^j) = \delta_{ij}$ (1 if $i = j$, 0 otherwise).

Remark 2.16 Point (i) of the above definition is actually unnecessary; indeed, according to Proposition 2.14, functions w^i defined by (ii) and (iii) are defined and continuous everywhere on Ω_h. □

One can easily understand, from Figure 2.6, the name 'hat function'; the figure also shows, on a crosshatched background, the support of the function w^i. These 'hat functions' form a basis of H_h^1; more precisely, we have the following result:

Proposition 2.17 *The set $\{w^i\}_{i \in \{1,...,n_v\}}$ forms a basis of H_h^1 and the set $\{w^i\}_{i \in I_\Omega}$, I_Ω being the set of indices i of all internal nodes of the mesh (i such that $q^i \notin \Gamma$), is a basis of H_{0h}^1.*

PROOF Since the space H_h^1 has dimension n_v, it is enough to prove that the n_v functions w^i are independent. However, this is trivially obvious since w^i is the only function that is non-zero at point q^i. The same argument is valid for H_{0h}^1. □

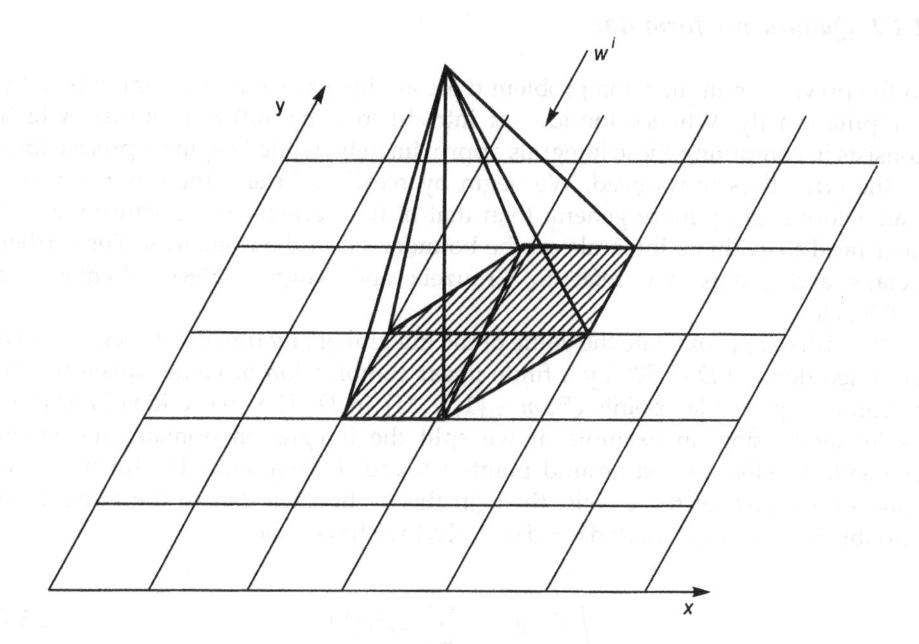

Figure 2.6 Support of the hat function associated with the ith node

In order to simplify the notation, we shall henceforth make an assumption that is certainly not necessary for programming: we shall assume that the first indices of the mesh vertices refer to internal vertices, the vertices located on the domain boundary being numbered after them. In this manner, the set I_Ω can be written as $I_\Omega = \{1, \dots, N_h\}$.

With this convention, the previous proposition enables us to look for the solution $\varphi_h \in H^1_{gh}$ of (2.25) in the form:

$$\varphi_h(\vec{x}) = \sum_{i=1}^{N_h} \varphi_i w^i(\vec{x}) + \sum_{i=N_h+1}^{n_v} g(q^i) w^i(\vec{x}). \tag{2.47}$$

In this expression, we have taken into account the boundary condition $\varphi_h(q^i) = g(q^i)$ for $q^i \in \Gamma$, $(i \in \{N_h + 1, \dots, n_v\})$ included in definition (2.41) of the space H^1_{gh}. The minimisation problem (2.25) is transformed as in the proof of Theorem 2.6 (formula (2.28)), and becomes:

$$\min_{\varphi_h \in H^1_{gh}} E(\varphi_h) = \min_{\Phi \in \mathbb{R}^{N_h}} E_h(\Phi), \tag{2.48}$$

where

$$\Phi = (\varphi_1, \dots, \varphi_{N_h}) \in \mathbb{R}^{N_h}, \quad \text{and} \quad E_h(\Phi) = E(\varphi_h). \tag{2.49}$$

2.4.2 *Quadrature formulae*

In the previous minimisation problem there are integrals that one cannot usually compute exactly. Whence the idea of introducing quadrature formulae, which consists in computing these integrals approximately, while keeping a precise idea of the error thus introduced. We begin by recalling briefly these notions in a framework slightly more general than that of two dimensions, because we will later need to evaluate integrals on the boundary Γ of the domain Ω. For further details and proofs, we refer to Crouzeix and Mignot (1984), Zienkiewicz (1971), etc.

We wish to approximate the value of the integral of any function f, defined on a bounded domain D of $\mathbb{R}^n$, by a finite linear combination of values taken by the function at particular points ξ^m, $m \in \{1,...,M\}$ in D. The basic idea is easy to understand using an example: if we split the integration domain into small elementary cells, centred around points ξ^m, and if we assume the function f is constant on each of these cells, then (in this section we denote the integration variable by $x = (x_1, \dots, x_n)$; $\mathrm{d}x = \mathrm{d}x_1 \dots \mathrm{d}x_n$) we have exactly

$$\int_D f(x)\mathrm{d}x = \sum_{m=1}^{M} \omega_m f(\xi^m), \tag{2.50}$$

where ω_m is the measure (area for $n = 2$, length for $n = 1$) of the small cell surrounding point ξ^m. If we generalise to an arbitrary function f, we have only:

$$\int_D f(x)\mathrm{d}x \simeq \sum_{m=1}^{M} \omega_m f(\xi^m). \tag{2.51}$$

This relation is called a Gaussian *quadrature formula*; the points ξ^m are the *quadrature points* and the real numbers ω_m are the *weights* of the quadrature formula (2.51). When we have equality in (2.51) for a particular function f, (i.e. (2.50) is true), we say the quadrature formula is *exact* for f. Let us give an estimate of the error introduced by this approximation, by recalling the following result, whose proof can be found in Crouzeix and Mignot (1984).

Proposition 2.18 *If the quadrature formula (2.51) is exact for polynomials of degree r, then there exists a strictly positive constant C such that, for any function f, $r+1$ times continuously differentiable in D, we have:*

$$\left| \int_D f(x)\mathrm{d}x - \sum_{m=1}^{M} \omega_m f(\xi^m) \right| \leq C|D|\, h^{r+1}, \tag{2.52}$$

where h is the diameter of D and $|D|$ its measure.

Let us now give a few examples of quadrature formulae, beginning in one dimension. Let $[a, b]$ be an interval in $\mathbb{R}$; the most common quadrature formulae on $[a, b]$ are:

the rectangle formula

$$\int_a^b f(x)\mathrm{d}x \simeq (b-a)f(a) \quad (\text{or } (b-a)f(b)); \tag{2.53}$$

the midpoint formula

$$\int_a^b f(x)\mathrm{d}x \simeq (b-a)f\left(\frac{a+b}{2}\right); \tag{2.54}$$

the trapezoidal formula

$$\int_a^b f(x)\mathrm{d}x \simeq (b-a)\frac{f(a)+f(b)}{2}; \tag{2.55}$$

another two points formula, $\xi^1 = a + \lambda(b-a)$ and $\xi^2 = b - \lambda(b-a)$, with $\lambda = (\sqrt{3}-1)/(2\sqrt{3})$

$$\int_a^b f(x)\mathrm{d}x \simeq (b-a)\frac{f(\xi^1)+f(\xi^2)}{2}; \tag{2.56}$$

Simpson's formula

$$\int_a^b f(x)\mathrm{d}x \simeq (b-a)\frac{f(a)+4f\left(\frac{a+b}{2}\right)+f(b)}{6}. \tag{2.57}$$

It is easy to show that the quadrature formula (2.53) is exact for constant polynomials, that (2.54) and (2.55) are exact for first degree polynomials, and last that (2.56) is exact for degree 2 polynomials, whereas (2.57) is exact for those of degree 3; moreover, in each case these results are optimal. Using Proposition 2.16 with $D = [a, b]$, we obtain the following result:

Proposition 2.19 (one-dimensional case) *The conclusions of Proposition 2.18 are valid with* $r = 0$ *for (2.53),* $r = 1$ *for (2.54) and (2.55),* $r = 2$ *for (2.56) and* $r = 3$ *for (2.57).*

Let us give a few examples of quadrature formulae in two dimensions, limiting ourselves to the case of triangles. Let T be a triangle with vertices q^1, q^2, q^3; the most common quadrature formulae are the following ($|T|$ denotes the area of T):

the formula centered at the center of mass $G = (q^1 + q^2 + q^3)/3$ of the triangle

$$\int_T f(x)\mathrm{d}x \simeq |T| f(G); \tag{2.58}$$

a formula involving the vertices of the triangle

$$\int_T f(x)\mathrm{d}x \simeq \frac{|T|}{3}\left(f(q^1)+f(q^2)+f(q^3)\right); \tag{2.59}$$

a formula involving the midpoints $\bar{q}^k$ of the triangle edges

$$\int_T f(x)\mathrm{d}x \simeq \frac{|T|}{3}\,(f(\bar{q}^1)+f(\bar{q}^2)+f(\bar{q}^3)). \tag{2.60}$$

To prove that a quadrature formula in two dimensions is exact for a degree r polynomial, we begin by a change of variables that transforms the triangle T into a triangle called the 'reference triangle' or the 'unit triangle': it is the isoceles triangle with length 1 on each of the coordinate axes of $\mathbb{R}^2$. One then shows that it suffices to prove the result for this triangle, which makes the computations easier. In this way, we observe that quadrature formulae (2.58) and (2.59) are exact for first degree polynomials, and that (2.60) is exact for second degree polynomials, and that these results are optimal; we then have the following result:

Proposition 2.20 (Two-dimensional case) *The conclusions of Proposition 2.18 are valid with $r = 1$ for (2.58) and (2.59) and $r = 2$ for (2.60).*

Remark 2.21 If we apply formula (2.59) when computing the matrix coefficients $I = ((I_{ij}))$, with

$$I_{ij} = \int_\Omega w^i(x)\, w^j(x)\, \mathrm{d}x,$$

we find:

$$I_{ij} \simeq \delta_{ij}\,\frac{1}{3} \sum_{\{k, q^i \in T_k\}} |T_k|. \tag{2.61}$$

The matrix I appears in finite element methods after discretising the constant term in partial differential equations. Formula (2.61), which is not exact, is called mass lumping; it allows the approximation of I by a diagonal matrix. We will have the opportunity to meet in Chapter 9 (in Section 9.3.2), an example where this approximation is used. □

2.4.3 *Interpolation*

The numerical integration formulae above are tightly linked with the the notion of interpolation of a function by polynomials. Let us return to our bounded domain Ω in $\mathbb{R}^2$, whose boundary we assume to be polygonal (so as to have $\Omega_h = \bar{\Omega}$), and provided with a triangulation $\mathcal{T}_h$.

Definition 2.22 The P^1 interpolant of a function f defined on Ω is the continuous function on Ω_h, linear on each triangle of the triangulation, that coincides with f at each of the mesh vertices.

Since quadrature formulae (2.58) and (2.59) are both exact for the P^1 interpolant of f, using such formulae means we have actually approximated f by its P^1 interpolant. More generally, we can locally define (i.e. on each triangle T) the

P^M interpolant of a function f, P^M being the set of polynomials of degree less than or equal to M in all variables $x_1, ...x_n$. For example, the function defined on T that coincides with f at the edge midpoints of T is a P^2 interpolant of f; using quadrature formula (2.60) then amounts to replacing f by this interpolant.

2.5 *PROGRAMMING THE METHOD*

2.5.1 *Orientation*

Let us now summarise what is left to do; we must:

- construct the triangulation of Ω, which we have assumed to have a polygonal boundary (we then have $\Omega_h = \bar{\Omega}$),
- construct the 'hat functions' w^i,
- find the solution of (2.48) and (2.49).

As far as the first point is concerned, we give an example of a triangulation in Section 2.5.2; there, the data structure is specified in a general way. The second point will not be detailed, since it is not necessary to store the basis functions w^i in an array; indeed, the basis functions only occur in the computation of the integrals, which will be carried out directly. For the third point, we have to choose a numerical method to solve problem (2.48). The methods commonly used for optimisation problems are *gradient methods*, all of which require the following two computations:

(a) Given $\Phi = (\varphi_1, \ldots, \varphi_{N_h})$, compute $E_h(\Phi)$ (i.e. compute E_h);

(b) Given $\Phi = (\varphi_1, \ldots, \varphi_{N_h})$, compute $\dfrac{\partial E_h}{\partial \varphi_i}(\Phi)$ (i.e. compute $Grad_\Phi E_h$).

For instance, the *optimal step gradient method* [Ciarlet (1989), Lascaux and Théodor (1987), Schatzman (1991)] to solve (2.48) translates, on a computer, into an iteration over all integers m of the following type:

Algorithm 1 (Optimal step gradient method):

0 choose an initialisation $\Phi^0 = (\Phi_1^0, ..., \Phi_{N_h}^0)$ for Φ and the accuracy ε; set $m = 0$;

1. set $G_i^m = \dfrac{\partial E_h}{\partial \varphi_i}(\Phi^m)$, for $i \in \{1, ..., N_h\}$, where $\Phi^m = (\Phi_1^m, ..., \Phi_{N_h}^m)$,
2. determine the real number ρ^m that minimizes, with respect to the variable $\rho \in \mathbb{R}$, the function $E_h(\Phi^m - \rho G^m)$ (cf. (2.77)), where $G^m = (G_1^m, ..., G_{N_h}^m)$,
3. set $\Phi^{m+1} = \Phi^m - \rho^m G^m$,
4. perform the stopping test: if $\sum_{i=1}^{N_h} |G_i^m|^2 < \varepsilon$, then stop; otherwise, increment m by 1 and go back to step 1.

Since the functional E_h has 'good properties' (it is a quadratic functional whose second derivative matrix is positive definite), we can easily show that the scalar

ρ^m is always positive; the terminology *descent method* comes from this property, since Φ^{m+1} is on the halfline starting from Φ^m in the direction opposite to that of the gradient.

Let us now describe the data structure we shall use for the triangulation.

2.5.2 Data structure for a triangulation

From a computational point of view, we shall use three Fortran arrays (a real array q, two integer arrays me and ng) and two integers (nv and nt) to store the information pertaining to the mesh; this information is:

nv = total number of mesh vertices;
nt = number of triangles;
$q(i,j)$ = ith coordinate of jth vertex;
$me(j,k)$ = storage index, in array q, of the jth vertex of the triangle with index k (this assumes a 'local' numbering of the vertices in each triangle);
$ng(i)$ = indicates whether the ith triangle is on the domain boundary ($ng(i) \neq 0$) or not ($ng(i) = 0$);
$ngT(k)$ = index of the region in the domain to which the kth triangle belongs; in fact, in some problems, there may be different materials, which makes it necessary to know rapidly the region where a given triangle is located. For the moment, we shall not use this array.

In C we can adopt the following data structure:

```
typedef struct {float x,y}; point;
typedef int triangle[3];
typedef int edge[2]; /* optional */
typedef struct{
int nv, nt;             /* number of vertices and triangles */
point *q;                 /* vertex coordinates */
triangle *me;             /* triangles connectivity */
int* ngt;                 /* triangle kind index */
int* ng;                /* vertex kind index */
                          /* the following is optional */
```

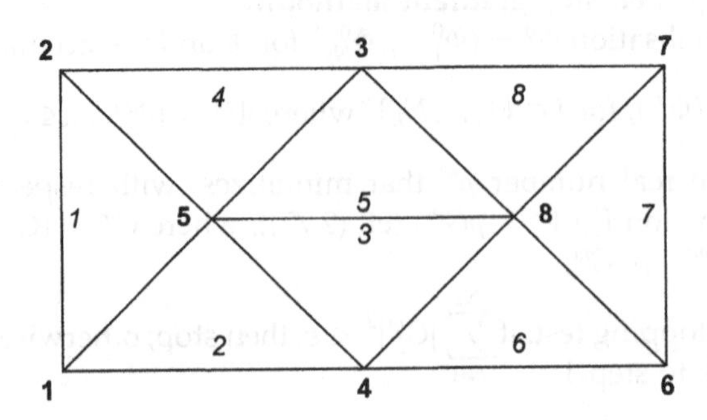

Figure 2.7 A tiny triangulation

```
  int ne;      /* number of edges */
  int* nde;    /* edge array */
  edge* edtr; /* indices of triangles neighbouring edge */
  triangle* tred; /* indices of 3 edges of triangle */
                                /* end optional */
}Triangulation;
```

We can, for instance, represent the triangulation of Figure 2.7, called *'tiny triangulation'*, by the following Fortran program. On the figure, bold indices are for the mesh vertices, and italic indices are for numbering the triangles. For readability's sake, we have not shown the 'local' numbering, i.e. the numbering of the three vertices within each triangle, as it can be deduced from the array *me* defined in the main program.

The names of the files for storing the triangulation or the problem data f and g are defined by the user at run time. Subroutine 'writeT' is used to store the triangulation in a file, whereas subroutine 'writeF' enables us to store an array of size *nv* in a file; it will either be the right hand side f of equation (2.1), assumed to be linear on each triangle, or the Dirichlet boundary condition g stored in array *phi*, which will also be used, at the end of the computation, to store the values of ϕ_h at each vertex of the mesh (we assume here that g is defined at all nodes of the mesh, whether they are internal or not).

```
C file: TinyTri.for
C
      PROGRAM MeshTiny
C     ----------------
C
C This program builds the tiny triangulation mesh and
C stores the model problem data
C
C     arrays:
C     -------
C     mesh:
C       q:coordinates of the mesh vertices,
C       me:numbering of the triangles vertices
C       ng:array indicating whether a vertex is
C         internal (ng(i)=0) or on the boundary.
C
C     data:
C       f: Laplacian right hand side
C       phi:Dirichlet boundary conditions
C
C     files:
C     ------
C       TinyTri.dta:name of file for storing the
C         triangulation,
C       filnam:name of file for storing f or phi
C             (the name of this file is given by the
C             user at program run time)
C
```

```
C       subroutines:
C       ------------
C         writeT:writes the mesh to a file whose
C           name is provided by the user at run time;
C           we suggest to call it TinyTri.dta.
C         writeF:writes array tab to a file whose
C           name is provided by the user at run time.
C
C
        PARAMETER (nvMax=30,ntMax=60)
        DIMENSION f(nvMax),phi(nvMax)
        CHARACTER*20 filnam
C
        nv=8
        nt=8
        DATA q/3*0.,4*1.,0.,2*.5,2.,0.,2.,1.,1.5,.5/
        DATA me/1,5,2, 1,4,5, 5,4,8, 3,2,5,
     >          8,3,5, 4,6,8, 6,7,8, 8,7,3/
        DATA ng/4*1,0,2*1,0/
        DATA f/nv*0./
        DATA phi/nv*0./
C
C
        write(*,*)'file name for mesh'
        write(*,*)'(example treated: TinyTri.dta)'
        read(*,'(a)') filnam
        CALL writeT (filnam,nv,nt,q,ng,me,ngt)
        write(*,*)'file name for storing f'
        write(*,*)'(Laplacian right hand side)'
        read(*,'(a') filnam
        CALL writeF (filnam,nv,f)
        write(*,*)'file name for storing phi'
        write(*,*)'(Dirichlet boundary conditions)'
        read(*,'(a)') filnam
        CALL writeF(filnam,nv,phi)
C
        END
C
C----------------------------------------------------------
        SUBROUTINE writeT(filnam,nv,nt,q,ng,me,ngt)
C----------------------------------------------------------
C
C  Writes mesh to file filnam
C
C
        CHARACTER*20 filnam
        DIMENSION q(2,nv),me(3,nt),ng(nv)
C
        OPEN(UNIT=13, file=filnam)
        ngt=0
        write(13,*) nv,nt
```

```
      DO 1 i=1,nv
        write(13,*) q(1,i),q(2,i),ng(i)
1     CONTINUE
      DO 2 i=1,nt
        write(13,*) me(1,i),me(2,i),me(3,i),ngt
2     CONTINUE
      CLOSE(13)
C
      END
C
C-----------------------------------------------------------
      SUBROUTINE writeF(filnam,nv,tab)
C-----------------------------------------------------------
C
C  Writes array tab to file filnam
C
C
      CHARACTER*20 filnam
      DIMENSION tab(nv)
C
      OPEN(UNIT=13, file=filnam)
      write(13,*) nv
      DO 1 i=1,nv
        write(13,*) tab(i)
1     CONTINUE
      CLOSE(13)
C
      END
```

Results on the screen when this program is run:

```
8 8
1 .000000   .000000   1
2 .000000   1.00000   1
3 1.00000   1.00000   1
4 1.00000   .000000   1
5 .500000   .500000   0
6 2.00000   .000000   1
7 2.00000   1.00000   1
8 1.50000   .500000   0
1 1  5  2  0
2 1  4  5  0
3 5  4  8  0
4 3  2  5  0
5 8  3  5  0
6 4  6  8  0
7 6  7  8  0
8 8  7  3  0
```

In C the same program works in the following way:

```
/*  file: tinytri.c   */
/*  This program builds the tiny triangulation mesh
         and writes the model problem data                        */
#include <math.h>
#include <stdio.h>
#include <stdlib.h>

char* tfile = "tinyt.dta";
char* ffile = "tinyf.dta";
char* gfile = "tinyg.dta";

typedef struct{ float x,y; }point;
typedef long triangle[3];
typedef struct{ int nv, nt;
  point *q;
  triangle *me;
  int *ngt;
  int* ng;
    }triangulation;
int    writeT(char* filnam,triangulation* t);
int    writeF(char* filnam,int nv,float* tab);

/*-------------------------------------------------*/
int writeT(char* filnam,triangulation* t)
/*-------------------------------------------------*/
{int i,a,b,c; FILE *f;
  if(f=fopen(filnam, "w"), !f) return 1;
  fprintf(f,"%d %d\n", t->nv,t->nt);
  for(i=0; i<t->nv; i++)
    fprintf(f,"%e %e %d\n",
      t->q[i].x, t->q[i].y, t->ng[i]);
  for(i=0; i<t->nt; i++)
    {  a = 1+t->me[i][0]; b = 1+t->me[i][1];
       c = 1+t->me[i][2];
      fprintf(f,"%d %d% d%d\n",a,b,c, t->ngt[i]);
    }
  fclose(f);
  return 0;
}
/*-------------------------------------------------*/
int writeF(char* filnam,int nv,float* tab)
/*-------------------------------------------------*/
/* Writes array tab to file filnam */
{  int i; FILE* f;
   if(f=fopen(filnam, "w"), !f) return 1;
   fprintf(f,"%d\n", nv);
   for(i=0; i<nv; i++) fprintf(f,"%e\n", tab[i]);
   fclose(f);
   return 0;
}
```

```
/*-----------------------------------------------*/
void main(void)
/*-----------------------------------------------*/
{
  triangulation t;
  float *f;   /* source function */
  float *phi; /* boundary conditions & initialisations */
  int i;
t.nv = 8;
t.nt = 8;
t.q= (point*)calloc(t.nv,sizeof(point));
t.q[0].x = 0;  t.q[0].y = 0;
t.q[1].x = 0;  t.q[1].y = 1;
t.q[2].x = 1;  t.q[2].y = 1;
t.q[3].x = 1;  t.q[3].y = 0;
t.q[4].x =.5;   t.q[4].y =.5;
t.q[5].x = 2;  t.q[5].y = 0;
t.q[6].x = 2;  t.q[6].y = 1;
t.q[7].x =1.5; t.q[7].y =.5;
t.me = (triangle*)calloc(t.nt,sizeof(triangle));
t.me[0][0] = 0; t.me[0][1] = 4; t.me[0][2] = 1;
t.me[1][0] = 0; t.me[1][1] = 3; t.me[1][2] = 4;
t.me[2][0] = 4; t.me[2][1] = 3; t.me[1][2] = 7;
t.me[3][0] = 2; t.me[3][1] = 1; t.me[3][2] = 4;
t.me[4][0] = 7; t.me[4][1] = 2; t.me[4][2] = 4;
t.me[5][0] = 3; t.me[5][1] = 5; t.me[5][2] = 7;
t.me[6][0] = 5; t.me[6][1] = 6; t.me[6][2] = 7;
t.me[7][0] = 7; t.me[7][1] = 6; t.me[7][2]= 2;
t.ng= (int*)calloc(t.nv,sizeof(int));
t.ng[0] = 1; t.ng[1] = 1; t.ng[2] = 1; t.ng[3] = 1;
t.ng[4] = 0; t.ng[5] = 1; t.ng[6] = 1; t.ng[7] = 0;
t.ngt = (int*)calloc(t.nt,sizeof(int));
f = (float*)calloc(t.nv,sizeof(float));
phi = (float*)calloc(t.nv,sizeof(float));
for(i=0;i<t.nv;i++)f[i]=1;
writeT(tfile,&t);
writeF(ffile,t.nv,f) ;
writeF(gfile,t.nv,phi);
printf("3 files generated: %s, %s, %s \n",
tfile,ffile,gfile);
}
```

2.5.3 *Computing the discrete energy*

From the known values φ_i of φ_h at each of the vertices q^i we must compute the real number $E_h(\Phi)$ given by

$$E_h(\Phi) = E(\varphi_h) = \frac{1}{2}\int_{\Omega_h} |(\nabla\varphi_h)(\vec{x})|^2 \mathrm{d}\vec{x} - \int_{\Omega_h} (f\varphi_h)(\vec{x})\mathrm{d}\vec{x}. \tag{2.62}$$

Let us recall that we assumed that Ω has a polygonal boundary, so that $\bar{\Omega} = \Omega_h$. Since $\varphi_h|_{T_k}$ is linear, $|\nabla\varphi_h|$ is constant on T_k and we have

$$E_h(\Phi) = \frac{1}{2}\sum_{k=1}^{n_t} |T_k|\,|\nabla\varphi_h|^2_{T_k} - \sum_{k=1}^{n_t}\int_{T_k}(f\varphi_h)(\vec{x})\mathrm{d}\vec{x}. \tag{2.63}$$

Now, on triangle T_k, φ_h has the form (2.42), where the coefficients α, β, γ are given explicitly by (2.43), that is (abs is short for absolute value):

$$\alpha = \frac{\begin{vmatrix} \varphi_1 & q_y^1 & 1 \\ \varphi_2 & q_y^2 & 1 \\ \varphi_3 & q_y^3 & 1 \end{vmatrix}}{\begin{vmatrix} q_x^1 & q_y^1 & 1 \\ q_x^2 & q_y^2 & 1 \\ q_x^3 & q_y^3 & 1 \end{vmatrix}}, \quad \beta = \frac{\begin{vmatrix} q_x^1 & \varphi_1 & 1 \\ q_x^2 & \varphi_2 & 1 \\ q_x^3 & \varphi_3 & 1 \end{vmatrix}}{\begin{vmatrix} q_x^1 & q_y^1 & 1 \\ q_x^2 & q_y^2 & 1 \\ q_x^3 & q_y^3 & 1 \end{vmatrix}}, \quad \gamma = \frac{\begin{vmatrix} q_x^1 & q_y^1 & \varphi_1 \\ q_x^2 & q_y^2 & \varphi_2 \\ q_x^3 & q_y^3 & \varphi_3 \end{vmatrix}}{\begin{vmatrix} q_x^1 & q_y^1 & 1 \\ q_x^2 & q_y^2 & 1 \\ q_x^3 & q_y^3 & 1 \end{vmatrix}}, \tag{2.64}$$

and

$$|T_k| = \frac{1}{2}\mathrm{abs}\left(\begin{vmatrix} q_x^1 & q_y^1 & 1 \\ q_x^2 & q_y^2 & 1 \\ q_x^3 & q_y^3 & 1 \end{vmatrix}\right). \tag{2.65}$$

The notation is that of Figure 2.8 where, to simplify notation, the vertices of triangle T_k are denoted by q^1, q^2, q^3, and vertex q^i has coordinates (q_x^i, q_y^i), and $\varphi_i = \varphi_h(q^i)$.

Last, all determinants in the previous formulae are of the type

$$\begin{vmatrix} a & b & 1 \\ c & d & 1 \\ e & f & 1 \end{vmatrix} = \begin{vmatrix} a-e & b-f & 0 \\ c-e & d-f & 0 \\ e & f & 1 \end{vmatrix} = (a-e)(d-f) - (b-f)(c-e). \tag{2.66}$$

This is easily programmed in Fortran in the following way

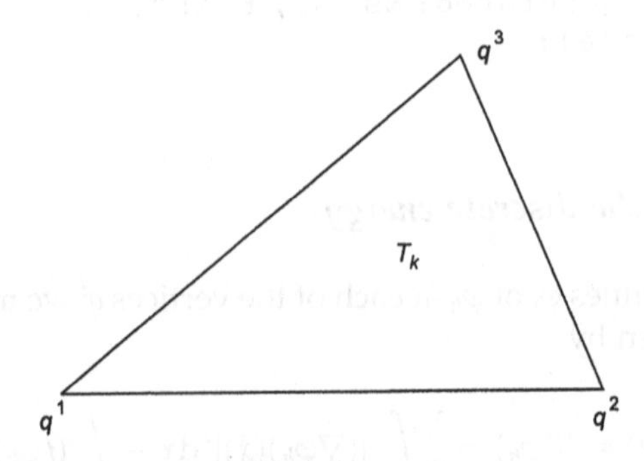

Figure 2.8

```
c
      FUNCTION deter (a,c,e,b,d,f)
c              -----
      deter=(a-e)*(d-f) (b-f)*(c-e)
      END
```

To compute the second integral in (2.62), we shall use the Gaussian quadrature formula (2.58) centred at only one point, namely the centre of mass of each of the triangles. We first have:

$$\int_{\Omega_h} (f\varphi_h)(\vec{x})\mathrm{d}\vec{x} = \sum_{k=1}^{n_t} \int_{T_k} (f\varphi_h)(\vec{x})\mathrm{d}\vec{x}, \tag{2.67}$$

then, with the notation of Figure 2.8,

$$\int_{T_k} (f\varphi_h)(\vec{x})\mathrm{d}\vec{x} \simeq f\left(\frac{q^1+q^2+q^3}{3}\right)\varphi_h\left(\frac{q^1+q^2+q^3}{3}\right)|T_k| \tag{2.68}$$

$$\simeq f\left(\frac{q^1+q^2+q^3}{3}\right)\frac{\varphi_1+\varphi_2+\varphi_3}{3}|T_k|. \tag{2.69}$$

This quadrature formula is exact if f is assumed to be constant on T_k, which is not the case here. If f is linear on triangle T_k, we can approximate it by a constant function on each triangle, the value of the constant being one-third of the sum of the values taken by f at each vertex of the triangle; starting from the quadrature formula (2.69), which is no longer exact (the error is, according to Proposition 2.20, of order h), we obtain the following approximation of the sought integral:

$$\int_{T_k} (f\varphi_h)(\vec{x})\mathrm{d}\vec{x} \simeq \frac{f(q^1)+f(q^2)+f(q^3)}{3}\,\frac{\varphi_1+\varphi_2+\varphi_3}{3}|T_k|. \tag{2.70}$$

Remark 2.23 We can improve the previous approximation by considering the quadrature formula (2.60). The integrals occurring in (2.67) will then be computed exactly and, denoting by f_i the value of f at node q^i of triangle T_k, we have on this triangle:

$$\int_{T_k} (f\varphi_h)(\vec{x})\mathrm{d}\vec{x} = \frac{|T_k|}{12}\;[(f_1+f_2)(\varphi_1+\varphi_2)$$

$$+(f_2+f_3)(\varphi_2+\varphi_3)+(f_3+f_1)(\varphi_3+\varphi_1)]. \tag{2.71}$$

Since the quadrature formula (2.60) is exact for second order polynomials in both variables x and y, formula (2.71) gives the exact value of integral (2.67), if f is linear on each triangle. If f is not linear, it gives an approximate value, the error being of order h^2, again according to Proposition 2.20. We have chosen to program the first approximation (2.70), rather than (2.71), only to simplify the program. □

We can now program the computation of $E_h(\Phi) = E(\varphi_h)$.

```
C file: minEe.for
C
C-----------------------------------------------------------
      FUNCTION E(q,me,phi,f,nv,nt)
C-----------------------------------------------------------
C
C Computes energy E. This function can be useful to
C check that the energy actually decreases along the
C iterations of the method used to minimise it.
C
C If the array f on entry to this subroutine is
C identically zero, this function computes
C ∫Ω ½|∇φ|², for any array phi(nv), which
C gives another useful purpose to this function.
C
C     arrays:
C     -------
C     mesh:
C       q(2,nv):coordinates of the mesh vertices,
C       me(3,nt):numbering of the triangles vertices
C
C     data:
C       f(nv):Laplacian right hand side,
C       phi(nv):at the beginning, Dirichlet boundary
C           conditions.
C
C     scalars:
C     --------
C     fk:value of f at the centre of mass of the index k
C       triangle, or equivalently, f being assumed linear
C       on each triangle, one-third of the sum of the
C       values taken by f at each of the 3 vertices of
C       the index k triangle.
C     area2Tk:twice the area of the index k triangle
C
      DIMENSION phi(nv),f(nv),me(3,nt),q(2,nv)
C
      E = 0.
      DO 1 k = 1,nt
         i1 = me(1,k)
         i2 = me(2,k)
         i3 = me(3,k)
         xi1=q(1,i1)
         xi2=q(1,i2)
         xi3=q(1,i3)
         yi1=q(2,i1)
         yi2=q(2,i2)
         yi3=q(2,i3)
         phii1=phi(i1)
```

```
            phii2=phi(i2)
            phii3=phi(i3)
C
            fk=(f(i1)+f(i2)+f(i3))/3.
            area2Tk = deter(xi1,xi2,xi3,yi1,yi2,yi3)
            gradFx = deter(phii1,phii2,phii3,yi1,yi2,yi3)
     >                                   /area2Tk
            gradFy = deter(xi1,xi2,xi3,phii1,phii2,phii3)
     >                                   /area2Tk
C
            E = E + 0.5*abs(area2Tk)*(
     >                  (gradFx*gradFx + gradFy*gradFy)/2.
     >                  -fk*(phii1+phii2+phii3)/3.)
1       CONTINUE
C
        RETURN
        END
```

Remark 2.24 This Fortran function will be useful later (for computing 'ρ^m' in Section 2.5.5). Indeed, we can use it to compute, for a given function φ, the following integral: $\int_{\Omega_h} |(\nabla\varphi)(\vec{x})|^2 d\vec{x}$; we just have to take, on input of the Fortran function, an array f that is identically zero. □

2.5.4 *Computing the gradient of the discrete energy*

It follows from (2.63) and (2.47) that

$$E_h(\Phi) = \frac{1}{2}\int_{\Omega_h}\left|\nabla\left(\sum_{i=1}^{N_h}\varphi_i w^i + \sum_{i=N_h+1}^{n_v} g_i w^i\right)(\vec{x})\right|^2 d\vec{x} - \int_{\Omega_h} f(\vec{x})\left(\sum_{i=1}^{N_h}\varphi_i w^i(\vec{x}) + \sum_{i=N_h+1}^{n_v} g_i w^i(\vec{x})\right)d\vec{x}, \tag{2.72}$$

still with the same notation $\Phi = (\varphi_1, \ldots, \varphi_{N_h})$. Then, for any $k \in \{1, \ldots, N_h\}$, we have:

$$\begin{aligned}\frac{\partial E_h}{\partial\varphi_k}(\Phi) &= \int_{\Omega_h}\nabla\left(\sum_{i=1}^{N_h}\varphi_i w^i + \sum_{i=N_h+1}^{n_v} g_i w^i\right)(\vec{x})\cdot\nabla w^k(\vec{x})d\vec{x} - \int_{\Omega_h}(fw^k)(\vec{x})d\vec{x}\\ &= \int_{\Omega_h}\nabla\varphi_h(\vec{x})\cdot\nabla w^k(\vec{x})d\vec{x} - \int_{\Omega_h}(fw^k)(\vec{x})d\vec{x}.\end{aligned} \tag{2.73}$$

Let $G_k = \partial E_h/\partial\varphi_k(\Phi)$. Then

$$G_k = \sum_{l=1}^{n_t}\left[\int_{T_l}\nabla\varphi_h(\vec{x})\cdot\nabla w^k(\vec{x})d\vec{x} - \int_{T_l}(fw^k)(\vec{x})d\vec{x}\right].$$

On each triangle T_l, we use the form (2.42) of φ_h, completed by (2.64), which gives us, still denoting by $q^{i_1}, q^{i_2}, q^{i_3}$ the three vertices of T_l:

$$\forall \vec{x} \in T_l, \quad \nabla\varphi_h(\vec{x}) \cdot \nabla w^k(\vec{x}) = \frac{\partial \varphi_h}{\partial x}(\vec{x})\frac{\partial w^k}{\partial x}(\vec{x}) + \frac{\partial \varphi_h}{\partial y}(\vec{x})\frac{\partial w^k}{\partial y(\vec{x})} \tag{2.74}$$

$$= \frac{\begin{vmatrix} \varphi_{i_1} & q_y^{i_1} & 1 \\ \varphi_{i_2} & q_y^{i_2} & 1 \\ \varphi_{i_3} & q_y^{i_3} & 1 \end{vmatrix}}{\begin{vmatrix} q_x^{i_1} & q_y^{i_1} & 1 \\ q_x^{i_2} & q_y^{i_2} & 1 \\ q_x^{i_3} & q_y^{i_3} & 1 \end{vmatrix}} \frac{\begin{vmatrix} \delta_{i_1,k} & q_y^{i_1} & 1 \\ \delta_{i_2,k} & q_y^{i_2} & 1 \\ \delta_{i_3,k} & q_y^{i_3} & 1 \end{vmatrix}}{\begin{vmatrix} q_x^{i_1} & q_y^{i_1} & 1 \\ q_x^{i_2} & q_y^{i_2} & 1 \\ q_x^{i_3} & q_y^{i_3} & 1 \end{vmatrix}} + \frac{\begin{vmatrix} q_x^{i_1} & \varphi_{i_1} & 1 \\ q_x^{i_2} & \varphi_{i_2} & 1 \\ q_x^{i_3} & \varphi_{i_3} & 1 \end{vmatrix}}{\begin{vmatrix} q_x^{i_1} & q_y^{i_1} & 1 \\ q_x^{i_2} & q_y^{i_2} & 1 \\ q_x^{i_3} & q_y^{i_3} & 1 \end{vmatrix}} \frac{\begin{vmatrix} q_x^{i_1} & \delta_{i_1,k} & 1 \\ q_x^{i_2} & \delta_{i_2,k} & 1 \\ q_x^{i_3} & \delta_{i_3,k} & 1 \end{vmatrix}}{\begin{vmatrix} q_x^{i_1} & q_y^{i_1} & 1 \\ q_x^{i_2} & q_y^{i_2} & 1 \\ q_x^{i_3} & q_y^{i_3} & 1 \end{vmatrix}}.$$

EXERCISE Show that $\partial w^k/\partial x = (q_y^{k_1} - q_y^{k_2})/|T_l|$, $\partial w^k/\partial y = (q_x^{k_1} - q_x^{k_2})/\ |T_l|$ where k_1 is next to k and k_2 is next to $k_1$in T_l.

As for the integral of fw^h on T_l, it can be approximated by

$$\int_{T_l} (fw^k)(\vec{x})\mathrm{d}\vec{x} \simeq \begin{cases} f\left(\dfrac{q^{i_1}+q^{i_2}+q^{i_3}}{3}\right)\dfrac{|T_l|}{3}, \text{if } k \in \{i_1, i_2, i_3\} \\ 0 \text{ otherwise,} \end{cases} \tag{2.75}$$

because

$$w^k\left(\frac{q^{i_1}+q^{i_2}+q^{i_3}}{3}\right) = \frac{1}{3}(\text{if } k \in \{i_1, i_2, i_3\}),\ 0\ (\text{otherwise}). \tag{2.76}$$

From a programming point of view, an important idea is simultaneously to compute *all* the G_k, which can be done by a loop over *all triangles* of the triangulation:

```
      DO 1 k = 1,nv
        G(k)=0.
1     CONTINUE
      DO 2 l = 1,nt
        DO 2 j = 1,3
          k = me(j,l)
          IF (ng(k).EQ.0) THEN
            G(k) = G(k) + ∫T1 (∇φh∇wk - fwk)
          ENDIF
2     CONTINUE
```

To understand how this works, let us look at what would happen if we computed each G_k separately:

```
      DO 1 k = 1,ns
        G(k) = 0
        DO 2 l = 1,nt
```

```
             DO 2 j = 1,3
               IF ((k.EQ.me(j,l)).AND.(ng(k).EQ.0)) THEN
                 G(k) = G(k) + ∫_{T_l}(∇φ_h∇w^k − fw^k)
               ENDIF
2            CONTINUE
1          CONTINUE
```

We thus have two nested 'DO loops', that is $O(nv * nt)$ operations, whereas the previous program only has $O(nt)$ operations. The first method is based on the fact that Ω is the union of all the triangles, and that on the index l triangle, the integral $\int_{T_l}(\nabla\varphi_h\nabla w^k - fw^k)$ is zero if q^k is not a vertex of T_l. But the number of such T_l is small, and independent of nv. One says that G_k has been computed by *assembly* over the triangles.

Thus, we obtain the following subroutine to compute $\nabla_\Phi E_h(\Phi) = \left(\frac{\partial E_h}{\partial\varphi_1},\ldots,\frac{\partial E_h}{\partial\varphi_{N_h}}\right)(\Phi)$.

```
C file: minEdE.for
C
C------------------------------------------------------------
      SUBROUTINE dEdPhi(q,me,ng,phi,f,nv,nt,gradE,g2h)
C------------------------------------------------------------
C Computes the gradient of the energy functional at each
C vertex of the mesh, and stores it in array gradE.
C
C Also computes the square of the Euclidean norm
C of the gradient of E and stores it in scalar g2h.
C
C      arrays:
C      -------
C      mesh:
C        q(2,nv):coordinates of the mesh vertices,
C        me(3,nt):numbering of the triangles vertices
C        ng(nv):array to indicate whether a vertex is
C        internal (ng(i)=0) or not.
C
C      data:
C        f(nv):Laplacian right hand side,
C        phi(nv):at the beginning, approximate solution
C
C      output:
C        gradE(nv):gradient of the functional E;
C
C      other arrays:
C        w(3):basis function
C
      DIMENSION q(2,nv),me(3,nt), ng(nv)
      DIMENSION phi(nv),f(nv),gradE(nv),w(3)
C
      DO 1 i = 1, nv
```

```
            gradE(i)=0
1         CONTINUE
          DO 2 k=1,nt
            i1=me(1,k)
            i2=me(2,k)
            i3=me(3,k)
            xi1=q(1,i1)
            xi2=q(1,i2)
            xi3=q(1,i3)
            yi1=q(2,i1)
            yi2=q(2,i2)
            yi3=q(2,i3)
            phii1=phi(i1)
            phii2=phi(i2)
            phii3=phi(i3)
C
            fk=(f(i1)+f(i2)+f(i3))/3.
            area2Tk=deter(xi1,xi2,xi3,yi1,yi2,yi3)
            gradFx=deter(phii1,phii2,phii3,yi1,yi2,yi3)
     >                          /area2Tk
            gradFy=deter(xi1,xi2,xi3,phii1,phii2,phii3)
     >                          /area2Tk
C
            DO 3 j=1,3
              i=me(j,k)
              IF ( ng(i).EQ.0 ) THEN
                DO 4 l=1,3
                  w(l)=0.0
4               CONTINUE
                w(j)=1.0
                gradWix=deter(w(1),w(2),w(3),yi1,yi2,yi3)
     >                          /area2Tk
                gradWiy=deter(xi1,xi2,xi3,w(1),w(2),w(3))
     >                          /area2Tk
                gradE(i)=gradE(i)+(gradFx*gradWix
     >                   +gradFy*gradWiy-fk/3)
     >                    *abs(area2Tk)/2.
              ENDIF
3           CONTINUE
2         CONTINUE
          g2h=0.0
          DO 5 i=1,nv
            gradEi=gradE(i)
            g2h=g2h+gradEi*gradEi
5         CONTINUE
          RETURN
          END
```

Remark 2.25 The value of $\partial E_h/\partial\varphi_k(\Phi)$ is stored in the array 'gradE' at index k. Only values of $\partial E_h/\partial\varphi_k(\Phi)$ for $k \in \{1,\dots N_h\}$ need to be computed; however, to make programming simpler, the array 'gradE' is filled for *all indices,* with the convention that gradE$(k) = 0$ if $k \in \{N_h + 1, \ldots, n_v\}$. In fact, we recall that *actually* the internal nodes of the mesh are not necessarily numbered first ($I_\Omega \neq \{1, \ldots, N_h\}$), so that if we wanted to set the dimension of the array gradE only to N_h, we would have to number the internal nodes through an array 'qpos', and for each index $k \in I_\Omega$, the value of $\partial E_h / \partial \varphi_k(\Phi)$ would be stored at index $qpos(k)$ in the array 'gradE', which complicates memory accesses.

Let us note that subroutine 'dEdPhi' also computes the Euclidean norm in $\mathbb{R}^{N_h}$, denoted by $||.||_{\mathbb{R}^{N_h}}$, of the previous vector, that is

$$||\nabla_\Phi E_h(\Phi)||^2_{\mathbb{R}^N_h} = \sum_{k=1}^{N_h} \left| \frac{\partial E_h}{\partial \varphi_k}(\Phi) \right|^2,$$

which will be useful in the sequel (conjugate gradient method in Section 2.5.6). □

2.5.5 *A first minimisation algorithm*

We detail here the programming of Algorithm1 proposed in Section 2.5.1 to solve the optimisation problem (2.48); it is the optimal step gradient method.

Algorithm 1 (optimal step gradient method):

0 choose an accuracy ϵ, choose an initialization $\Phi^0 = (\varphi_1^0, \ldots, \varphi_{N_h}^0)$ for Φ, set $m = 0$;

1 compute, for all indices $k \in \{1, ..., N_h\}$,

$$G_k^m = \frac{\partial E_h}{\partial \varphi_k}(\Phi^m), \quad \Phi^m = (\varphi_1^m, \ldots, \varphi_{N_h}^m);$$

2 compute the scalar ρ that minimises the function $E_h(\Phi^m - \rho G^m)$ with respect to the variable ρ, where $G^m = (G_1^m, \ldots, G_{N_h}^m)$;

3 set $\Phi^{m+1} = \Phi^m - \rho^m G^m$. Increase m by 1, and go back to step 1 until

$$\sum_k (G_k^m)^2 < \epsilon \sum_k (G_k^0)^2.$$

Let us describe in detail the computation of step 2. We choose a slightly more general framework, where we must find the value of ρ that achieves the minimum of $E_h(\Phi^m - \rho H^m)$, where $H^m \in \mathbb{R}^{N_h}$. Since E_h is a quadratic function of Φ, its Taylor expansion stops at the second order term, and we obtain:

$$E_h(\Phi^m - \rho H^m) = E_h(\Phi^m) - \rho \sum_{k=1}^{N_h} \frac{\partial E_h}{\partial \varphi_k}(\Phi^m) H_k^m + \frac{\rho^2}{2} \sum_{k,l=1}^{N_h} \frac{\partial^2 E_h}{\partial \varphi_k \partial \varphi_l}(\Phi^m) H_k^m H_l^m. \tag{2.77}$$

The minimal value ρ^m sought is then

$$\rho^m = \frac{\sum_{k=1}^{N_h} \frac{\partial E_h}{\partial \varphi_k}(\Phi^m) H_k^m}{\sum_{k,l=1}^{N_h} \frac{\partial^2 E_h}{\partial \varphi_k \partial \varphi_l}(\Phi^m) H_k^m H_l^m} = \frac{(G^m)^t H^m}{(H^m)^t A H^m}, \tag{2.78}$$

where A is the matrix of second derivatives of E_h (that already came up in the proof of Theorem 2.6), whose kl element equals (here $\Omega = \Omega_h$)

$$A_{kl} = \frac{\partial^2 E_h}{\partial \varphi_k \partial \varphi_l}(\Phi) = \int_{\Omega_h} \nabla w^k(\vec{x}) \cdot \nabla w^l(\vec{x})\, d\vec{x}. \tag{2.79}$$

If we denote by H_h^m the continuous and piecewise linear function on Ω_h equal to H_k^m at each index k node of the mesh, then we have, according to (2.79):

$$(H^m)^t A H^m = \int_{\Omega_h} |\nabla H_h^m|^2 d\vec{x} = ||\nabla H_h^m||^2_{L^2(\Omega_h)} \tag{2.80}$$

this relation enables us to use Remark 2.24 to compute the denominator of ρ^m.

The following program computes ρ^m in a slightly more general framework than ours, which will allow us to use it also in the next algorithm. To be able to use it here, the array 'hConj' must contain values exactly opposite to those in array 'gradE' (that contains values of G_k at each node of index k of the mesh), which we can write symbolically by 'hConj = − gradE'. Also, according to Remark 2.25, *all nodes* of the mesh, whether or not they are internal, must be taken into account in the program, which requires, in order for the computation to be exact, that values in the array 'gradE' be zero for all indices corresponding to boundary nodes of the mesh.

```
C file: minErho.for
C
C-----------------------------------------------------------
      FUNCTION rho(q,me,nv,nt,phi,gradE,hConj,hConjm)
C-----------------------------------------------------------
C This function computes the scalar rho that minimises
C the energy functional on the line going through phi
C and along vector hConj.
C
C     arrays:
C     -------
C     mesh:
C       q(2,nv):coordinates of the mesh vertices,
C       me(3,nt):numbering of the triangles vertices
C     input:
C       gradE(nv):gradient, at each node of the mesh,
C             of functional E,
C       phi(nv):array containing the approximate values
C          of the solution at each vertex of the mesh,
```

```
C          hConj(nv):descent direction.
C        other array:
C          hConjm(nv):zero array (use of function E to
C            compute the denominator of rho)
C
C
         DIMENSION q(2,nv),me(3,nt)
          DIMENSION phi(nv),gradE(nv),hConj(nv),hConjm(nv)
C
         DEhConj=0.
         DO 1 i=1,nv
           hConjm(i)=0.
1        CONTINUE
         DO 2 i=1,nv
           DEhConj=DEhConj+gradE(i)*hConj(i)
2        CONTINUE
C
         rho = -0.5*DEhConj/E(q,me,hConj,hConjm,nv,nt)
C
         RETURN
         END
```

Remark 2.26 If Algorithm 1 is used for *all nodes* of the mesh, we must take care to initialise Φ correctly, that is to choose a Φ^0 satisfying the Dirichlet boundary conditions. In this case, since G^m is always chosen to be zero at all boundary nodes, the solution at each step m always satisfies the right boundary conditions. □

At the end of this chapter, we shall give numerical results pertaining to this method; before that we shall propose a second algorithm for solving (2.48) that is little different at the programming level, but that will prove to be much more efficient than the one above.

2.5.6 *The conjugate gradient method*

The conjugate gradient method is an improvement on the previous method, mostly because the choice of the descent direction is now optimised. Indeed, in the previous method, the gradients at two successive iterations are orthogonal, whereas in the conjugate gradient method, the descent direction at iteration m is orthogonal to all the previous ones. In particular, this implies that the method converges in N_h iterations at most, since the gradients thus constructed are linearly independent.

We give the algorithm in the framework of our model optimisation problem (2.48).

Algorithm 2 (conjugate gradient method):

0 choose accuracy ϵ, the maximum number of iterations *mMax* (as a precaution), choose an initialisation $\Phi^0 = (\varphi_1^0, \dots, \varphi_{N_h}^0)$ for Φ, set $m = 0$, $H^{-1} = (H_1^{-1}, \dots, H_{N_h}^1)$, with $H_k^{-1} = 0, \forall k \in \{1, \dots, N_h\}$ and define $l^{-1} = 1$;

1 compute, for all indices $k \in \{1, \ldots, N_h\}$,

$$G_k^m = \frac{\partial E}{\partial \varphi_k}(\Phi^m), \quad \Phi^m = (\varphi_1^m, \ldots, \varphi_{N_h}^m);$$

set $G^m = (G_1^m, ..., G_{N_h}^m)$ and define

$$l^m = ||G^m||_{\mathbb{R}^{N_h}}^2 = \sum_{k=1}^{N_h} (G_k^m)^2;$$

2 compute $H^m = -G^m + \gamma H^{m-1}$, where $H^m = (H_1^m, ..., H_{N_h}^m)$, with $\gamma = \frac{l^m}{l^{m-1}}$;

3 set $\Phi(\rho) = \Phi^m + \rho H^m$, then compute the scalar ρ^m that minimises the function $E(\Phi(\rho))$ over $\mathbb{R}$;

4 set $\Phi^{m+1} = \Phi^m(\rho^m)$, increment m by 1 and go back to step 1 until $l^m < \epsilon$ or $m > mMax$ (in this case, write that the program does not work).

The minimal value ρ^m is computed as in the case of the optimal step gradient algorithm ((2.77), (2.78)), and we obtain, still with the same notation as in (2.80),

$$\rho^m = -\frac{\sum_{k=1}^{N_h} G_k^m H_k^m}{\sum_{k,l=1}^{N_h} \frac{\partial^2 E_h}{\partial \varphi_k \partial \varphi_l}(\Phi^m) H_k^m H_l^m} = -\frac{(G^m, H^m)_{\mathbb{R}^{N_h}}}{||\nabla H^m||_{L^2(\Omega_h)}^2}, \tag{2.81}$$

where the symbol $(.,.)_{\mathbb{R}^{N_h}}$ denotes the inner product associated with the Euclidean norm in $\mathbb{R}^{N_h}$.

The value of the parameter γ ensures orthogonality, in the inner product defined by A, between two successive descent directions, (i.e. $(h^m)^T A h^{m-1} = 0$). We shall come back to this point in Chapter 4, where we shall recall the general properties of the conjugate gradient algorithm.

Remark 2.27 This algorithm is very efficient for minimising a continuously differentiable functional bounded from below. For more details, we refer to Polak (1971), Strang (1986) and Ciarlet (1989), for example. Vector H^m is called the *conjugate direction* to that of G at iteration m. Let us remark that it is not necessary to store all vectors H^m: at one iteration, only the previous value H^{m-1}, as well as the value of H^m, need to be known. Two arrays *hConj*, *hConjm* are thus sufficient: the first one allows us to store values of H^m, the second one the values of H^{m-1}; at the end of iteration m, the values of array *hConj* are written to array *hConjm*. □

We give the modifications required by the previous algorithm, in order to take into account *all the nodes* of the mesh. We still denote by $\Phi = (\varphi_1, \ldots, \varphi_{n_v})$ the 'unknown' of the problem, knowing that actually values of this vector are perfectly well known at those nodes of the mesh that are located on the boundary of the domain.

Modified Algorithm 2 (conjugate gradient method on all nodes of the mesh)

0 choose accuracy ϵ, the maximum number of iterations *mMax* (as a precaution), choose an initialisation $\Phi^0 = (\varphi_1^0, \ldots, \varphi_{n_v}^0)$ for Φ *satisfying the Dirichlet boundary conditions*, set $m = 0$, $H^{-1} = (H_1^{-1}, \ldots, H_{n_v}^{-1})$, with $H_k^{-1} = 0, \forall k \in \{1, \ldots, n_v\}$ and define $l^{-1} = 1$;

1 compute, for all indices $k \in \{1, \ldots, N_h\}$,

$$G_k^m = \frac{\partial E}{\partial \varphi_k}(\Phi^m), \quad \Phi^m = (\varphi_1^m, \ldots, \varphi_{n_v}^m);$$

$G_k^m = 0, \forall k \in \{N_h + 1, \ldots, n_v\}$, then $G^m = (G_1^m, \ldots, G_{n_v}^m)$ and define

$$l^m = ||G^m||_{\mathbb{R}^{n_v}}^2 = \sum_{k=1}^{n_v} (G_k^m)^2;$$

2 compute $H^m = -G^m + \gamma H^{m-1}$, where $H^m = (H_1^m, \ldots, H_{n_v}^m)$, with $\gamma = \frac{l^m}{l^{m-1}}$;

3 set $\Phi(\rho) = \Phi^m + \rho H^m$, then compute the scalar ρ^m that minimises over $\mathbb{R}$ the function $E(\Phi(\rho))$;

4 set $\Phi^{m+1} = \Phi^m(\rho^m)$, increase m by 1 and go back to step **1** until $l^m < \epsilon$ or $m > mMax$ (in this case, write that the program does not work).

We notice that since the values of G_k^m are zero at the boundary nodes, (i.e. $k \in \{N_h + 1, \ldots, n_v\}$), this is also true for the corresponding values of H_k^m; thus for all m, Φ^m satisfies the Dirichlet boundary condition. Also, one can easily check that the formulae defining γ and ρ^m remain valid, provided one replaces sums over indices $k \in \{1, \ldots, N_h\}$ by the same sums over indices $k \in \{1, \ldots, n_v\}$, which assumes we have computed the k, l element of matrix A, for all $k, l \in \{1, \ldots, n_v\}$. The previous Fortran function 'rhoE' can be used, if array 'hConj' contains the values of vector H^m.

2.5.7 *Some comments on the program in the Appendix*

The data structure is analogous to that defined in Section 2.5.2, but for any triangulation (composed of, at most, *ntMax* triangles and *nvMax* vertices) of any bounded domain in $\mathbb{R}^2$. The program starts by reading, in subroutine 'readT', the mesh data, (i.e. actual number of vertices *nv* and of triangles *nt*, arrays *q*, *me*, *ng* and possibly *ngt*) that have been stored in a file whose name is asked at run time.

The source term f is assumed to be linear on each triangle, the value of f at each mesh vertex is stored in array 'f' of size *nvMax*, and the Dirichlet boundary condition is stored in array 'phi', also of size *nvMax*. These values are assumed to be stored in a file whose name is given by the user at the beginning of the program execution.

Working with *all the nodes* of the mesh has advantages for a possible program vectorisation: indeed, for some 'Do loops' (DO 2 or DO 3), it is not necessary to test whether or not the node is internal.

Let us note that the program computes the area of each triangle several times. To save time, we can store these values once and for all at the beginning of the program, and place them in an array.

2.6 *EFFICIENCY OF THE METHOD*

2.6.1 *A Simple Test Case*

To check the program, it is enough to take a simple example, for which we know the exact solution.

Let us consider the unit square $\Omega = [0,1] \times [0,1]$ and the function defined on this domain by: $\varphi_0(x,y) = \sin(ax)\cos(by)$, where a and b are two real parameters. The $-\Delta\varphi_0 = (a^2+b^2)\varphi_0$, so that if we let $f = (a^2+b^2)\sin(ax)\cos(by)$, then the exact solution of the problem

$$-\Delta\varphi = f \text{ in } \Omega \quad \varphi|_\Gamma = \varphi_0, \tag{2.82}$$

is nothing other than φ_0.

The triangulation of Ω is obtained by decomposing each of the small elementary squares formed after we have partitioned the interval $[0,1]$ into N intervals of the same length (cf. Figure 2.9) into two triangles. The parameter h is here $\sqrt{2}/N$. The numerical results obtained both by Algorithm 1 and by (modified) Algorithm 2

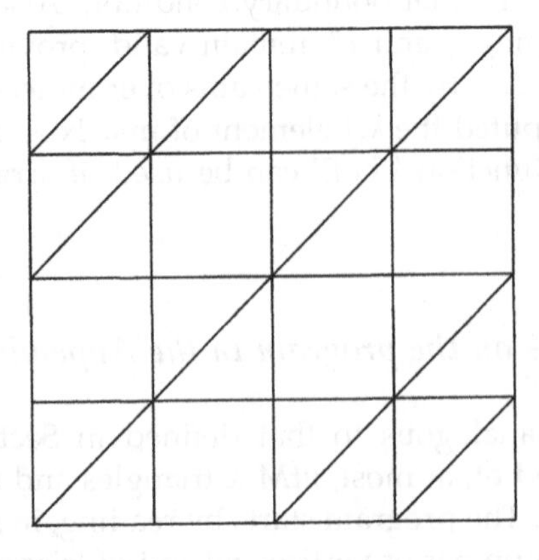

Figure 2.9 Triangulation of the unit square

will be given in Section 2.6.3. Before we do that, we shall give a precise estimate for the error made in the approximation described above.

2.6.2 Error Estimates

Theorem 2.8 gives an estimate for the error between the exact solution φ of (2.8) and that φ_h of (2.25) in the homogeneous (i.e. $g = 0$) case, in the general framework of approximation by a Galerkin method. We shall now make this result explicit in the case of the $P1$ finite element approximation described in Section 2.5. To do this, we need a condition on the angles of the triangulation, which can be expressed in the following way. Let us denote by $h(T)$, the diameter of the circle going through the three vertices of triangle T and by $\delta(T)$, the diameter of the smallest circle contained within the triangle. We consider a family $\mathcal{T}_h$, $h \to 0$ of triangulations indexed by h that is the largest of all sides of the triangles in $\mathcal{T}_h$. We assume henceforth that there exists a positive constant C_T such that

$$\forall h, \quad \sup_{T \in \mathcal{T}_h} \left(\frac{h(T)}{\delta(T)} \right) \leq C_T. \tag{2.83}$$

In practice, this forbids triangulations in which some triangles have too 'obtuse' angles, since we assume $\delta(T)$ cannot go to zero.

Theorem 2.28 If the exact solution φ of the homogeneous model problem (i.e. (2.1)–(2.2) with $g = 0$) is sufficiently smooth, the error between φ and the approximate solution φ_h computed with a $P1$ finite element approximation satisfies, with the same notation as in (2.21):

$$||\varphi - \varphi_h||_1 \leq Ch||\nabla\nabla\varphi||_\infty, \tag{2.84}$$

where C is a positive constant independent of h and of φ, and where $\nabla\nabla\varphi$ is the matrix of second derivatives of φ. We have also used the following notation:

$$||\nabla\nabla\varphi||_\infty = \max\left(\sup_{\vec{x}\in\Omega} \left|\frac{\partial^2}{\partial x^2}\varphi(\vec{x})\right|, \sup_{\vec{x}\in\Omega} \left|\frac{\partial^2}{\partial x \partial y}\varphi(\vec{x})\right|, \sup_{\vec{x}\in\Omega} \left|\frac{\partial^2}{\partial y^2}\varphi(\vec{x})\right| \right). \tag{2.85}$$

To prove this result, we must first introduce the *barycentric coordinates* of a point x in the plane (to simplify writing, $\vec{x}$ is now replaced by x) located within a triangle T_k with vertices q^{i_j}, $j \in \{1, 2, 3\}$, and we assume this triangle is oriented counterclockwise: they are 'local' coordinates of point x with respect to the three vertices of T_k.

Lemma 2.29 Let x be any point in triangle T_k with vertices q^{i_j}, $j \in \{1, 2, 3\}$; there exists a unique triple $(\lambda_1(x), \lambda_2(x), \lambda_3(x))$ of positive real numbers such that:

$$x = \sum_{j=1}^{3} \lambda_j(x)\, q^{i_j}, \quad \text{with :} \quad \sum_{j=1}^{3} \lambda_j(x) = 1. \tag{2.86}$$

These numbers $\lambda_j(x), j \in \{1,2,3\}$ are linear functions of the coordinates of point x and we have, denoting the identity matrix by I:

$$I = \nabla_x x = \sum_{j=1}^{3} \nabla_x \lambda_j(x)\, q^{i_j}; \quad \sum_{j=1}^{3} \nabla_x \lambda_j(x) = 0. \tag{2.87}$$

Also, under hypothesis (2.83), we have:

$$\forall j \in \{1,2,3\},\ |\nabla \lambda_j| \leq \frac{1}{\delta(T_k)}. \tag{2.88}$$

Definition 2.30 With the same notation as in the above lemma, the numbers $(\lambda_1(x), \lambda_2(x), \lambda_3(x))$ defined by (2.86) are called the barycentric coordinates of point x.

PROOF OF LEMMA 2.29 For each fixed x in triangle T_k, relations (2.86) provide a system of three equations in three unknowns $(\lambda_1(x), \lambda_2(x), \lambda_3(x))$, and the matrix of this system is invertible, since it is nothing but the transpose of the matrix of system (2.43). The barycentric coordinates $(\lambda_1(x), \lambda_2(x), \lambda_3(x))$ are then uniquely defined as functions of the coordinates of point x, and furthermore they are linear functions of these coordinates. Also, each of the numbers $\lambda_j(x)$ is non negative, since if x is internal to triangle T_k, each of the three subtriangles formed with point x and the two vertices of triangle T_k (for instance the triangle with vertices x, q^{i_2}, q^{i_3} followed along x, q^{i_2} then q^{i_3}) is positively oriented. From the relation $\nabla_x \lambda_j(x) = 0$, it follows that each of the numbers $\lambda_i(x)$ lies between 0 and 1.

Relation (2.87) follows simply from (2.86) by differentiation. We shall now prove an estimate on the gradients of the barycentric coordinates. On each triangle T_k, since the barycentric coordinates $\lambda_j(x), j \in \{1,2,3\}$ are linear functions of the coordinates of point x, their gradients $\nabla_x \lambda_j$ are constants. For all points x and y in triangle T_k, we then have $\forall j \in \{1,2,3\}$, $\nabla_x \lambda_j \cdot (x-y) = \lambda_j(x) - \lambda_j(y) \in [-1,1]$ from what is above, so that:

$$|\nabla_x \lambda_j \cdot (x-y)| \leq 1.$$

However, if we denote by S the unit sphere in $\mathbb{R}^2$, we have:

$$|\nabla_x \lambda_j| = \max_{u \in S} |\nabla \lambda_j \cdot u|,$$

and for all $u \in S$, there exists two points x and y in triangle T_k such that $x - y = \delta(T_k) u$; we deduce that

$$\forall j \in \{1,2,3\},\ |\nabla \lambda_j| \le \max_{x,y \in T_k} \frac{|\nabla \lambda_j \cdot (x-y)|}{\delta(T_k)} \le \frac{1}{\delta(T_k)},$$

which proves relation (2.88) and ends the proof of the lemma. □

We shall now prove Theorem 2.28.

PROOF OF THEOREM 2.28 According to estimate (2.34), it is enough to show that we have the estimate

$$||\varphi - \psi_h||_1 \le Ch||\nabla\nabla\varphi||_\infty, \tag{2.89}$$

for a particular choice of the function ψ_h in the space H^1_{0h}. We shall see that the 'right choice' is the P^1 interpolant, at each of the internal mesh nodes, of the exact solution φ.

The proof is in two steps: a first step where we give an L^2 estimate of the error $\varphi - \psi_h$, and a second step where we estimate its gradient.

Step 1: L^2 estimates.
Let us write a Taylor expansion of function φ at point x in triangle T_k with vertices $q^{i_j}, j \in \{1,2,3\}$; for any $j \in \{1,2,3\}$, there exists a point $\xi_j \in [x, q^{i_j}]$ such that

$$\varphi(q^{i_j}) = \varphi(x) + \nabla_x\varphi(x) \cdot (q^{i_j} - x) + \tfrac{1}{2}(q^{i_j} - x)^T \nabla\nabla\varphi(\xi_j)(q^{i_j} - x). \tag{2.90}$$

With the same notation as in Lemma 2.29, let us then form the following linear combination:

$$\begin{aligned}\sum_{j=1}^{3} \lambda_j(x)\varphi(q^{i_j}) &= \varphi(x)\sum_{j=1}^{3}\lambda_j(x) + \nabla\varphi(x) \cdot \sum_{j=1}^{3}\lambda_j(x)(q^{i_j} - x)\\ &+ \tfrac{1}{2}\sum_{j=1}^{3}\lambda_j(x)(q^{i_j} - x)^T\nabla\nabla\varphi(\xi_j)(q^{i_j} - x).\end{aligned} \tag{2.91}$$

By virtue of (2.86) we rewrite this relation as:

$$\sum_{j=1}^{3} \lambda_j(x)\varphi(q^{i_j}) = \varphi(x) + \tfrac{1}{2}\sum_{j=1}^{3}\lambda_j(x)(q^{i_j} - x)^T\nabla\nabla\varphi(\xi_j)(q^{i_j} - x). \tag{2.92}$$

Thus we obtain the estimate

$$\begin{aligned}\left|\sum_{j=1}^{3} \lambda_j(x)\varphi(q^{i_j})\varphi(x)\right| &= \tfrac{1}{2}\left|\sum_{j=1}^{3}\lambda_j(x)(q^{i_j} - x)^T\nabla\nabla\varphi(\xi_j)(q^{i_j} - x)\right|\\ &\le \tfrac{1}{2}h^2 \sup_{\xi \in T_k} ||\nabla\nabla\varphi(\xi)||,\end{aligned} \tag{2.93}$$

since, for $x \in T_k$, $0 \leq \lambda_j(x) \leq 1$, for all $j \in \{1,2,3\}$, and:

$$|q^{i_j} - x| = |\sum_{k=1}^{3} \lambda_k(x)\,(q^{i_j} - q^{i_k})| \leq h(T_k) \sum_{k=1}^{3} \lambda_k(x) = h(T_k) \leq h. \tag{2.94}$$

Let us note that the function ψ_h defined on each triangle T_k by $\psi_h = \sum_{j=1}^{3} \lambda_j\, \varphi(q^{i_j})$ is the P^1 interpolant of φ at each of the nodes of the mesh. Integrating estimate (2.93) on Ω, ($|\Omega|$ is the area of Ω) we then obtain:

$$||\varphi - \psi_h||_{L^2(\Omega)} \leq \tfrac{1}{2} h^2 ||\nabla\nabla\varphi||_\infty |\Omega|^{1/2}. \tag{2.95}$$

Step 2: H^1 estimates.
Let us multiply (2.90) by $\nabla\lambda_j(x)$, then sum over all indices j: we obtain, thanks to (2.87):

$$\begin{aligned}\sum_j \nabla\lambda_j(x)\varphi(q^{i_j}) &= \varphi(x)\sum_j \nabla\lambda_j(x) + \nabla\varphi(x)\cdot \sum_j \nabla\lambda_j(x)(q^{i_j} - x)\\ &\quad + \tfrac{1}{2}\sum_j \nabla\lambda_j(x)(q^{i_j} - x)^T \nabla\nabla\varphi(\xi_j)(q^{i_j} - x)\\ &= \nabla\varphi(x) + \tfrac{1}{2}\sum_j \nabla\lambda_j(x)(q^{i_j} - x)^T \nabla\nabla\varphi(\xi_j)(q^{i_j} - x) \end{aligned} \tag{2.96}$$

Taking into account (2.94), (2.88) and condition (2.83), we deduce the following upper bound, at each point x of triangle T_k

$$|\nabla\varphi(x) - \nabla\psi_h(x)| \leq \tfrac{1}{2} ||\nabla\nabla\varphi||_\infty h^2(T_k)\frac{1}{\delta(T_k)} \leq \frac{C_T}{2} h ||\nabla\nabla\varphi||_\infty, \tag{2.97}$$

which gives, by integration, the required estimate for the gradient of $\varphi - \psi_h$. □

Remark 2.31 We can actually find a more precise estimate for this result, in the form:

$$||\varphi - \varphi_h||_1 \leq Ch ||\varphi||_{H^2(\Omega)}, \tag{2.98}$$

where the space $H^2(\Omega)$ is defined in (2.10), the norm in this space being naturally given by:

$$||\varphi||^2_{H^2(\Omega)} = ||\varphi||^2_{H^1(\Omega)} + \left\|\frac{\partial^2}{\partial x^2}\varphi\right\|^2_{L^2(\Omega)} + \left\|\frac{\partial^2}{\partial x \partial y}\varphi\right\|^2_{L^2(\Omega)} + \left\|\frac{\partial^2}{\partial y^2}\varphi\right\|^2_{L^2(\Omega)}.$$

Also, if the domain Ω is convex, we have:

$$||\varphi - \varphi_h||_{L^2(\Omega)} \leq Ch^2 ||\varphi||_{H^2(\Omega)}. \tag{2.99}$$

The proof of this kind of result can be found in Ciarlet (1978).

Let us also note that in the nonhomogeneous case, we obtain a result analogous to that of Theorem 2.28, g_h being simply the P^1 interpolant of g at each of the boundary nodes of the mesh. □

2.6.3 Numerical results

Figure 2.10 shows a 3D representation of the solution of problem (2.82) with $a = b = 6.28$, computed using the modified Algorithm 2. The array 'phi' is initialised by zero at the internal nodes. We have also plotted the level lines of this solution. The computations were performed with $N = 29$, which corresponds to a regular mesh (cf. figure 2.9) with 600 vertices and 1682 triangles.

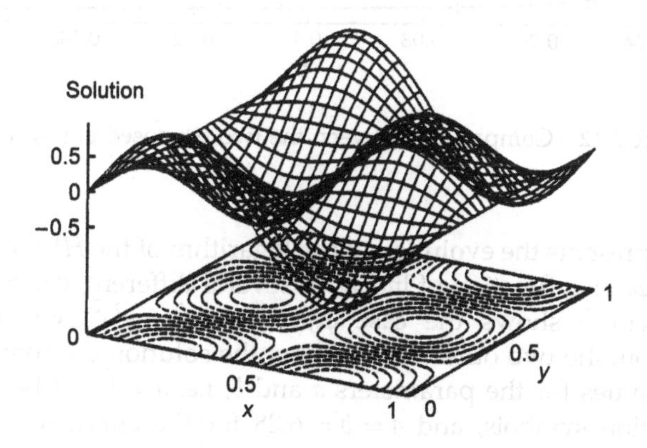

Figure 2.10 3D representation and level lines of the approximate solution

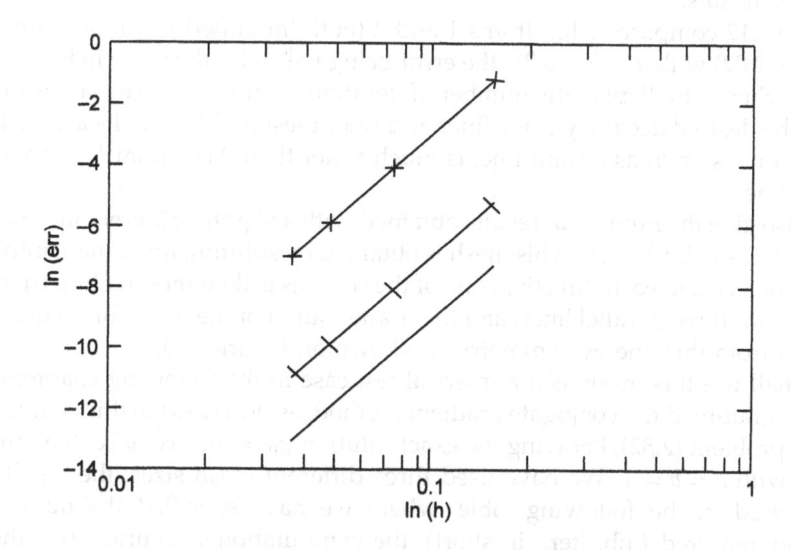

Figure 2.11 H^1 error between the exact solution and the computed solution

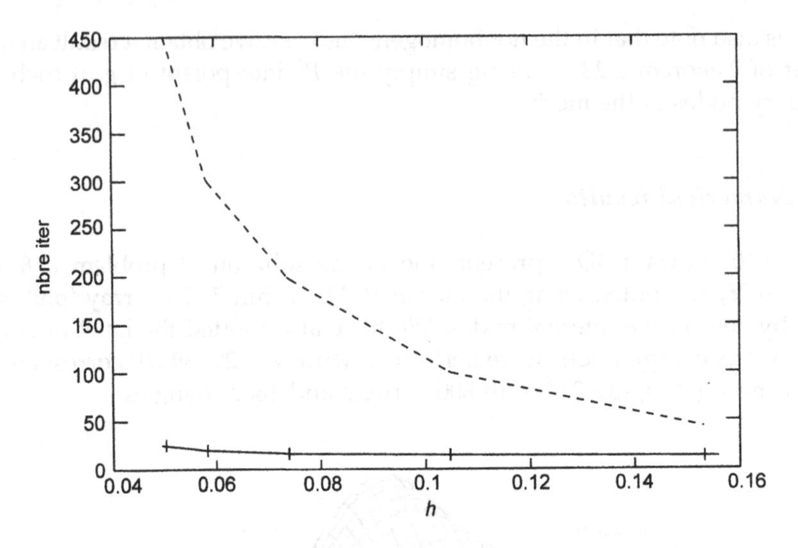

Figure 2.12 Comparison between the two proposed algorithms

Figure 2.11 represents the evolution of the logarithm of the H^1 norm of the error between φ and φ_h as a function of $\ln(h)$, for several different 'exact solutions' φ_0; the solid line curve shows the case $\varphi_0(x,y) = (x^2+y^2)^2$, which is a much smoother solution, the two other curves show the solution φ_0 from Section 2.6.1, with different values for the parameters a and b, i.e. $a = b = 3.14$ for the curves with multiplication symbols, and $a = b = 6.28$ for the curve with the addition symbols. We observe that, in all cases, they are actually straight lines, which confirms the results of Theorem 2.28, and that the 'smoother' the solution, the better the results.

Figure 2.12 compares Algorithms 1 and 2 (both 'modified'), still on numerical example (2.82) with $a = b = 6.28$; the error being naturally the same in both cases, we have chosen to display the number of iterations required by each algorithm to obtain the desired accuracy ε, for finer and finer meshes. The results are striking: Algorithm 2, shown as a solid line, is much faster than Algorithm 1, shown as a dashed line.

We also give the numerical results obtained with a slightly different mesh of the square $\Omega =]-L, L[\times]-L, L[$. This mesh is obtained by splitting this square into four squares of size L in each direction; one of the squares is then meshed into triangles formed with three parallel lines, and the triangulation of the other three squares is deduced from this one by symmetry, as shown on Figure 2.13.

We shall use this mesh as a numerical test case in the following chapters. We have programmed the conjugate gradient method, as described in this chapter, for solving problem (2.82), knowing the exact solution $\varphi_0 = \sin x \, \cos y$ i.e. the same as before, with $a = b = 1$. We have used three different mesh sizes; the results are summarised in the following table, where we have specified the number of iterations required ('nb. iter.' in short), the computational accuracy (i.e. the H^1 error between φ and φ_h), as well as the computer time (CPU time).

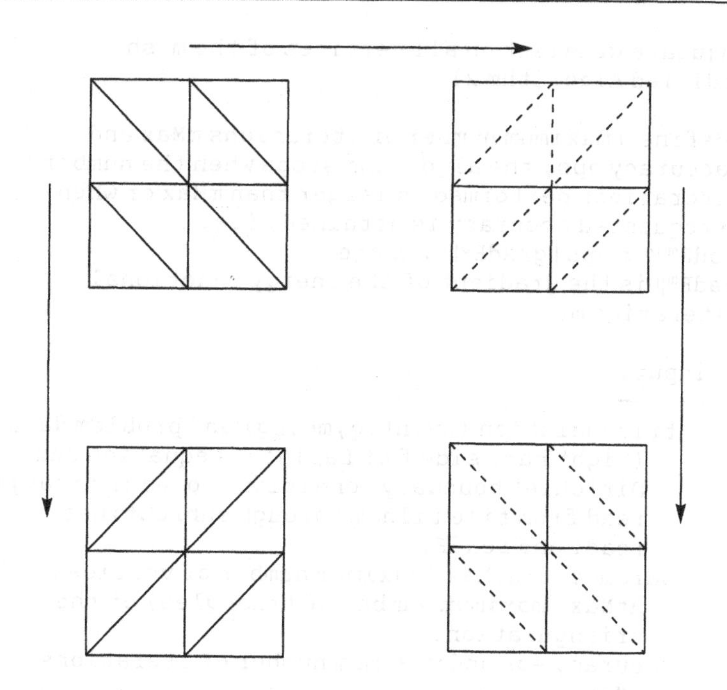

Figure 2.13 Triangulation of the square $]-L, L[\times]-L, L[$

h/L	nv	nt	nb. iter.	CPU time	Error
1/7	49	72	9	7″	3.210^{-2}
1/13	169	288	21	53″	4.810^{-4}
1/25	625	1152	40	420″	8.610^{-5}

APPENDIX 2.A

The Complete Fortran Program for the Solution of the Laplacian with Dirichlet Boundary Conditions by Energy Minimisation

```
C file LplcMinE.for
C
      PROGRAM Laplace
C     ---------------
C
C This program solves the model problem, i.e. the
C Laplacian with Dirichlet boundary conditions:
C -Δφ = f, φ given on the boundary.
C
C The method used is minimisation of the energy by
```

```
C  conjugate gradient on all vertices of the mesh
C  (modified Algorithm 2).
C
C  We define a maximum number of iterations mMax and
C  an accuracy eps; the algorithm stops when the number
C  of iterations performed is larger than mMax or when
C  the requested accuracy is attained, i.e.
C  ||gradE^m||^2 < eps ||gradE^0||^2 , where
C  ||gradE^m|| is the gradient of the energy functional
C  at iteration m.
C
C      input:
C      ------
C        triangulation (nv,nt,q,me,ng) and problem data
C          (right hand side f of Laplace's equation and
C          Dirichlet boundary condition stored in array phi)
C          read from file filnam through subroutines
C          readT and readF;
C        parameter:nvMax (maximum number of vertices and
C          ntMax (maximum number of triangles) of the
C          triangulation.
C        accuracy eps and maximum number of iterations
C          mMax,
C        initialization of arrays hConjm and g2hm.
C
C      output:
C      -------
C        phi:array containing values of the approximate
C          solution at each node of the mesh
C        g2h: norm squared of the gradient of
C          the functional.
C
C      arrays:
C      -------
C      mesh:
C        q(2,nvMax):coordinates of the vertices of the mesh,
C        me(3,ntMax):numbering of the vertices of the
C          triangles,
C        ng(nvMax):arrays indicating whether a vertex is
C          internal (ng(i)=0) or not.
C
C      data:
C        f(nvMax):Laplacian right hand side
C        phi(nvMax):on startup, Dirichlet boundary
C          conditions (later, values of the approximate
C          solution).
C
C      other arrays:
C        gradE(nvMax):gradient, at each node of the
C          mesh, of the functional E,
C        hConj(nvMax):new descent direction,
```

```
C         hConjm(nvMax):old descent direction.
C
C       scalars:
C       --------
C         g2h,g2hm:norm squared of the gradient of the
C           energy functional at current iteration
C           (g2h) and at previous iteration (g2m),
C         gamma:coefficient γ = ||gradE^m||^2 / ||gradE^(m-1)||^2,
C         ro:real number that minimises energy on the line
C           through phi and along hConj
C
C       files:
C       ------
C         filnam:name of file for storing the mesh,
C           the model problem data (Laplacian right hand side
C           and Dirichlet boundary condition), or the
C           approximate solution, for instance, for future
C           graphical exploitation.
C
C       subroutines:
C       ------------
C         readT:reads the mesh
C         readF:reads problem data
C         dEdPhi:computes gradient gradE and Euclidean
C           norm squared g2h of gradient of E,
C         writeF:writes approximate solution
C
C       functions:
C       ----------
C         rho:computes ρ,
C         E:computes energy
C         deter:computes a certain determinant
C
C       library (subroutines and functions):
C       ------------------------------------
C         minEe.for:file containing function E,
C         minEdE.for:file containing subroutine
C           dEdPhi,
C         minErho.for:file containing function rho,
C         utilP1.for:file containing function deter,
C           subroutines readT,readF and writeF.
C
C
      PARAMETER (nvMax=2000, ntMax=4000)
      DIMENSION q(2,nvMax),me(3,ntMax),ng(nvMax)
      DIMENSION phi(nvMax),f(nvMax)
      DIMENSION gradE(nvMax),hConj(nvMax),hConjm(nvMax)
      CHARACTER*20 filnam
C
      DATA eps/1.0e-14/
```

```
      DATA mMax/300/
C
      DATA hConjm/nvMax*0./
      g2hm=1.0
C
      write(*,*) 'name of file where mesh is stored'
      read(*,'(a)') filnam
      CALL readT(filnam,nv,nt,q,ng,me,ngt)
      write(*,*) 'name of file where f is stored'
      write(*,*) '(Laplacian right hand side)'
      read(*,'(a)') filnam
      CALL readF(filnam,nv,f)
      write(*,*) 'name of file where phi0 is stored'
      write(*,*) '(Dirichlet boundary conditions)'
      read(*,'(a)') filnam
      CALL readF(filnam,nv,phi)

      write(*,*)'iter energy             ro       g2h'
C
      DO 1 m = 1,mMax
        CALL dEdPhi(q,me,ng,phi,f,nv,nt,gradE,g2h)
        IF(m.EQ.1) eps = eps*g2h
        gamma=g2h/g2hm
        DO 2 i=1,nv
          hConj(i)=-gradE(i)+gamma*hConjm(i)
2       CONTINUE
        ro = rho(q,me,nv,nt,phi,gradE,hConj,hConjm)
        E1 = E(q,me,phi,f,nv,nt)
        write(*,*) m,E1,ro,g2h
        IF(g2h.LT.eps) GOTO 4
        DO 3 i = 1,nv
          phi(i) = phi(i)+ ro* hConj(i)
          hConjm(i)=hConj(i)
3       CONTINUE
        g2hm=g2h
1     CONTINUE
C
4     CONTINUE
C
      write(*,*) 'name of file for storing the '
      write(*,*) 'approximate solution'
      read(*,'(a)') filnam
      CALL writeF(filnam,nv,phi)
      PAUSE
      END
C
C**********************************
C
C     LIBRARY
C
C**********************************
```

```
C
C
      INCLUDE 'minEe.for'
      INCLUDE 'minEdE.for'
      INCLUDE 'minErho.for'
      INCLUDE 'utilP1.for'
```

We give the full contents of the file 'utilP1.for', even though some subroutines are not used for the moment ('getBand' and 'verif1' for instance).

```
C  file: utilP1.for
C
C---------------------------------------------------
C  SUBROUTINE writeT(filnam,nv,nt,q,ng,me,ngt)
C---------------------------------------------------
C
C  Writes mesh to file filnam
C
      CHARACTER*20 filnam
      DIMENSION q(2,nv),me(3,nt),ng(nv)
C
      OPEN(UNIT=13, file=filnam)
      ngt=0
      write(13,*) nv,nt
      DO 1 i=1,nv
        write(13,*) q(1,i),q(2,i),ng(i)
1     CONTINUE
      DO 2 i=1,nt
        write(13,*) me(1,i),me(2,i),me(3,i),ngt
2     CONTINUE
      CLOSE(13)
C
      END
C
C---------------------------------------------------
      SUBROUTINE writeF(filnam,nv,tab)
C---------------------------------------------------
C
C  Writes array tab to file filnam
C
C
C
      CHARACTER*20 filnam
      DIMENSION tab(nv)
C
      OPEN(UNIT=13, file=filnam)
      write(13,*) nv
      DO 1 i=1,nv
        write(13,*) tab(i)
1     CONTINUE
      CLOSE(13)
```

```
C
      END
C
C-----------------------------------------------------
      SUBROUTINE readT(filnam,nv,nt,q,ng,me,ngt)
C-----------------------------------------------------
C
C This subroutine reads the triangulation stored into
C   file filnam.
C If necessary, array ngt contains the index of the region
C where a triangle lies, and must then be declared as an
C array in main program;
C otherwise, ngt can just be a scalar (ngt=0).
C
C
      CHARACTER*20 filnam
      DIMENSION q(2,nv),me(3,nt),ng(nv),ngt(1)
C
      OPEN(UNIT=13,file=filnam,STATUS='OLD')
      read(13,*) nv,nt
      DO 1 i=1,nv
        read(13,*) q(1,i),q(2,i),ng(i)
1     CONTINUE
      DO 2 i=1,nt
        read(13,*) me(1,i), me(2,i), me(3,i), ngti
        IF (ngti.NE.0)ngt(i)=ngti
2     CONTINUE
      CLOSE(13)
C
      END
C
C-----------------------------------------------------
      SUBROUTINE readF(filnam,nv,tab)
C-----------------------------------------------------
C
C This subroutine reads the content of array tab
C of size nv, which is stored in file filnam.
C
C
      CHARACTER*20 filnam
      DIMENSION tab(nv)
C
      OPEN(UNIT=13,file=filnam,STATUS='OLD')
      read(13,*) nv
      DO 1 i=1,nv
        read(13,*) tab(i)
1     CONTINUE
      CLOSE(13)
C
      END
C
```

```
C-----------------------------------------------------
      SUBROUTINE getBand(me,ng,nt,nv,nBand)
C-----------------------------------------------------
C
C  This subroutine computes the bandwidth of the linear
C  system matrix: nBand = max|i - j| where q^i, q^j
C  are vertices of a common triangle.
C
C
      DIMENSION me(3,nt),ng(nv)
      nBand=0
      DO 2 k=1,nt
        DO 2 ir=1,3
          jr=mod(ir,3)+1
          i=me(ir,k)
          j=me(jr,k)
          nB1=abs(j-i)
        IF(nBand.LT.nB1) nBand=nB1
2     CONTINUE
      END
C
C-----------------------------------------------------
      SUBROUTINE verif1(nv,nBand,a)
C-----------------------------------------------------
C  This subroutine checks that the sum of the elements
C  of a line of the matrix is zero.
C  It should be called before applying
C  Dirichlet boundary conditions.
C
      DIMENSION a(*)
      write(*,*) nv,nBand
      DO 1 i=1,nv
        sum1=0.
        jmin=max(1,i-nBand)
        DO 2 j=jmin,i-1
          sum1=sum1 +a((i-1)*nBand+j)
2       CONTINUE
        jmax=min(nv,i+nBand)
        sum2=0.
        DO 3 j=i+1,jmax
          sum2=sum2 +a((j-1)*nBand+i)
3       CONTINUE
          s11=a((i-1)*nBand+i)
        IF((sum1+sum2+s11).NE.0.)
     >  write(*,*)i,j,jmin,jmax,sum1,sum2,s11
1     CONTINUE
      END
C
C -----------------------------------------------------
      FUNCTION errH10(q,me,phi,f,nv,nt)
C-----------------------------------------------------
```

```
C  This function computes the H_0^1 error between the
C  exact solution (stored in array f) and the computed
C  solution (stored in array phi) for a Laplacian with
C  Dirichlet boundary conditions
C       Error = ∫_Ω(|∇(φ - f)|^2)
C
C
       DIMENSION phi(nv),f(nv),me(3,nt),q(2,nv)
C
       E = 0.
       DO 1 k = 1,nt
       i1 = me(1,k)
       i2 = me(2,k)
       i3 = me(3,k)
         err1=phi(i1)-f(i1)
         err2=phi(i2)-f(i2)
         err3=phi(i3)-f(i3)
       a2=deter(q(1,i1),q(1,i2),q(1,i3),
     >   q(2,i1),q(2,i2),q(2,i3))
       gradFx=deter(err1,err2,err3,
     >   q(2,i1),q(2,i2),q(2,i3))/a2
       gradFy=deter(q(1,i1),q(1,i2),q(1,i3),
     >   err1,err2,err3)/a2
1      E = E + abs(a2)*(gradFx*gradFx + gradFy*gradFy)/2.
       errH10=E
C
       END
C
C-----------------------------------------------------
       FUNCTION deter (a,c,e,b,d,f)
C-----------------------------------------------------
C
C  Computes determinant (a,b,1|c,d,1|e,f,1)
C
C
         deter = (a-e)*(d-f) - (b-f)*(c-e)
       END
```

APPENDIX 2.B

The Complete C Program for the Solution of the Laplacian with Dirichlet Boundary Conditions by energy minimisation

```
/* file lplcmin.c */
#define eps 1.0e-14
#define mMax 10

#include <math.h>
#include <stdio.h>
```

```
#include <stdlib.h>
#define min(a,b) (a)>(b) ? (b) : (a)
#define max(a,b) (a)<(b) ? (b) : (a)

typedef struct{ float x,y; }point;
typedef long triangle[3];
typedef struct{ int nv, nt;
  point *q;
  triangle *me;
  int *ngt;
  int* ng;
    }triangulation;

int    readT(char* filnam,triangulation* t);
int    readF(char* filnam,int nv,float* tab);
int    writeF(char* filnam,int nv,float* tab);
/*void getBand(triangulation* t);
void verif1(int nv,int nBand,float* a);
*/float deter (float a,float c,float e,float b,
             float d,float f);
float E(triangulation* t, float* phi, float* f);
float dEdPhi(triangulation* t, float* phi, float* f,
             float* gradE);
float rho(triangulation* t,float* gradE, float* hConj,
             float* hConjm);

/*-----------------------------------------------*/
int readT(char* filnam,triangulation* t)
/*-----------------------------------------------*/
{  int i,a,b,c; FILE *f;
   if(f=fopen(filnam, "r"), !f) return 1;
   fscanf(f,"%d%d\n", &t->nv,&t->nt);
   t->q = (point*)calloc(t->nv,sizeof(point));
   t->me = (triangle*)calloc(t->nt,sizeof(triangle));
   t->ng = (int*)calloc(t->nv,sizeof(int));
   t->ngt = (int*)calloc(t->nt,sizeof(int));
   for(i=0; i<t->nv; i++)
     fscanf(f,"%e%e%d\n",
       &t->q[i].x, &t->q[i].y, &t->ng[i]);
for(i=0; i<t->nt; i++)
   {  fscanf(f,"%d%d%d%d\n",&a,&b,&c, &t->ngt[i]);
      t->me[i][0] = a-1; t->me[i][1] = b-1;
      t->me[i][2] = c-1;
   }
 fclose(f);
 return 0;
}

/*-----------------------------------------------*/
int readF(char* filnam,int nv,float* tab)
/*-----------------------------------------------*/
```

```
/*  Reads array tab from file filnam */
{  int i, nvv; FILE* f;
   if(f=fopen(filnam, "r"), !f) return 1;
   fscanf(f,"%d\n", &nvv);
   if(nv != nvv)
     { printf("dimension error\n"); return 1;}
   for(i=0; i<nv; i++) fscanf(f,"%e\n", &tab[i]);
   fclose(f);
   return 0;
}

/*-----------------------------------------------*/
int writeF(char* filnam,int nv,float* tab)
/*-----------------------------------------------*/
/* Writes array tab to file filnam */
{  int i; FILE* f;
   if(f=fopen(filnam, "w"), !f) return 1;
   fprintf(f,"%d\n", nv);
   for(i=0; i<nv; i++) fprintf(f,"%e\n", tab[i]);
   fclose(f);
   return 0;
}
/*-----------------------------------------------*/
float deter (float a,float c,float e,float b,
                   float d,float f)
/*-----------------------------------------------*/
{
        return (a-e)*(d-f) - (b-f)*(c-e);
}

/*-----------------------------------------------*/
float E(triangulation* t, float* phi, float* f)
/* ----------------------------------------------*/
{
 float xi1, xi2, xi3, yi1, yi2, yi3, phii1, phii2, phii3 ;
 float fk, area2Tk, gradFx, gradFy, ee;
 int k,i1,i2,i3;

   ee = 0.;
   for(k=0;k<t->nt;k++)
   {
    i1 = t->me[k][0]; i2 = t->me[k][1]; i3 = t->me[k][2];
    xi1=t->q[i1].x;    xi2=t->q[i2].x; xi3=t->q[i3].x;
    yi1=t->q[i1].y;    yi2=t->q[i2].y; yi3=t->q[i3].y;
    phii1=phi[i1];   phii2=phi[i2]; phii3=phi[i3];
    fk=(f[i1]+f[i2]+f[i3])/3.;
    area2Tk = deter(xi1,xi2,xi3,yi1,yi2,yi3) ;
    gradFx =deter(phii1,phii2,phii3,yi1,yi2,yi3)/area2Tk;
    gradFy =deter(xi1,xi2,xi3,phii1,phii2,phii3)/area2Tk;
    ee += 0.5*fabs(area2Tk)*((gradFx*gradFx +gradFy*gradFy)/2.
        -fk*(phii1+phii2+phii3)/3.);
```

```
    }
   return ee;
  }
  /*------------------------------------------------------*/
  float dEdPhi(triangulation* t,float* phi,float* f,
     float* gradE)
  /*------------------------------------------------------*/
  {
   float w[3], xi1, xi2, xi3, yi1, yi2, yi3;
   float phii1, phii2, phii3 ;
   float fk, area2Tk, gradFx, gradFy, g2h, gradWix, gradWiy;
   int i,k,j,l,i1,i2,i3;

   for(i=0;i<t->nv;i++) gradE[i] = 0;
    for(k=0;k<t->nt;k++)
    {
      i1 = t->me[k][0]; i2 = t->me[k][1]; i3 = t->me[k][2];
      xi1=t->q[i1].x; xi2=t->q[i2].x; xi3=t->q[i3].x;
      yi1=t->q[i1].y; yi2=t->q[i2].y; yi3=t->q[i3].y;
      phii1=phi[i1];  phii2=phi[i2];  phii3=phi[i3];

      fk=(f[i1]+f[i2]+f[i3])/3.;
      area2Tk = deter(xi1,xi2,xi3,yi1,yi2,yi3) ;

      gradFx = deter(phii1,phii2,phii3,yi1,yi2,yi3) area2Tk;
      gradFy = deter(xi1,xi2,xi3,phii1,phii2,phii3) / area2Tk;

      for(j=0;j<3;j++)
      {
        i = t->me[k][j];
        if ( t->ng[i]== 0 )
        {
          for(l=0;l<3;l++) w[l] = 0.0;
          w[j] = 1.0;
          gradWix = deter(w[0],w[1],w[2],yi1,yi2,yi3)/
    >           area2Tk;
          gradWiy = deter(xi1,xi2,xi3,w[0],w[1],w[2]) /
    >           area2Tk;
          gradE[i] += (gradFx*gradWix + gradFy*gradWiy
                           - fk/3) *fabs(area2Tk)/2;
          }
      }
    }
   g2h=0.0; for(i=0;i<t->nv;i++) g2h += gradE[i] * gradE[i];
   return g2h;
}
/*----------------------------------------------*/
float rho(triangulation* t,float* gradE, float* hConj,
            float* hConjm)
/*----------------------------------------------*/
{
```

```
   float rh, DEhConj=0.;
   int i;
    for(i=0; i<t->nv;i++)
    {
    hConjm[i]=0.;
    DEhConj += gradE[i]*hConj[i];
    }
   rh = -0.5*DEhConj/E(t, hConj,hConjm) ;
   return rh;
}

/*------------------------------------------*/
void main()
/*------------------------------------------*/
{
int i,m;
float g2h, eps1,gamma, g2hm=1, ro, E1;
float *phi, *f, *gradE, *hConj, *hConjm;
triangulation t;
char* filnam = calloc(20, sizeof(char));

printf("name of mesh file ");
scanf("%s",filnam);
readT(filnam,&t);

   f = (float*)calloc(t.nv,sizeof(float));
   phi = (float*)calloc(t.nv,sizeof(float));
   gradE = (float*)calloc(t.nv,sizeof(float));
   hConj = (float*)calloc(t.nv,sizeof(float));
   hConjm = (float*)calloc(t.nv,sizeof(float));

printf("name of file for f ");
scanf("%s",filnam);
readF(filnam,t.nv,f);
printf("name fo file for phi ");
scanf("%s",filnam);
readF(filnam,t.nv,phi);

printf("iter energy            ro        g2h\n");
for(m=0; m<mMax; m++)
{
g2h = dEdPhi(&t,phi,f,gradE);
      if(m==0) eps1 = eps*g2h;
        gamma = g2h / g2hm;
        for(i=0;i<t.nv;i++)
          hConj[i] =-gradE[i] + gamma*hConjm[i];
        ro = rho(&t,gradE,hConj,hConjm);
        E1 = E(&t,phi,f);
        printf("%d%e%e%e\n",m,E1,ro,g2h);
        if(g2h<eps1) break;
        for(i=0;i<t.nv;i++)
```

```
        {
          phi[i] += ro* hConj[i];
          hConjm[i] = hConj[i];
}
        g2hm = g2h;
    }
  printf("name of file for storing the approximate solution\n");
  scanf("%s", filnam);
  writeF(filnam,t.nv,phi) ;
  printf("end\n");
}
```

3 Finite Element Method: Variational Formulation and Direct Methods

Summary

In the previous chapter, we have proposed a program for solving the model problem, that is the Laplacian with Dirichlet boundary conditions on a bounded domain Ω in $\mathbb{R}^2$, using triangular finite elements. Even though it was very simple, this program was somewhat costly in computer time, because numerous operations were repeated several times, instead of having their results stored once and for all in the computer memory. We shall now propose a different method that reduces to the solution of a linear system whose matrix will be stored. As this matrix is very large (of the order of the square of the number of mesh vertices), a clever way of storing it is necessary in order to save on memory; we shall propose several different storage formats towards this aim, including band storage and CSR storage.

3.1 *STATEMENT OF THE PROBLEM*

Let us recall the data of the model problem, as well as the method we used in the previous chapter to solve it numerically. Given a bounded domain Ω in $\mathbb{R}^2$, with sufficiently smooth boundary Γ, our goal is to solve for the unknown φ, the following problem

$$-\Delta\varphi(\vec{x}) = f(\vec{x}), \quad \forall \vec{x} \in \Omega, \tag{3.1}$$

$$\varphi(\vec{x}) = g(\vec{x}), \quad \forall \vec{x} \in \Gamma, \tag{3.2}$$

where the source term f is a given square integrable function on Ω and where the Dirichlet boundary data g is a given function on Γ, assumed slightly smoother than simply square integrable ($g \in H^{1/2}(\Gamma)$, cf. Remark 2.3). We have seen (cf. Theorems 2.1 and 2.2) that, modulo some smoothness assumptions on φ, problem (3.1)–(3.2) was equivalent to the following minimisation problem

$$\min_{\varphi \in H^1_g(\Omega)} E(\varphi), \tag{3.3}$$

the energy functional E being given by:

$$E(\varphi) = \frac{1}{2}\int_\Omega |\nabla\varphi(\vec{x})|^2 \mathrm{d}\vec{x} - \int_\Omega f(\vec{x})\varphi(\vec{x})\,\mathrm{d}\vec{x}. \tag{3.4}$$

This functional E acts on the space $H^1_g(\Omega)$, defined in the general case (i.e. whether $g = 0$ or not) by

$$H^1_g(\Omega) = \{w : \Omega \to \mathbb{R}, \quad w \in H^1(\Omega), \text{ and } \quad \forall \vec{x} \in \Gamma, w(\vec{x}) = g(\vec{x})\}, \tag{3.5}$$

with:

$$H^1(\Omega) = \{w \in L^2(\Omega), \quad \frac{\partial w}{\partial x}, \frac{\partial w}{\partial y} \in L^2(\Omega)\}. \tag{3.6}$$

We have then discretised this problem, by defining a family $(\mathcal{T}_h)_h, h \to 0$ of admissible triangulations of the domain Ω, assumed to have a polygonal boundary, then by solving (3.3) in a finite dimensional affine subspace H^1_{gh} of $H^1_g(\Omega)$ defined by

$$H^1_{gh} = \{\varphi_h \in H^1_h \text{ and: } \forall q^i \text{ vertex of } \Gamma_h, \quad \varphi_h(q^i) = g(q^i)\}, \tag{3.7}$$

where: $H^1_h = \{\varphi_h : \Omega_h = \cup_{k=1}^{n_t} T_k \to \mathbb{R}, \varphi_h$ continuous on Ω_h, and

$$\forall k \in \{1, ..., n_t\}, \varphi_{h|T_k} \text{ affine function}\}. \tag{3.8}$$

We have then shown that functions in H^1_h were completely determined by their values at each of the n_v vertices q^j of the mesh (cf. Proposition 2.14), then we have constructed a basis $w^i, i \in \{1, ..., n_v\}$ of this space (cf. Proposition 2.17) by letting

$$\forall i \in \{1, ..., n_v\}, w_i \in H^1_h, \quad w^i(q^j) = \delta_{ij} = 1 \text{ if } j = i, 0 \text{ otherwise.} \tag{3.9}$$

Denoting I_Ω the set of indices of the internal nodes of the mesh, we have then sought the solution φ_h of the discrete problem

$$\min_{\varphi_h \in H^1_{gh}} E(\varphi_h) \tag{3.10}$$

in the form

$$\varphi_h(\vec{x}) = \sum_{j \in I_\Omega} \varphi_j w^j(\vec{x}) + \sum_{j \notin I_\Omega} g(q^j) w^j(\vec{x}), \tag{3.11}$$

thus reducing the problem to the solution of a simple optimisation problem in $\mathbb{R}^{N_h}$, N_h being the number of internal nodes of the mesh ($N_h = card(I_\Omega)$), and this problem can be written in the form

$$\min_{\Phi \in \mathbb{R}^{N_h}} E^h(\Phi), \tag{3.12}$$

where

$$\Phi = (\varphi_1, \ldots, \varphi_{N_h}) \in \mathbb{R}^{N_h}, \; E^h(\Phi) = E(\varphi_h), \text{ and } \varphi_h \text{ given by (3.11).} \tag{3.13}$$

The optimisation problem (3.12) was then solved by a conjugate gradient method that was simple and easy to program, yet costly in computer time, since some computations were repeated at each iteration, rather than being stored once and for all; this was the case, for instance, of the computation, at each node i of the mesh of the integral over Ω of fw^i, a computation that is required to evaluate the gradient G_i at this node of the functional.

To alleviate this prohibitive cost, we shall now propose a second method, based on a different formulation of the initial problem (3.1)–(3.2), and this method will lead us to store the results of some computations. To do this, remember that we have shown in the previous chapter that problem (3.1)–(3.2) could be written, with again the same smoothness assumptions, in the following *variational form*:

find $\varphi \in H^1_g(\Omega)$ such that

$$\forall w \in H^1_0(\Omega), \quad \int_\Omega \nabla\varphi(\vec{x}) \cdot \nabla w(\vec{x}) \mathrm{d}\vec{x} = \int_\Omega f(\vec{x}) w(\vec{x}) \mathrm{d}\vec{x}. \tag{3.14}$$

It is precisely this variational form that will be the starting point for the new method that we shall develop in this chapter.

3.2. SOLVING THE DISCRETE VARIATIONAL FORMULATION BY BAND STORAGE OF THE MATRIX

3.2.1 Discrete variational formulation

The procedure for discretising problem (3.14) is the same one that was used for problem (3.3). Given a family $(\mathcal{T}_h)_h$, $h \to 0$, of admissible triangulations of Ω, we set about solving the 'approximate' variational problem:

find $\varphi_h \in H^1_{gh}$ such that

$$\forall w_h \in H^1_{0h}, \quad \int_{\Omega_h} \nabla\varphi_h(\vec{x}) \cdot \nabla w_h(\vec{x}) \mathrm{d}\vec{x} = \int_{\Omega_h} f(\vec{x}) w_h(\vec{x}) \mathrm{d}\vec{x}, \tag{3.15}$$

where the functional spaces are those defined previously, in particular

$$H^1_{0h} = \{\varphi_h \in H^1_h \text{ and: } \forall q^i \text{ vertex of } \Gamma_h, \quad \varphi_h(q^i) = 0\}, \tag{3.16}$$

and where Ω_h is the union of all triangles in $\mathcal{T}_h$.

Thus formulated, and if we substitute expression (3.11) into (3.15), the problem reduces to the solution of a linear system in the unknown $\Phi = (\varphi_i)_{i \in I_\Omega}$, as the next proposition shows.

Proposition 3.1 *Problem (3.15) is equivalent to the solution of the linear system*

$$A\Phi = F, \tag{3.17}$$

where A is the matrix of size $N_h \times N_h$ with entries

$$A_{ij} = \int_{\Omega_h} \nabla w^i(\vec{x}) \cdot \nabla w^j(\vec{x}) d\vec{x}, \quad (i,j) \in I_\Omega^2, \tag{3.18}$$

and where F is the vector of dimension N_h with components

$$F_j = \int_{\Omega_h} (fw^j)(\vec{x}) d\vec{x} - \int_{\Omega_h} \nabla g_h(\vec{x}) \cdot \nabla w^j(\vec{x}) d\vec{x}, \quad j \in I_\Omega, \tag{3.19}$$

with: $g_h(x) = \sum_{i \notin I_\Omega} g_i w^i(x)$, $g_i = g(q^i)$.

PROOF Let us recall that the N_h 'hat' functions $w^i, i \in I_\Omega$, form a basis of the space H^1_{0h}. But, since relation (3.15) is linear in w_h, it holds for all functions $w_h \in H^1_{0h}$ if and only if it holds for all elements in the previous basis, which gives us, by using decomposition (3.11) of the unknown φ_h: $\forall i \in I_\Omega$,

$$\int_{\Omega_h} \Big(\sum_{j \in I_\Omega} \varphi_j \nabla w^j(\vec{x}) \cdot \nabla w^i(\vec{x}) + \sum_{j \notin I_\Omega} g_j \nabla w^j(\vec{x}) \cdot \nabla w^i(\vec{x}) \Big) \, d\vec{x} = \int_{\Omega_h} (fw^i)(\vec{x}) \, d\vec{x}.$$

The linear system thus obtained has precisely equation (3.17) as its matrix form, which ends the proof, since the converse is immediate. □

Remark 3.2 System (3.17) is uniquely solvable, since the matrix A is positive definite; indeed, this property was proved in Theorem 2.6. □

Remark 3.3 It is possible to simplify the computation of the right hand side of (3.19), by using a trick that consists in 'enlarging' the system and solving it on *all nodes of the mesh*, and not only on the internal nodes. Let us actually assume we have constructed the matrix A for all nodes of the mesh, while respecting definition (3.18) for the internal nodes, and that we solve the following system in the unknown $\tilde{\Phi} = (\varphi_1, ..., \varphi_{n_v})$,

$$A\tilde{\Phi} = \tilde{F}, \tag{3.20}$$

where the right hand side $\tilde{F} = (F_1, ..., F_{n_v})$ is simply defined by:

$$F_i = \int_{\Omega_h} (fw^i)(\vec{x}) d\vec{x}. \tag{3.21}$$

To simplify notation in this chapter, we prefer to keep notation A rather than $\tilde{A}$ to denote the matrix of the new system thus formed; we will denote by A this new matrix or the old one, the context allowing us to distinguish between them without ambiguity. The solution $\tilde{\Phi}$ of this system will only satisfy (3.17) if $\forall i \notin I_\Omega$, $\varphi_i = g(q^i)$, and this condition will be approximately fulfilled by replacing, for each $i \notin I_\Omega$, the diagonal element of the matrix A_{ii} by a very large constant e and replacing simultaneously the right hand side F_i by $eg(q^i)$. Indeed, the ith line of system (3.20) is then written, for $i \notin I_\Omega$:

$$e\varphi^i + \sum_{j \neq i} A_{ij}\varphi^j = eg(q^i) \quad \text{i.e.} \quad \varphi^i = g(q^i) + O(1/e). \tag{3.22}$$

In the remainder of the book, we shall constantly use this remark and, so as to simplify notation, we shall also suppress the tildes on the unknown $\tilde{\Phi}$ and on the right hand side $\tilde{F}$. In practice, matrix A will be defined by relation (3.18), extended to all vertices of the mesh. Last, let us note that if e is a sufficiently large constant, the matrix of the new system remains symmetric and positive definite.

The following subroutine 'RHS' allows us to compute the right hand side F_i defined by (3.21) at each index i node of the mesh. The computation uses quadrature formula (2.71) from the previous chapter; the result is stored in array 'ff'. The areas of the triangle have previously been stored in array 'area', constructed in subroutine 'getArea'.

```
c File: RHS.for
C
C ----------------------------------------------------
            SUBROUTINE getArea(q,me,nt,nv,area)
C ----------------------------------------------------
C This program computes the area of each triangle and
C stores it in array area. It warns if a triangle has
C a negative area.
C
C           input:   triangulation (nv,nt,q,me)
C           ------
C
C

            DIMENSION q(2,nv), me(3,nt),area(nt),qq(2,3)
            DO 1 k=1,nt
              DO 2 il=1,3
                i=me(il,k)
                qq(1,il)=q(1,i)
2               qq(2,il)=q(2,i)
              area(k)=( (qq(1,2)-qq(1,1))*(qq(2,3)- qq(2,1))
     >          - (qq(2,2)-qq(2,1))*(qq(1,3)-
     >          qq(1,1)) )/2
              IF(area(k).LE.0) write(*,*)'Area of triangle ',k,
     >          ' is <=0'
1           CONTINUE
            END
C
C
```

```
C ----------------------------------------------------
              SUBROUTINE RHS(me,f,nv,nt,area,ff)
C ----------------------------------------------------
C
C Computes the right hand side of the linear system
C associated with Laplace's equation and stores the
C result in array ff.
C
C             ff(i) = ∫Ω fw^i
C
C
C
              DIMENSION f(nv),me(3,nt),area(nt),ff(nv)
              DIMENSION inext(4)
              DATA inext/2,3,1,2/
C
              DO 1 j=1,nv
                ff(j)=0.
1             CONTINUE
              DO 2 k = 1,nt
                DO 2 il=1,3
                  i=me(il,k)
                  ip=me(inext(il),k)
                  ipp=me(inext(il+1),k)
                  ff(i)=ff(i)+(2*f(i)+f(ip)+f(ipp))*area(k)/12.
2             CONTINUE
                END
C
```

3.2.2 *Construction and band storage of the linear system matrix*

According to Remark 3.3 we shall compute all A_{ij} coefficients of matrix A after definition (3.18) extended to all indices. Also, the principle driving the computation is exactly the same as in the previous chapter (as it was above for the right hand side computation), that is the computation proceeds by going over the triangles of the mesh (which, on the computer, means a 'loop' on all triangle indices k), rather than by going over the mesh vertices (which would be a much longer computation, since it would correspond to a double 'loop' over all vertices indices i, j). We shall now specify this computation. Let us denote by

$$A_{ij}(T_k) = \int_{T_k} \nabla w^i(\vec{x}) \cdot \nabla w^j(\vec{x}) \mathrm{d}\vec{x}, \tag{3.23}$$

the contribution from each given triangle T_k to element A_{ij} of the system matrix (3.20); we have the following result:

Lemma 3.4 (i) $A_{ij}(T_k) \neq 0$ if and only if q^i and q^j are vertices of triangle T_k ; (ii) $A_{ij} = \sum_{k=1}^{n_t} A_{ij}(T_k) \neq 0$ if there exists a triangle having nodes q^i and q^j as vertices.

PROOF The proof is immediate; it rests on the fact that, on the one hand, the support of function w^i is the union of all triangles having point q^i as a vertex (cf. Figure 2.6 in the previous chapter), and on the other hand, the coefficients $A_{ij}(T_k)$ are zero as soon as the support of functions w^i and w^j are disjoint. □

Eventually, to determine the coefficients A_{ij} of matrix A, it is enough to compute, for each index k, the contributions $A_{ij}(T_k)$ corresponding to vertices q^i and q^j of triangle T_k, then to sum all these contributions for $k \in \{1, ..., n_t\}$. Let us now specify the actual computation of $A_{ij}(T_k)$; the i and j indices are indices of mesh vertices corresponding to "local' indices in triangle T_k, denoted respectively by il and jl. In other words, by using the data structure introduced in section 2.5.2, we have: $i = me(il, k)$, $j = me(jl, k)$. Since functions w^i and w^j are affine functions on T_k, their gradient is constant there, so that according to formulae (2.64) and (2.65) from chapter 2, we have:

$$A_{ij}(T_k) = |T_k|\nabla w^i \nabla w^j$$

$$= \left(\begin{vmatrix} 1 & q_2^{il} & 1 \\ 0 & q_2^{il_+} & 1 \\ 0 & q_2^{il_{++}} & 1 \end{vmatrix} \begin{vmatrix} 1 & q_2^{jl} & 1 \\ 0 & q_2^{jl_+} & 1 \\ 0 & q_2^{jl_{++}} & 1 \end{vmatrix} + \begin{vmatrix} q_1^{il} & 1 & 1 \\ q_1^{il_+} & 0 & 1 \\ q_1^{il_{++}} & 0 & 1 \end{vmatrix} \begin{vmatrix} q_1^{jl} & 1 & 1 \\ q_1^{jl_+} & 0 & 1 \\ q_1^{jl_{++}} & 0 & 1 \end{vmatrix} \right) \frac{1}{4|T_k|}$$

$$= \frac{1}{4|T_k|} [(q_2^{il_+} - q_2^{il_{++}})(q_2^{jl_+} - q_2^{jl_{++}}) - (q_1^{il_+} - q_1^{il_{++}})(q_1^{jl_+} - q_1^{jl_{++}})],$$

where we have set $il_+ = il \bmod 3 + 1$ and $il_{++} = il_+ \bmod 3 + 1$, and where $il \bmod jl$ denotes the remainder in the Euclidean division of il by jl. With this convention, $i_+ = me(il_+, k)$ is the index of the point following the point of index $i = me(il, k)$ in the local orientation of the triangle, and similarly, $i_{++} = me(il_{++}, k)$ is the index of the point following the point of index i_+. Let us note that, in this computation, we have used the invariance of the determinant under circular permutation of its lines, and the fact that the determinant of expression (2.65) is equal to twice the area of triangle T_k, if the later is positively oriented.

Now that we have finished computing the coefficients of matrix A, we set ourselves the problem of its storage on the computer. Obviously, because we are concerned with computer memory, it is out of the question to declare this matrix as a bi-dimensional array; such an array would have a size equal to the square of the number of mesh vertices, which could quickly become very large. Even more, according to Lemma 3.4, matrix A has a *sparse* structure i.e. it has many zero coefficients; we notice it is also symmetric. We shall thus take advantage of this particular structure to try to minimise the storage of A. In this section we describe a storage that is simple to implement, called *band storage* of the matrix; we complete this study in Section 3.4 by that of another storage, more economical in memory, but more complicated to implement.

First, by taking into account the symmetry of the matrix, we only store its lower triangular part, i.e. elements A_{ij} with $i \geq j$. Let us call *bandwidth* the number μ defined by

$$\mu = \max_{\{i,j\,:i>j,\,A_{ij}\neq 0\}} |i-j|; \tag{3.24}$$

we have $A_{ij} = 0$ if $i > \mu + j$. Moreover, on each line of the matrix, there are at most $\mu + 1$ non-zero terms. In order to minimise memory space, we store the matrix A in a 1D array a of size $n_a = (n_v - 1)\mu + n_v$. More precisely, element A_{ij} is stored in array a at index $k = (i-1)\mu + j$ (in the program, we will never actually perform the multiplication by μ, since all computations on the coefficients A_{ij} are done inside 'loops' in which i and j are incremented by 1). This storage format is particularly simple, if we are careful to respect the following bounds on i and j

$$\max\{1, i-\mu\} \le j \le i \le \min\{n_v, j+\mu\}, \tag{3.25}$$

so as never to access an index that overflows the size n_a of array a. We also notice that the space allocated to array a is slightly larger than what it actually needs, since on each line i with $i < \mu + 1$, it is only necessary to store i terms; the memory a *priori* required is thus only $n_a - \mu(\mu - 1)/2$. Respecting the above constraints (3.25) allows us not to access these 'additional' indices, which do not correspond to any stored element of the original matrix A.

Having made these remarks, we can ask the following question: is the mapping $\{i,j\} \to k = (i-1)\mu + j$ a bijection ? The answer is yes. Indeed, for a given k, two values of i (and thus also of the couple (i,j), since j is linked to i by the relation $j = k - (i-1)\mu$) are possible; these i values are given by

$$i = k \text{ div } (\mu + 2) \quad \text{or} \quad i = k \text{ div } (\mu + 1),$$

where i div j means the quotient in the Euclidean division of i by j. But among these two possible values of the couple (i,j), only one satisfies the constraints (3.25), which means that under those constraints, the mapping $\{i,j\} \to k = (i-1)\mu + j$ is actually a bijection. Let us note that in practice, we shall never have to 'reconstruct' the couple (i,j) from index k, as programming always produces indices into a in the form $(i-1)\mu + j$.

Computing the contribution $A_{ij}(T_k)$ is the topic of the following function 'elemMat', whereas constructing and storing matrix A is done in subroutine 'BgetMat'. The latter assumes that the bandwidth μ is known; to compute it, use subroutine getBand from the Appendix of the previous chapter. Last, we give subroutine 'BDirCL', which allows us to recover the right boundary conditions, as explained in Remark 3.3.

Remark 3.5 According to Lemma 3.4, if μ' denotes

$$\mu' = \max_{\{k,i,j:>j,q^i,q^j\in T_k\}} |i-j|,$$

then $\mu < \mu'$. Since μ' is easier to compute than μ, we often use this 'modified bandwidth', and abuse notation by identifying it with the real bandwidth μ. □

```
C File: elem.for
C ---------------------------------------------------
C             FUNCTION elemMat(k,il,jl,areak,q,me)
C ---------------------------------------------------
C Computes the elemental contribution over triangle
C of index k to element of index i,j in matrix A,
C i.e. Aij(Tk), where i and j are the indices
C of the vertices with local indices il and jl in
C the triangle
C (i.e. i=me(il,k), j=me(jl,k)).
C The triangle area is given as input (areak).
C
C
C N.B. This subroutine assumes that the triangles
C    are positively oriented.
C
            DIMENSION q(2,*), me(3,*), inext(4)
            DATA inext/2,3,1,2/
            ip=me(inext(il),k)
            ipp=me(inext(il+1),k)
            jp=me(inext(jl),k)
            jpp=me(inext(jl+1),k)
            elemMat= ( (q(2,ip)-q(2,ipp))*(q(2,jp)- q(2,jpp))
     >      + (q(1,ip)-q(1,ipp))*(q(1,jp)-q(1,jpp)))
     >       /(4*areak)
            END
C
C
C File: BgetMat.for
C
C ---------------------------------------------------
            SUBROUTINE BgetMat(q,me,area,nt,nv,a,nBand)
C ---------------------------------------------------
C
C Computes and stores in band format in 1D array a
C the lower triangular part of the linear system
C matrix A associated with solving the model problem
C (Laplacian with Dirichlet boundary conditions)
C written in variational form. The bandwidth nBand
C is assumed known.
C
C Computing the matrix elements Aij is done for all
C indices i and j, even if they are boundary vertices.
C One must, after calling this subroutine, call
C subroutine BDirCL so as to retrieve the right Dirichlet
C boundary conditions on the domain boundary.
C
C The actual computation of the Aij coefficients is
C done triangle by triangle, by only computing, on
C each triangle Tk, the elemental contributions from
```

```
C two local consecutive indices il and jl of vertices
C in this triangle (then we have: i=me(il,k), j=me(jl,k))
C as well as diagonal elements (jl=il). This follows
C from the symmetry of A.
C For i different from j, the result of this computation
C is stored in array a at index ij=(i-1)*nBand+j or
C ji=(j-1)*nBand+i, depending on whether i>j or j>i.
C
C
C          input:
C          -------
C            triangulation: nv,nt,q(2,nv),me(3,nt) and
C              area(nt);
C            nBand: matrix bandwidth, computed previously
C              by subroutine getBand from file utilP1.for.
C          output:
C          -------
C          a(nMat), nMat=(nv-1)*nBand+nv: array containing
C              the elements of matrix A; more precisely
C              element Aij is stored in array a at index
C              ij=(i-1)*nBand+j.
C
C
C
           DIMENSION q(2,nv), me(3,nt), area(nt),a(*)
C
           nMat=nBand*(nv-1)+nv
           DO 1 i=1,nMat
   1       a(i)=0.0
C
           DO 2 k=1,nt
             DO 2 il=1,3
               i=me(il,k)
               jl=mod(il,3)+1
               j=me(jl,k)
               xmatElem=elemMat(k,il,jl, area(k),q,me)
               IF (i.GT.j)THEN
                 ij=(i-1)*nBand+j
                 a(ij)=a(ij)+xmatElem
               ELSE
                 ji=(j-1)*nBand+i
                 a(ji)=a(ji)+xmatElem
               ENDIF
               ii=(i-1)*nBand+i
               xmatElem= elemMat(k,il,il,area(k),q,me)
               a(ii)=a(ii)+ xmatElem
  2        CONTINUE
           END
C
C
C
```

```
C
C-----------------------------------------------------------
      SUBROUTINE BDirCL(a,phi0,nv,ng,ff,cte, nBand)
C -----------------------------------------------------------
C This subroutine lets us recover the right
C Dirichlet boundary conditions for the boundary
C vertices.
C
C For each index i of a boundary vertex (ng(i)≠0),
C we modify at the same time the ith diagonal
C element of matrix A, replacing it by a very large
C constant cte, and the ith component of the right
C hand side f, replacing it by the product cte*phi0,
C where phi0 is the Dirichlet boundary condition. In
C this way, the ith line of the linear sytem is written,
C after simplifying by the constant cte
C   phi(i) + small terms = phi0(i),
C and the boundary condition is satisfied up to
C order (1/cte) terms.
C
C
C             input:
C             -------
C               triangulation: nv,ng(nv);
C               matrix a(*) and its bandwidth nBand;
C               Dirichlet boundary condition stored in array
C                 phi0 (N.B. this array is assumed filled for
C                 all nodes);
C               linear system right hand side ff(nv);
C               very large constant cte.
C
C             output:
C             -------
C               modified matrix a and right hand side ff.
C
C
      DIMENSION a(*),phi0(nv),ng(nv),ff(nv)
      DO 1 i=1,nv
      IF(ng(i).NE.0)THEN
        a((i-1)*nBand+i)=cte
        ff(i)=cte*phi0(i)
      ENDIF
1     CONTINUE
      END

C
C
C
```

We shall now describe a *direct* method for solving linear system (3.20).

3.2.3 *Solution of the linear system by Choleski factorisation of matrix A*

The solution of linear system (3.20) is in two steps; in the first step we factor the matrix A by the Choleski method, which consists in determining a lower triangular matrix L ($L_{ij} = 0$ if $j > i$) of size $n_v \times n_v$, with strictly positive diagonal elements, such that

$$A = LL^T. \tag{3.26}$$

The theoretical justification for the existence of such a matrix L, A being symmetric and positive definite, can, for instance, be found in Ciarlet (1989). Let us expand relation (3.26); we obtain

$$A_{ij} = \sum_{k \le \min[i,j]} L_{ik} L_{jk}, \tag{3.27}$$

which suggests that we compute L column by column, beginning with the computation of the diagonal element. From a programming point of view, this leads to the following scheme:

$$\text{DO j} = 1, \text{nv}$$

$$L_{jj} = \sqrt{A_{jj} - \sum_{k<j} L_{jk}^2} \tag{3.28}$$

$$\text{DO } i = \text{j} + 1, \text{nv}$$

$$L_{ij} = (A_{ij} - \sum_{k<j} L_{ik} L_{jk}) / L_{jj} \tag{3.29}$$

END i loop

END j loop

The second step consists in decomposing the solution of (3.20) into that of two successive linear systems. The right hand side $F \in R^{n_v}$ being given, we first determine the solution $\psi \in R^{n_v}$ of the linear system

$$L\psi = F, \tag{3.30}$$

then we solve the equation in R^{n_v}:

$$L^T \Phi = \psi. \tag{3.31}$$

Vector Φ is the sought solution since it satisfies $A\Phi = F$. This step is usually called *forward–backward solve*, the forward solve being the one associated with solving the lower triangular system (3.30), whereas the backward solve is linked to the solution of the upper triangular system (3.31).

Let us now detail the structure of the programs solving these two steps. We first note that the Choleski factor L of A also has a band structure, with a bandwidth at most equal to that of A. This enables us to use the same storage structure for L as for A. Also, algorithm (3.28)–(3.29) shows that there is no need to use two arrays l and a to store the values of A and those of L. Indeed, as the computation proceeds, we can 'overwrite' the values of array a (that correspond to storage of an element A_{ij} of A) with the corresponding values of matrix L (i.e. by the element L_{ij} with the same index). Subroutine 'Choleski' below performs the Choleski factorisation of a matrix A, stored in band format in an array a; on input, array a contains the coefficients of A and on output it contains the corresponding coefficients of the Choleski factor of A. As for subroutine 'BSolvSys', it solves the linear system via 'forward-backward' solves. The triangular structure of each of the linear systems (3.30) and (3.31) enables us here again to use for each one, only one array to store both the right hand side of the system to be solved and its solution.

```
C File: BCholes.for
C -----------------------------------------------------
C SUBROUTINE BCholeski(n,a,nBand)
C -----------------------------------------------------
C This subroutine computes the Choleski factor L of
C a symmetric positive definite matrix A of size n*n,
C of which only the lower triangular part is stored in
C band format in an array a of size nMat=(n-1)*nBand+n.
C We assume the bandwidth nBand is known. The Choleski
C factor L has the same band structure as matrix A.
C
C The factorisation is performed column by column,
C beginning with the computation of the diagonal
C element. Matrix L is stored at the same location
C as A in array a
C
C
C             input:
C             -------
C             bandwidth nBand of matrix A(n,n), i.e.
C               nBand = max{|i - j| : i > j,A(i,j) ≠ 0};
C             array a(*) for storing matrix A: A(i, j) is
C               stored in a((i-1)*nBand +j).
C
C
C             N.B. the constraints on indices induced by
C               the band storage must be taken into account:
C               max(1, i - nBand) ≤ j ≤ i ≤ min(n, j + nBand)
C
C
C             output:
C             -------
C               array a(*) contains the elements of matrix L i.e.
C                 a((i-1)*nBand +j) now contains element L(i,j).
```

```
C
C
          DIMENSION a(*)
          DO 1 j=1,n
            sum=0.0
            j1=(j-1)*nBand
            kMin=max(1,j-nBand)
            DO 2 k=kMin,j-1
              jk=j1+k
              sum=sum+a(jk)**2
2           CONTINUE
              jj=j1+j
              a(jj)=sqrt(a(jj)-sum)
C
            iMax=min(n,j+nBand)
            DO 4 i=j+1,iMax
              sum=0.0
              i1=(i-1)*nBand
              kMin=max(1,i-nBand)
              DO 3 k=kMin,j-1
                ik=i1+k
                jk=j1+k
                sum=sum+a(ik)*a(jk)
3             CONTINUE
            ij=i1+j
            a(ij)=(a(ij)-sum)/a(jj)
4           CONTINUE
1         CONTINUE
          END
C
C
C-------------------------------------------------------
C         SUBROUTINE BSolvSys(n,a,g,nBand)
C-------------------------------------------------------
C This subroutine computes the solution x to system
C Ax = g, where A is a symmetric positive definite matrix
C of size n*n factored with the Choleski algorithm, i.e.
C A = LLT, with L a lower triangular matrix with positive
C diagonal.
C Matrix L is stored in band format in array a of length
C nMat=(n-1)*nBand+n, where nBand is the bandwidth
C (assumed previously computed) of matrices A and L.
C
C Computation is in two successive stages:
C - a forward solve stage, where we solve system Ly = g,
C - a back solve step, where we solve system LTx = y.
C It is not necessary to use two additional arrays
C to store x and y; array g(n) represents the system
C right hand side at the beginning, then, as the
C computation proceeds, the solutions y, then x.
C
```

```
C
         DIMENSION a(*),g(n)
         DO 1 i=1,n
           sum=g(i)
           i1=(i-1)*nBand
           jMin=max(1,i-nBand)
           DO 2 j=jMin,i-1
             sum=sum-a(i1+j)*g(j)
2          CONTINUE
           g(i)=sum/a(i1+i)
1        CONTINUE
C
         DO 11 i1=1,n
           i=n-i1+1
           sum=g(i)
           jMax=min(n,i+nBand)
           DO 12 j=i+1,jMax
             j1=(j-1)*nBand
             sum=sum-a(j1+i)*g(j)
12           CONTINUE
           g(i)=sum/a((i-1)*nBand+i)
11       CONTINUE
         END
```

The complete program for solving the model problem with band storage of the matrix obtained by P^1 finite element discretisation of the variational formulation is given in Appendix 3.A.

EXERCISE Modify this program by using the Crout factorisation of A i.e. $A = LDL^T$ with L a unit (diagonal elements equal to 1) lower triangular matrix, and D a diagonal matrix. Store D where the diagonal of L would have been stored. □

3.2.4 *How to Check that the Program Works.*

There are essentially two types of tests we can make to check that a program works: partial tests, and global tests. Let us give an example. Among partial tests, we can for instance notice that the function φ identically equal to 1 on Ω satisfies problem (3.1)–(3.2) with $f = 0$ and $g = 1$, so that the decomposition (3.11) for φ writes:

$$1 = \sum_{j=1}^{n_v} w^j(x), \quad \forall x \in \Omega. \tag{3.32}$$

Thus matrix A has the following property

$$\forall i \in \{1, ..., n_v\}, \quad \sum_{j=1}^{n_v} A_{ij} = 0, \tag{3.33}$$

which can be checked, *a posteriori*, thanks to subroutine "verif1' from file 'utilP1.-for' located in Appendix 2.A. In this subroutine, we have taken into account the fact that only the lower triangular part of A is stored, which means that elements of the form A_{ij}, with $j > i$, are actually in array a at index $k = (j-1) * \mu + i$, corresponding to the pair (j, i).

An example of a global test is to compare the computed solution with the exact solution of problem (3.1)–(3.2), which assumes that we know the latter. To do this we proceed as in the previous chapter: we choose *a priori* a function φ_0, we compute its Laplacian in Ω, and we set $f = -\Delta\varphi_0$. Then we compute, using the above program, the approximate solution φ of problem

$$-\Delta\varphi = f, \quad \varphi = \varphi_0 \text{ on } \Gamma,$$

and estimate the error $\varphi - \varphi_0$ in the H^1_{0h} norm. This error computation is done in subroutine 'errH10', also stored in file 'utilP1.for'. We can also perform the previous computation on finer and finer meshes and check how the error decreases as a function of the discretisation step h, as was done in chapter 2. It is better, for this task, not to take too 'smooth' (i.e. polynomial) a function φ_0, as the numerical results obtained in this way are much better than those predicted by the theory !

3.2.5 *Optimising the Matrix Band Storage*

The smaller the bandwidth μ the better the method from a 'memory' standpoint, but also from a 'computing time' standpoint, as the number of non-zero elements in the Choleski factor of A is proportional to μ. Thus the problem is to optimise the numbering of the nodes, in such a way as to minimise the bandwidth, which is a particularly difficult graph theoretical problem. We can, however, propose the following 'frontal numbering' algorithm; first, we shall introduce a natural definition that we will find useful later.

Definition 3.6 We say that two vertices of the triangulation $\mathcal{T}_h$ are neighbours if they are the end points of a single edge of a triangle in this triangulation.

The Cuthill and Mc Kee Algorithm

0 Choose a vertex and assign it number 1; determine the set S^1 of vertices which are neighbours of vertex q^1; set $m = 1$.

1 Assign a number to each vertex in S^m.

2 For each vertex q^i in S^m, find the set V_i of its neighbouring vertices that have not yet been numbered.

3 Set $S^{m+1} = \cup_{\{i:q^i \in S^m\}} V_i$. If $S^{m+1} \neq \emptyset$, set $m = m+1$ and go back to step **1**.

EXERCISE Program this algorithm (to do this you will have to know, for each vertex in the mesh, the list of its neighbouring vertices. We will build such a list in the next section.) □

Another posible improvement is to define, instead of a global bandwidth, a bandwidth for each row of matrix A. We leave it as an exercise to the reader to define the new storage for A, as well as to program the corresponding Choleski factorisation.

3.3 NUMBERING OF EDGES AND NEIGHBOURING POINTS

We have just seen that the Cuthill and McKee algorithm required the knowledge, for each vertex of the triangulation, of its neighbouring vertices or, equivalently, for each edge of the mesh, the knowledge of its extremal points. Such a direct search appears difficult with just the knowledge of nv, nt and the array *me*. This is why we shall define additional arrays linked to the edges of the mesh.

3.3.1 Data Structure for the Edges

For each edge with index *ie*, we shall store:

— the indices of the two extremal vertices *infEd* and *isupEd*, with the convention $infEd(ie) < isupEd(ie)$

— the indices of the two triangles that share this edge; if we choose to orient the edge from the point with smallest index ($infEd(ie)$) to the point with largest index ($isupEd(ie)$), the index of the triangle to the left of this edge is stored in $iTLeft(ie)$, whereas $iTRight(ie)$ contains the index of the triangle to the right of this edge. If one of these triangles does not exist, which occurs when the edge is on the boundary, then the corresponding value in *iTLeft* or *iTRight* is taken to be zero.

In order to be able to define easily the edges of a given triangle, we also construct a bi-dimensional array $nEdT(3, nt)$ that gives, for each index k triangle, the indices of the three edges of this triangle, with the following convention: the edge with local index $iel \in \{1, 2, 3\}$ is the one opposite the vertex with local index *iel*. The edge with index $ie = nArT(iel, k)$ thus has the vertices $me(iel_{+}, k)$ and $me(iel_{++}, k)$ as its extremities, with a notation we have already met in Section 3.2.2, that is: $i_{+} = i \bmod 3 + 1$, $i_{++} = i_{+} \bmod 3 + 1$.

We recall Euler's formula, which is easy to prove by induction, and which links the number *ne* of edges to the number of vertices (*nv*) and of triangles (*nt*)

$$nt - ne + nv = ncc, \tag{3.34}$$

where *ncc* is the number of connected components of Ω.

After computing the number of edges *ne* by the previous formula, we define five new arrays:

```
DIMENSION infEd(na),isupEd(na), iTLeft(na), iTRight(na), nEdT(3,nt)
```

Even if these arrays are known, constructing the neighbours of a vertex remains an open problem: how can we store the result ? We answer this question in the next section.

3.3.2 *Construction of the Edge Arrays*

We could find all the edges with a loop over the triangles, followed by another loop over the three local indices of each vertex of each triangle:

```
            DO 1 k=1,nt
              DO 1 il=1,3
                i=me(il,k)
1               ip=me( mod(il,3)+1,k)
```

But in such a procedure, each internal edge is counted twice, whereas the boundary edges are counted only once. If we want to store an edge, we must first check that it has not been already stored; this is the main difficulty when programming the storage of the edges.

By convention, the edges are oriented from the vertex with lowest index (in the global numbering of all the edges of the mesh) to the one with highest index. We say the edge with index *ie* 'starts' at the vertex with index *infEd*(*ie*). We shall use a list structure, with the help of two arrays list and listHead, to store the indices of the edges starting at a given vertex. For each index *i*, listHead(*i*) holds the highest index of any edge starting at the point with index *i*; the edge starting at the point with index *i* that has the largest index among the remaining edges has its index stored in list(listHead(*i*)), and so on until we find a zero index, in which case there are no other edges starting from this vertex. It is then possible to find all edges starting from the vertex with index *i*, by going over the array in the following way:

```
            ie = listHead(i)
1           IF( ie.GT.0 ) THEN
C
C [infEd(ie),isupEd(ie)] is an edge
C
              ie = list(ie)
C
C ie is the index of the next edge
C
              GOTO 1
            ENDIF
```

The following program gives the implementation of the method to build the various edge arrays referred to in paragraph 3.3.1. Then we apply it to the tiny triangulation example of chapter 2.

```
C File: Edge.for
C
```

```
C -----------------------------------------------------
      SUBROUTINE getEdge(nv,nt,me,list, listHead,
     >      iTLeft, iTRight,infEd, isupEd,nEdT,ne)
C -----------------------------------------------------
C This subroutine numbers edges and builds the
C following arrays: iTLeft,iTRight, infEd,isupEd,nEdT.
C It computes the total number ne of edges.
C This construction uses auxiliary address arrays:
C list and listHead.
C
C           input: triangulation
C           ------
C           nv:   number of vertices,
C           nt:   number of triangles,
C           me(3,nt): local vertex numbering on
C             each triangle (me(il,k))=index of
C             il th vertex of index k triangle);
C
C           N.B. The triangles are assumed to be positively
C                oriented. Array ng is not required here.
C
C           output:
C           -------
C           ne: total number of edges
C           infEd(ne), isupEd(ne): indices of extremal
C             points of edge with the convention:
C             infEd(ie) < isupEd(ie);
C           iTLeft(ne), iTRight(ne): indices of the
C             triangles located to the left and to the right
C             of each edge, if we agree to orient edge ie
C             from point with index infEd(ie) to point with
C             index isupEd(ie);
C           nEdT(3,nt): index of the 3 edges of each triangle,
C             with the convention:
C             the edge with index nEdT(iel,k) is opposite
C             point with index me(iel,k), in triangle with
C             index k.
C
C           auxiliary arrays:
C           ---------------------
C           list(*): contains all indices for the edges
C             starting from a given point
C           listHead(*): for each vertex with index i,
C             listHead(i) contains the index of the edge with
C             highest index starting at vertex with index i;
C             the next edge has
C             list(listHead(i)) as index,...
C
C           N.B.1 The values of iTLeft(ie) and iTRight(ie)
C           are constructed at two different times:
C           one at the time where the edge is constructed,
```

```
C           the other at the time where we meet this edge
C           for the second time.
C
C           N.B.2 By Euler's formula, we have:
C               ne < nv+nt
C
C
            DIMENSION me(3,nt)
            DIMENSION infEd(*),isupEd(*),iTLeft(*),iTRight(*)
            DIMENSION nEdT(3,nt)
            DIMENSION list(*),listHead(*)
C
            ne=0
            DO 4 ie=1,nv+nt
4           iTRight(ie)=0
            DO 1 i=1,nv
1           listHead(i)=0
            DO 2 k=1,nt
              DO 2 il=1,3
                ilp=mod(il,3)+1
                i=me(ilp,k)
                j=me(mod(ilp,3)+1,k)
                ilow=min(i,j)
                ihigh=max(i,j)
                ie=listHead(ilow)
C
C We go over the list array so as to know whether the edge
C has already been created.
C
3               IF( ie.NE.0) THEN
                  IF( (ihigh.EQ.isupEd(ie) ).AND.
     >              ( ilow.EQ.infEd(ie) ) )THEN
                    nEdT(il,k)= ie
                    IF (ilow.EQ.i) THEN
                      iTLeft(ie) = k
                    ELSE
                      iTRight(ie)=k
                    ENDIF
                    GOTO 2
                  ENDIF
                  ie=list(ie)
                  GOTO 3
                ENDIF
C
C This is a new edge
                ne=ne+1
                infEd(ne)=ilow
                isupEd(ne)=ihigh
                list(ne)=listHead(ilow)
                listHead(ilow)=ne
                IF (ilow.EQ.i) THEN
```

```
                  iTLeft(ne)=k
                ELSE
                  iTRight(ne) = k
                ENDIF
                nedT(il,k)=ne
2               CONTINUE
                END
C
C
C PROGRAM listEdge
C -----------------
C This program aims at testing the previous
C subroutine in the case of the tiny triangulation.
C The triangulation is assumed to be stored in file
C TinyTri.Dta, created in the previous chapter
C (section 2.5.2); it is thus read via subroutine
C readT stored in file utilP1.for.

                PARAMETER(nvMax=8,ntMax=8, neMax=nvMax+ntMax)
                INTEGER list(3*ntMax), listHead(nvMax)
                INTEGER infEd(neMax),isupEd(neMax)
                INTEGER iTLeft(neMax),iTRight(neMax)
                INTEGER nEdT(3,ntMax)
C               DIMENSION q(2,nvMax),me(3,ntMax),ng(nvMax)
                CALL readT('TinyTri.dta',nv,nt,q,me,ng,ngt)
                CALL getEdge(nv,nt,me,ng,list,listHead,
     >              iTLeft,iTRight,infEd,isupEd,nEdT,ne)
                Do 16 k=1,nt
16                write(*,*)(nEdT(iel,k),iel=1,3)
                write(*,*)
                write(*,*) list
                write(*,*)listHead
                write(*,*)
                DO 15 ie=1,ne
15                write(*,*)ie,infEd(ie),isupEd(ie)
     >              ,iTRight(ie), iTLeft(ie)
                PAUSE
                END
C
C ************************************
C
C               LIBRARY
C
C ************************************
C
                INCLUDE 'utilP1.for'
C
C
```

Results for the tiny tirangulation are as follows:

```
 1  2  3
 4  3  5
 6  7  4
 1  8  9
 8  7 10
11  6 12
13 11 14
15 10 13

0 0  2 0 3 4  0 0 1 8 0 6 0 11 10
5 9 15  12 7 14  13 0

 1 2 5 1 4
 2 1 2 1 0
 3 1 5 2 1
 4 4 5 3 2
 5 1 4 0 2
 6 4 8 6 3
 7 5 8 3 5
 8 3 5 4 5
 9 2 3 4 0
10 3 8 5 8
11 6 8 7 6
12 4 6 0 6
13 7 8 8 7
14 6 7 0 7
15 3 7 8 0
```

The same program in C gives:

```
#include <stdlib.h>
/*---------------------------------------------*/
/* Numbers the edges and builds
 Tleft,Tright,lowV,highV,edgeT */
 typedef int triangle[3];
/* OUTPUT */
extern int ne; /* ne =ns+nt- nb holes (=nb edges) */
extern int* lowV; /* lowV[ne] number of vertex at edge start */
extern int* highV; /* highV[ne] number of vertex at edge end */
extern int* Tleft; /* Tleft[ne] number of triangle left of edge */
extern int* Tright; /* Tright[ne] number of triangle right of edge */
extern triangle* edgeT;/* edgeT[nt][3] number of 3 edges of each tr */
extern int* listHead; /* listHead[ns] number of tr having i as a vertex */
/*---------------------------------------------*/
int doedge(int nv, int nt, triangle* me)
/*---------------------------------------------*/
 {
 int found, i,k,j,l,lp, low, high,ip;
 int* list=NULL; /* auxiliary array */
 int next[4]={1,2,0,1};
```

```
ne=-1;
list = (int*)calloc(50+nv+nt,sizeof(int)); /* 50 holes max */
lowV = (int*)calloc(50+nv+nt,sizeof(int));
highV = (int*)calloc(50+nv+nt,sizeof(int));
Tleft = (int*)calloc(50+nv+nt,sizeof(int));
Tright = (int*)calloc(50+nv+nt,sizeof(int));
edgeT = (triangle*)calloc(nt,sizeof(triangle));
listHead = (int*)calloc(nv, sizeof(int));
for(k=0;k<nt+nv+50;k++){Tright[k]=Tleft[k]=-1;}
for(k=0;k<nt;k++)
   for(l=0;l<=2;l++)
     {
        lp=next[l];
        i=me[k][lp] ; /* starting vertex of edge */
        j=me[k][next[lp]]; /* end vertex of edge */
        low= i>j ? j : i; /* smallest in low */
        high= i>j ? i : j;
        ip=listHead[low]; /* begin edg list */
         found=0;
         while(( ip!= 0)&&(!found))
     {
         if(ip>49+nv+nt)
         printf("bug in doedge\n");
         if( (high==highV[ip] )&& ( low == lowV[ip] ) )
           { /* edge is already built */
           edgeT[k][l]= ip;/* store edge */
           if (low==i) Tleft[ip] = k;
           else Tright[ip]=k;/* triangle right of ip */
           found=1;
        }
        ip=list[ip] ; /* next edge in list */
      }
     if(!found) /* found a new edge */
         {
           lowV[++ne]=low ; /* store starting vertex */
           highV[ne]=high; /* store end vertex */
           list[ne]=listHead[low]; /* add to list */
           listHead[low]=ne; /* put this as start */
           if (low == i)
           Tleft[ne]=k; /* left triangle */
           else Tright[ne] = k;
           edgeT[k][l]=ne;
         }
     }
     for(k=0;k<nt;k++)
     for(i=0;i<=2;i++) /* vertex i is in tr listHead[j] */
     listHead[me[k][i]]=k;
     free((void**)&list);
       return 0;
     }
```

3.3.3 A List Structure in Fortran

We shall study a slightly simpler problem than the previous one: *find all triangles having point with index i as vertex.*

In a natural way, we introduce the following definition:

Definition 3.7 We shall say a triangle in triangulation $\mathcal{T}_h$ is *a neighbour of the point with index i* if this point is one of its vertices.

We shall introduce two arrays:

*list*1(*na*) contains all the indices of the triangles neighbouring all vertices in the mesh,

listHead(ns) contains, for each index *i*, the position in array *list*1 of a neighbouring triangle of the vertex with index *i*,

so that all the triangles neighbouring the vertex with index *i* have their indices stored in array *list*1 between indices *listHead*(i) and *listHead*($i+1$) $-$ 1.

Building the *list*1 array is not easy since we do not know in advance the number of triangles that are neighbours of each vertex. We shall thus consider a slightly different structure, where the triangles neighbouring a vertex will be *linked together* by a pointer stored in an array named *list*2, rather than being simply stored sequentially in array *list*1. We can find the indices k of all triangles that are neighbours of a vertex with index i with the following procedure:

```
          kptr = listHead(i)
1           IF ( kptr.GT. 0 ) THEN
              k = ineighbour(kptr)
              kptr = list2(kptr)
              GOTO 1
          ENDIF
```

In this way, *list*2(*kptr*) contains the location, in array *ineighbour*, of the next triangle that is a neighbour of the vertex with index *i*; when we reach the last triangle, array *list*2 must point to 0, so as to show there are no more neighbouring triangles. This operation is costly, as it requires an additional array *ineighbour*(*ne*). But the advantage is that now the triangles can be stored in any order, which is easier to implement. We shall take advantage of this to get rid of array *ineighbour*, by coding the values it contains in the form of an address in an array. To make this clearer, let us consider the following program:

```
    INTEGER list(3*nt), listHead(nv)
C
C This program aims at building the list and listHead
C arrays. They are address arrays.
C
C For each vertex with index i, listHead(i) contains
C the address kptr in array list of the index k of the
```

```
C neighbouring triangle of this vertex with the highest
C index (since the indices k below are swept in
C increasing order).
C
C The address of the next triangle (i.e. the one with
C highest index among other triangles having this point as
C vertex) is stored in list(listHead(i)), and so on...
C (kptr is address of index k of triangle in array list).
C
C
            DO 1 i=1,nv
1           listHead(i)=0
            DO 2 k=1,nt
              DO 2 il=1,3
                i=me(il,k)
                kptr = 3*(k-1) + il
                list(kptr)=listHead(i)
2               listHead(i)=kptr
C
C Conversely, to know the indices of all triangles having
C point with index i as a vertex, we look for
C their addresses starting with the highest one
C (stored in listHead(i)), then we go along the
C address array list
C (list(listHead(i)), list(list(listHead(i))),...)
C so as to know the addresses of the next indices, until
C we meet a zero value, in which case we stop, since
C we have the adresses of all the triangles neighbouring
C the vertex with index i.
C
C Once we have an address kptr, we find the index k of
C the corresponding triangle by noting that, in the
C previous program, k-1 is the quotient in the Euclidean
C division of kptr-1 by 3. We can also retrieve the
C local index il of the vertex with index i in triangle
C with index k, by noting that il-1 is the remainder in
C the previous Euclidean division
C
            kptr=listHead(i)
10            IF (kptr.NE.0) THEN
              k = (kptr-1)/3 +1
              il = mod(kptr-1,3)+1
              kptr = list(kptr)
              GOTO 10
            ENDIF
```

The result for the tiny triangulation from chapter 2 (cf. Figure 2.7) is the following:

```
list : 0 0 0 1 0 2 6 5 0 0 3 7 9 10 12 8 0 13 17 0 18 21 20 14
listHead : 4 11 24 16 15 19 23 22
```

```
i kptr k il -->list(kptr) k il...

1 --  4 2 1  -->  1 1 1
2 -- 11 4 2  -->  3 1 3
3 -- 24 8 3  --> 14 5 2 --> 10 4 1
4 -- 16 6 1  -->  8 3 2 -->  5 2 2
5 -- 15 5 3  --> 12 4 3 -->  7 3 1 --> 6 2 3 --> 2 1 2
6 -- 19 7 1  --> 17 6 2
7 -- 23 8 2  --> 20 7 2
8 -- 22 8 1  --> 21 7 3 → 18 6 3→ 13 5 1 → 9 3 3
```

Let us recall that the me array for this triangulation is given by:

```
k me(1,k) me(2,k),me(3,k)
1  1 5 2
2  1 4 5
3  5 4 8
4  3 2 5
5  8 3 5
6  4 6 8
7  6 7 8
8  8 7 3
```

EXERCISE (skyline storage) Propose and study a storage where the bandwidth varies with the row of the matrix; i.e. each row i has its own bandwidth $\mu(i)$ and we make use of it to reduce storage space.

3.4 COMPRESSED SPARSE ROW STORAGE*

We observe that even if we try, as we did above, to optimise the matrix storage (for instance by defining a per-row bandwidth), there still remains a large number of zeros to be stored (of the order of 90%), and this is both unnecessary and costly of memory space. Let us consider, for instance, the solution of the model problem in the unit square, with the triangulation described in Section 2.6.1, using $N+1$ discretisation points on both coordinate axes (cf. Figure 2.9). We assume the vertices are numbered by increasing y-coordinates, and for equal y-coordinates by increasing x-coordinates (or the opposite choice). As $nv = (N+1)^2$ and $nt = 2N^2$, the number of non-zero elements of A is less than the number of edges, that is $2N^2 + (N+1)^2 - 1$, whereas band storage requires $(N+1)^2 * (N+2)$ memory locations, since here $\mu = N+1$.

The idea is thus to store only the non-zero elements of A, in a one-dimensional array *aa* of size *nelemA*. To do this, we shall go over the elements of this matrix using a double loop over the i and j indices, and when we find an element $A_{ij} \neq 0$, we store it in array *aa* at an index that gets incremented by 1. Naturally, if we want to be able to retrieve the indices i and j corresponding to the element thus stored, we must keep in memory some more information; for instance, we can store the

* This storage is called 'Morse' storage in the French text, but we have chosen to use the American term.

column index j in an array $ja(nElemA)$ and the row index i in another array $ia(nElemA)$. Actually there is a more astute solution: instead of storing in array $ia(nElemA)$ the row index i, we shall store into a smaller array $ia(nv)$ the location in array aa of the first non-zero element of the ith line of A (i.e. the first element of this line to be actually stored in array aa). For convenience, we define element $ia(nv+1)$ as being equal to $nElemA+1$, as this enables us to locate easily the index in array aa at which the last diagonal element of matrix A is stored: it is index $ia(nv+1)-1$. This CSR storage format thus requires the following arrays:

```
REAL aa(nElemA),
INTEGER ja(nElemA), ia(nv+1)
```

3.4.1 Construction of the CSR Storage

Let us first assume that matrix A is already stored in an array $A(nv, nv)$. To store it in CSR format, we just need to operate in the following way:

```
      k=1
        DO 1 i=1,nv
          ia(i)=k
          DO 1 j=1,nv
            IF (A(i,j).NE. 0.) THEN
              aa(k)= A(i,j)
              ja(k)= j
              k=k+1
            ENDIF
1       CONTINUE
      ia(nv+1)=k
```

EXAMPLE

$$A = \begin{pmatrix} 4 & -1 & 0 & 2 \\ -1 & 4 & -1 & 0 \\ 0 & -1 & 4 & -1 \\ 2 & 0 & -1 & 4 \end{pmatrix}$$

$$\begin{array}{lcccccccccccc} k & 1 & 2 & 3 & 4 & 5 & 6 & 7 & 8 & 9 & 10 & 11 & 12 \\ aa = & (\mathbf{4} & -1 & 2 & \mathbf{-1} & 4 & -1 & \mathbf{-1} & 4 & -1 & \mathbf{2} & -1 & 4) \\ ja = & (1 & 2 & 4 & 1 & 2 & 3 & 2 & 3 & 4 & 1 & 3 & 4) \\ ia = & (1 & 4 & 7 & 10 &) \end{array} \tag{3.35}$$

In array aa, we have written in boldface the first non-zero element of each line of A. Array ia gives their location in array aa. Since the matrix is symmetric, if we only wish to store its lower triangular part, i.e. if we add a test like

```
... IF (A(i,j).NE. 0.).AND.(i.GE.j) THEN ...
```

then the arrays can be written as:

$$aa = (4, -1, 4, -1, 4, 2, -1, 4),$$
$$ja = (1, 1, 2, 2, 3, 1, 3, 4), \quad ia = (1, 2, 4, 6). \tag{3.36}$$

From now on we shall work with symmetric matrices, and we shall only store their lower triangular part.

Obviously the above situation is 'academic', and we shall never have to transform a 'dense' storage of A into a 'reduced' storage, for the very good reason that matrix A will never be stored as a two-dimensional array. In practice we shall have to compute and store elements A_{ij} simultaneously.

To do this, let us note that according to Lemma 3.4, element A_{ij} is non-zero only if $q^i q^j$ is an edge of the triangulation. However, the converse is not necessarily true. But that does not matter much, even if it entails storing a few more elements, and we shall use this criterion to select the elements of A that are to be actually stored.

The construction of array *ia* then becomes particularly simple. First, this array is used to store, for each vertex with index i, the number of its neighbours. This computation can be done by counting the number of edges that have point q^i as a vertex, and on the computer this becomes a loop over all triangles followed by a loop over the three vertices of each triangle. By doing this, each internal edge is counted twice, whereas boundary edges are only counted once. To remedy this, when we meet an internal edge, we increment by 1 only the value in array *ia* associated with the point with largest index, whereas if we are on a boundary edge, we increment by 1 both values in array *ia* associated with both ends of the edge. Let us make an important remark in connection with the program that follows.

Boundary edges are *characterised* through array *ng*; more precisely, we shall say that *an edge is on the boundary if and only if its extremal points are boundary vertices of the mesh.* Thus, we can only use this program for triangulations that satisfy this property. This excludes, in particular, the triangulation in Figure 2.9 of chapter 2; indeed, the lower right triangle and the upper left triangle do not meet this condition. For this example it is easy to modify locally the triangulation (by swapping the diagonals in the lower right square and the upper left square) so that the above condition becomes satisfied.

```
C (extract from subroutine: CgetMat.for)
C
C We first compute, for each vertex with index i,
C the number of its "strict" neighbours ( i.e. itself
C excluded), and we store the result in ia(i).
C This construction assumes that each triangle is
C positively oriented.
C Boundary edges are characterised with array ng.
C
C
          DO 110 i=1,nv
110         ia(i)=0
          DO 111 k = 1,nt
```

```
                 DO 111 il = 1,3
                   ilp = mod(il,3)+1
                   i=me(il,k)
                   ip=me(ilp,k)
                   IF (ip.LE.i) THEN
                     ia(i) = ia(i)+1
                   ELSEIF ((ng(i).NE.0).AND.
                     (ng(ip).NE.0)) THEN
                     ia(ip) = ia(ip)+1
                   ENDIF
111              CONTINUE
```

At the end of this double loop *ia*(i) contains the number of vertices that form, together with point q^i, an 'actual' edge (that is one not reduced to a point) of the triangulation.

We now put array *ia* in the desired form, i.e. *ia*(i) contains the index in array *aa* of the first element on the ith row of matrix A that is actually stored (note that we no longer say 'first non-zero' element of the ith row of matrix A, why ?). This construction is particularly simple to implement, as the following lines of code show

```
               nElemA=0
               DO 211 i = 1,nv
                 ineighb=ia(i)
                 ia(i)= nElemA + 1
211                nElemA = nElemA + ineighb + 1
               ia(nv+1)=nElemA+1
```

With this procedure, we have also computed the number *nElemA* of elements of A to be stored.

We construct *ja* and *aa* simultaneously. The difficulty in programming lies in the fact that, as the construction of *aa* is done by looping over triangles, the computation of *aa*(k) is not done once: it is done by adding elementary contributions from each triangle. This means that for each couple (i, j), we must be able to retrieve the index k at which A_{ij} is stored in array *aa*. We show how to program this computation.

```
C
C Construction of arrays ja and aa
C (array ia is known)
C Aii is stored in array aa
C at index ia(i+1)-1, with, by convention,
C ia(nv+1)=nElemA+1.
C
C
               DO 1 k = 1, nElemA
               aa(k)=0.
1              ja(k)=0
               DO 2 kt = 1,nt
```

```
              DO 2 il=1,3
              i = me(il,kt)
              DO 2 jl=1,3
                j=me(jl,kt)
                k=0
                IF (i.EQ.j) THEN
                  k=ia(i+1)-1
                  ja(k)=i
                ELSEIF (i.GT.j) THEN
                  DO 3 l= ia(i), ia(i+1)-2
C
C j cannot be elsewhere
C
                    IF (ja(l).EQ.0) GOTO 4
                    IF ( ja(l) .EQ. j )THEN
                      k=l
                      GOTO 4
                    ENDIF
3                 CONTINUE
             write(*,*)'error in CcalMat.for at point i,j=',i,j
4                 IF (k.EQ.0) THEN
                    k = l
                    ja(l) = j
                  ENDIF
                ENDIF
                IF(k.NE.0)aa(k)=aa(k)+
     >            elemMat(kt,il,jl,area(kt),q,me)
2            CONTINUE
                  END
```

Exercise There are still some zero values in the above array. Write a program to get rid of them. □

Exercise Build an analogous program for the case of non-symmetric matrices. □

We note that in the above construction, the elements of the ith row of A stored in array aa between indices $ia(i)$ and $ia(i+1)-1$ are stored in an arbitrary order, and not in increasing column order. This is unfortunate because much time is subsequently lost searching for the index k at which the newly computed elementary contribution will be stored.

Exercise Directly modify the above program, in such a way that we have, in array aa: □

$$ja(k+1) > ja(k)\ , \quad \forall k \in [ia(i), ia(i+1)-1]. \tag{3.37}$$

It is also possible to modify, *a posteriori*, the order of these elements, thanks to the following sorting subroutine.

```
        DO 3 i=1,nv
3         CALL Sort(ja(ia(i)),aa(ia(i)), ia(i+1)-ia(i))
```

We give the sorting subroutine in a particularly compact form [Knuth (1973)].

```
C File sort.for
C           SUBROUTINE Sort(criter,record,n)
C
C This subroutine is written in a very compact form,
C which makes it difficult to understand.
C
C It sorts elements in array criter in increasing
C order, and simultaneously modifies elements
C of array record.
C
C
            INTEGER crit,criter(n)
            INTEGER n,record(n)
            INTEGER l,r,j,i, rec
            IF(n.le.1) RETURN
            l=n/2+1
            r=n
2           IF(l.gt.1)THEN
              l=l-1
              rec=record(l)
              crit=criter(l)
            ELSE
              rec=record(r)
              crit=criter(r)
              record(r)=record(1)
              criter(r)=criter(1)
              r=r-1
              IF(r.eq.1) THEN
                record(1)=rec
                criter(1)=crit
                RETURN
              ENDIF
            ENDIF
3           j=l
4           i=j
            j=2*j
            IF(j-r.lt.0)THEN
              GOTO 5
            ELSEIF(j.eq.r)THEN
              GOTO 6
            ELSE
              GOTO 8
            ENDIF
5           IF(criter(j).lt.criter(j+1))j=j+1
6           IF(crit.ge.criter(j))GOTO 8
            record(i)=record(j)
```

```
          criter(i)=criter(j)
          GOTO 4
8         record(i)=rec
          criter(i)=crit
          GOTO 2
          END
```

Once we have constructed and stored matrix A in CSR format, we must specify how to modify the programs in all the steps leading to the solution of linear system (3.20). In particular, we must program, for this new storage format, the Choleski factorisation of the matrix, as well as the forward and backward solves linked to the solutions of the triangular systems (3.30) and (3.31). This is what we shall now do, beginning with the simple computation of the product of matrix A, stored in CSR format, by a vector Φ.

3.4.2 *Computing $A\phi$*

If A has no particular symmetry and both its lower triangular and upper triangular parts are stored in CSR format in array aa, computing the product of A by a vector Φ is programmed in the following way:

$$(A\Phi)_i = \sum_{j=1}^{n_v} A_{ij}\varphi_j = \sum_{k=ia(i)}^{ia(i+1)-1} aa(k)\varphi(ja(k)). \tag{3.38}$$

If A is symmetric, we can formally decompose it as $A = L_1 + L_2^T$, where L_1 is the lower triangular part, including the diagonal, of A (i.e. $L_{ij} = A_{ij}$ for $i \geq j$, 0 otherwise) and L_2^T is its strictly upper triangular part. So to compute $A\Phi$ it is enough to know how to compute $L\Phi$ and $L^T\Phi$ for any triangular matrix L stored in CSR format, and this is what we shall now study.

3.4.3 *Computing $L\Phi$*

Let us denote, in all that follows and with obvious notation, by $[al, il, jl]$ the arrays for storing L in CSR format. We have:

$$(L\Phi)_i = \sum_{j=1}^{n_v} L_{ij}\varphi_j = \sum_{k=il(i)}^{il(i+1)-1} al(k)\varphi(jl(k)) \tag{3.39}$$

3.4.4 *Computing $r = L^T\Phi$*

The index j entry of vector r is equal to

$$r_j = \sum_{i\geq j} L_{ij} z_i, \tag{3.40}$$

which requires a different program from above. We can proceed in the following way:

- we initialize array r to 0;
- we loop over the i indices, then we loop over the j indices and we increment each r_j by $L_{ij}z_i$.

If we now loop over indices $k \in \{1, ..., nElemL\}$, then $al(k)z_i$ actually contributes to the value of r_j if $j = jl(k)$ and $k \in [il(i), il(i+1) - 1]$. The loop over k decomposes into a double loop, first over $i \in [1, ..., nv]$, then over $k \in [il(i), il(i+1) - 1]$, and (3.40) is programmed as follows:

```
          DO 1 j=1,nv
1           r(j)=0.
          DO 2 i=1,nv
            DO 2 k=il(i), il(i+1)-1
2             r(jl(k))=r(jl(k)) + al(k)*z(i)
```

3.4.5 *Solution of* $Lz = r$

We must program the forward solve (3.30) with CSR storage $[al, il, jl]$ for L. Since L is a lower triangular matrix, the solution of system $Lz = r$ is programmed by a loop over the i indices increasing from 1 to nv as follows:

$$z_i = \frac{1}{L_{ii}}[r_i - \sum_{j=1}^{i-1} L_{ij}z_j]. \tag{3.41}$$

Since matrix L is stored in CSR format, the index i entry of Lz in this formula is computed in the same way as in expression (3.39), which leads to the following program sketch for computing z_i:

$$z_i = \frac{1}{al(il(i+1)-1)}[r_i - \sum_{k=il(i)}^{il(i+1)-2} al(k)z_{jl(k)}] \tag{3.42}$$

3.4.6 *Solution of* $L^T z = r$

In the backward solve step (3.31), the matrix involved is now upper triangular so that solving system $L^T z = r$ is done by a 'reverse' loop over indices i decreasing from nv to 1:

$$z_i = \frac{1}{L_{ii}}[r_i - \sum_{j=i+1}^{n_v} L_{ji}z_j]. \tag{3.43}$$

The actual computation of the index i entry of $L^T z$ involved in this computation is carried out in a manner identical to the explanations given in paragraph 3.4.4, which gives the following program:

```
        DO 1 i=1,nv
1         z(i)=r(i)/al(il(i+1)-1)
        DO 2 i=nv,1,-1
        DO 2 k=il(i), il(i+1)-2
          jlk=jl(k)
2         z(jlk) = z(jlk) - al(k)*z(i)/ al (il(jlk+1)-1)
```

Here, as in (3.42), we have taken into account the fact that the diagonal element L_{ii} is stored in array al at index $il(i+1)-1$, with the convention $il(nv+1)-1 = nElemL$. Let us also note that in each of the previous two steps (forward and backward solves), only one array is required to store both the equation right hand side at the beginning of the program, and its solution on output.

3.4.7 Choleski factorisation of A

Starting from matrix A stored in CSR format, thanks to arrays aa, ia, ja, we shall now construct its Choleksi factor L stored in CSR format in arrays al, il, jl. Let us note that although A has a 'very sparse' band structure (cf. the example alluded to at the beginning of section 3.4), this is *a priori* not true of matrix L. Thus we could be content to use band storage, as described in section 3.2, for the latter matrix. We shall nevertheless show how to program this factorisation for the case where matrix L is stored in CSR format.

Let us briefly recall the **algorithm**:

$$\begin{aligned}
&\mathrm{DO\,j} = 1, \mathrm{nv} \\
&\quad L_{jj} = \sqrt{A_{jj} - \sum_{k<j} L_{jk}^2} \\
&\quad \mathrm{DO\,i} = j+1, \mathrm{nv} \\
&\quad\quad L_{ij} = (A_{ij} - \sum_{k<j} L_{ik}L_{jk})/L_{jj} \\
&\quad\quad \text{End i loop} \\
&\quad\quad \text{End j loop}
\end{aligned}$$

Let us note that

$$\sum_{k<j} L_{jk}^2 = \sum_{l=il(j)}^{il(j+1)-2} al(l)^2, \tag{3.44}$$

and that

$$\sum_{k<j} L_{ik}L_{jk} = \sum_{k=il(j)}^{il(j+1)-2} al(k_{ik})al(k), \tag{3.45}$$

if we denote by k_{ik} the index, in array *al*, where element L_{ik} is stored. To compute this index, we must sweep over all indices $l \in [il(i), il(i+1)-2]$ in array *al* until we find l such that $jl(l) = k$. We note that this procedure is faster if, within each interval $[il(i), il(i+1)-2]$ the elements of *al* are stored in increasing column order, i.e. $ja(l+1) > ja(l)$. Once (3.45) is computed, we must then find the index where the result is stored in array *al*. Schematically the algorithm is as follows:

```
C (extract from CCholes.for)
C
C Choleski factorisation of matrix A stored in CSR
C format thanks to arrays aa, ia, ja;
C The factor L is stored in CSR format thanks to
C arrays al, il, jl.
C This factorisation assumes that
C jl(k+1) > jl(k), ∀k ∈ [il(i), il(i+1) - 2],  ∀i
C
C
      DO 1 j=1,nv
      al(il(j+1)-1) = sqrt(aa(ia(j+1)-1) - Σ_{k=il(j)}^{il(j+1)-2} al(k)^2)
      DO 1 i=j+1,nv
        DO 2 l=il(i),il(i+1)-2
          IF( jl(l).EQ.j ) THEN
          al(l) = 1/al(il(j+1)-1) [A_ij - Σ_{k=il(j)}^{il(j+1)-2} al(k_ik)al(k)]
          ELSEIF( jl(l).GT.j) GOTO 3
          ENDIF
2       CONTINUE
3       CONTINUE
1     CONTINUE
```

In the above, computing the quantity under the square root is easy, however computing $al(l)$ is less so, for two reasons:

we do not know where A_{ij} is stored,
the index k_{ik} is unknown.

A possible solution to overcome the first point is to loop over the k indices (see loop 'DO 32' below) rather than over the l indices (previous 'DO 2' loop), and to increment k and l simultaneously.

```
      DO 34 j=1,nv
      DO 34 i=j+1,min(j+nband,ns)
```

Here we limit the search for indices i to the bandwidth of matrix A. To find the index where element A_{ij} is stored, we just have to loop over the row with index i and look for the index k such that $ja(k) = j$; the search for this index is made easier if the elements are stored within this row according to the order: $ja(k+1) > ja(k)$. The program runs in the following way:

```
      DO 32 k=ia(i),ia(i+1)-2
```

```
          IF(ja(k).GE.j) GOTO 37
32        CONTINUE
37        IF(ja(k).NE.j) THEN
            alij=0.
          ELSE
            alij=aa(k)
34        ENDIF
```

As for the second point alluded to above, we shall write the sum in a different way, by incrementing indices i_k and j_k until $jl(i_k) = jl(j_k)$

$$\sum_{k=il(j)}^{il(j+1)-2} al(k)al(k_{ik}) = \sum_{i_k \in I_k(i), j_k \in I_k(j): jl(i_k)=jl(j_k)} al(i_k)al(j_k),$$

where $I_k(i) = \{i_k : il(i) \leq i_k \leq il(i_k + 1) - 2\}$.

The result of this sum is to be stored at index i_k such that $jl(i_k) = j$. To make the program clearer, we give line by line comments.

```
          lk=0                                  ! index to store alij
          ik=il(i)                                   !! multiply row i
          jk=il(j)                                           !! by row j
          maxik=il(i+1)-2                       ! maximum index for ik
          maxjk=il(j+1)-2                       ! maximum index for jk
35        IF(ik.LE.maxik) THEN
            jlik=jl(ik)                              ! couple i,jlik
            IF(jlik.EQ.j) lk=ik                    !index to store Lij
36          IF(jk.LE.maxjk) THEN
            jljk=jl(jk)                              ! couple j,jljk
              IF(jlik.GT.jljk)THEN
                jk=jk+1                            ! jljk is too small
                GOTO 36
              ELSEIF(jlik.LT.jljk)THEN
                ik=ik+1                            ! jlik is too small
                GOTO 35
              ELSE                                     ! jlik=jljk
                alij=alij-al(ik)*al(jk) ! compute alij
                ik=ik+1
                jk=jk+1
                GOTO 35
              ENDIF
            ENDIF                                      !! jk<maxjk?
          ENDIF                                        !! ik<maxik?
          IF(lk.NE.0) al(lk)=alij/al(il(j+1)-1)
34     CONTINUE
```

The last difficulty is to construct arrays *il* and *jl*. We assume that all the elements L_{ij} located between the first non-zero element of row i of A (its index in array *aa* is $ia(i)$) and the diagonals are all non-zero. It is then easy to construct these arrays.

```
          k=0
          DO 41 i=1,nv
            il(i)=k+1
            IF(ja(ia(i)).GT.i) write(*,*) i,ia(i),ja(ia(i)),k
            DO 41 j=ja(ia(i)),i
              k=k+1
41            jl(k)=j
```

The complete program is displayed in Appendix 3.B.

3.5 EFFICIENCY OF THE METHOD

The following table gives the computer timings necessary when programming the problem defined in the previous chapter (end of Section 2.6.3) with the triangulation of Figure 2.13, and this by the Choleski method with both storage formats envisioned: band storage (the corresponding column is labeled 'band Chol.') and CSR storage format ('CSR Chol.'). Four increasingly fine meshes are treated, for which we give the H^1 error between the exact solution φ and the computed solution φ_h ('Error'). The timings are compared with those obtained with the iterative method of the previous chapter ('Energy').

nv	Band Chol.	CSR Chol.	Energy	Error
49	3.3"	3.4"	7"	$3.2 10^{-2}$
170	11"	12.4"	53"	$4.8 10^{-4}$
625	86"	96"	420"	$8.6 10^{-5}$
1741	405"	480"	**	$3.4 10^{-5}$

APPENDIX 3.A

The Fortran program for solving the Laplacian with Dirichlet boundary conditions by Variational Formulation and Choleski Factorisation of the Linear System Matrix Stored in Band Format

```
C File: BLaplace.for
C
C
C            PROGRAM BLaplace
C            ----------------
C This program solves the model problem, i.e. the
C Laplacian with Dirichlet boundary conditions:
C -ΔΦ = f, Φ given on the boundary.
C
C The problem is written in variational form, then
C discretised by P¹ finite elements. The approximate
```

```
C solution at each of the vertices of the triangulation
C is determined by solving a linear system, which is
C solved by a direct method, by factoring the matrix
C in Choleski form.
C The matrix is stored in band format.
C Matrix A, then its Choleski factor L are
C stored in the same array a.
C
C
C
C            input:
C            ------
C            triangulation (nv,nt,q,me,ng)
C              and problem data
C              (right hand side f of Laplace's equation and
C              Dirichlet boundary condition stored in array
C              phi0) read from file filnam through
C              subroutines readT and readF;
C            parameter:nvMax (maximum number of vertices)
C              ntMax (maximum number of triangles) of the
C              triangulation, nBandMax (maximum bandwidth)
C              and nMatMax=(nvMax-1)*nBandMax + nvMax
C              (maximum size of array a).
C
C            output:
C            -------
C            ff:array containing the values of the
C              approximate solution at each of the mesh
C              vertices, stored in a file (whose
C              name is given by the user) in view of a possible
C              graphical exploitation
C
C            arrays:
C            -------
C            mesh:
C              q(2,nvMax):coordinates of the vertices of the mesh,
C              me(3,ntMax):numbering of the vertices of the
C                triangles,
C              ng(nvMax):array indicating whether a vertex is
C                on the boundary (ng(i)≠0) or not (ng(i)=0).
C
C            data:
C              f(nvMax):Laplacian right hand side
C              phi0(nvMax):Dirichlet boundary conditions
C
C            other arrays:
C              a(nMatMax):array for storing matrix A in band
C                format at the beginning of the program, then
C                its Choleski factor,
C              area(ntMax):area of each triangle
C              ff(nvMax):on startup, right hand side of the
```

```
C              linear system to solve, then solution of
C              this system.
C
C          scalars:
C          --------
C            nBand: matrix bandwidth
C            nMat: dimension of array a
C            cte: value of the very large constant
C              enabling to satisfy the Dirichlet boundary
C              condition
C            err: error in H10 norm between the exact
C              solution phi0 and the approximate
C              solution ff.
C
C          files:
C          ------
C          filnam:name of file for storing the mesh,
C            the model problem data (Laplacian right
C            hand side and Dirichlet boundary condition),
C            or the approximate solution, for instance,
C            for future graphical exploitation.
C
C          subroutines:
C          ------------
C            readT:reads the mesh
C            readF:reads problem data;
C            getBand:computes bandwidth of matrix A;
C            getArea:computes the area of each triangle;
C            BgetMat:computes matrix A and stores it in
C              band format, in array a;
C            RHS:computes, and stores in array ff, the
C              system right hand side;
C            BDirCL:program to satisfy Dirichlet
C              boundary conditions;
C            BCholeski:Choleski factorization of
C              matrix A stored in band format;
C            BSolvSys:solution of linear system
C              by forward and backward solves; uses the
C              Choleski factor L of A
C              stored in array a in band format;
C            writeF:writes approximate solution.
C
C          functions:
C          ----------
C            errH10:computes H10 error between
C              exact solution phi0 and approximate
C              solution ff.
C
C          library (subroutines and functions):
C          ------------------------------------
C            utilP1.for:file containing functions deter
```

```
C             (to compute areas), errH10 and subroutines
C             readT, readF, writeF and getBand;
C           RHS:file containing subroutines getArea
C             and RHS;
C           elem.for:file containing function elemMat;
C           BgetMat.for:file containing subroutines
C             BgetMat and BDirCL;
C           BCholes.for:file containing subroutines
C             BCholeski and BSolvSys.
C
C
      PARAMETER (nvMax=1800, ntMax=3400, nBandMax=70)
      PARAMETER (nMatMax=(nvMax-1)*nBandMax+nvMax)
      DIMENSION q(2,nvMax),me(3,ntMax),ng(nvMax)
      DIMENSION area(ntMax)
      DIMENSION phi0(nvMax),f(nvMax),ff(nvMax)
      DIMENSION a(nMatMax)
      CHARACTER*20 filnam
C
      DATA cte/1.0e25/
C
      write(*,*) 'name of file where mesh is stored'
      read(*,'(a)') filnam
      CALL readT(filnam,nv,nt,q,ng,me,ngt)
      write(*,*) 'name of file where f is stored'
      write(*,*) '(Laplacian right hand side)'
      read(*,'(a)') filnam
      CALL readF(filnam,nv,f)
      write(*,*) 'name of file where phi0 is stored'
      write(*,*) '(Dirichlet boundary conditions)'
      read(*,'(a)') filnam
      CALL readF(filnam,nv,phi0)
C
      CALL getBand(me,ng,nt,nv,nBand)
      CALL getArea(q,me,nt,nv,area)
      CALL BgetMat(q,me,area,nt,nv,a,nBand)
      CALL RHS(me,f,nv,nt,area,ff)
      CALL BDirCL(a,phi0,nv,ng,ff,cte,nBand)
      CALL BCholeski(nv,a,nBand)
      CALL BSolvSys(nv,a,ff,nBand)
C
      err=errH10(q,me,ff,phi0,nv,nt)
      write(*,*) 'error=',err
      write(*,*) 'name of file for storing the '
      write(*,*) 'approximate solution'
      read(*,'(a)') filnam
      CALL writeF(filnam,nv,ff)
      PAUSE
      END
C
C
```

```
C****************************
C           LIBRARY
C****************************
C
            INCLUDE 'elem.for'
            INCLUDE 'BgetMat.for'
            INCLUDE 'BCholes.for'
            INCLUDE 'utilP1.for'
            INCLUDE 'RHS.for'
```

APPENDIX 3.B

The Fortran Program for Solving the Laplacian with Dirichlet Boundary Conditions by Variational Formulation and Choleski Factorisation of the Linear System Matrix Stored in CSR Format

```
C   PROGRAM CLaplace
C This program solves the model problem, i.e. the
C Laplacian with Dirichlet boundary conditions:
C -ΔΦ = f,Φ given on the boundary.
C
C The problem is written in variational form, then
C discretised by P1 finite elements. The approximate
C solution at each of the vertices of the triangulation
C is determined by solving a linear system, which is
C solved by a direct method, by factoring the matrix
C in Choleski form.
C The matrix is stored in CSR format.
C Matrix A and its Choleski factor L are
C not stored in the same array
C
C
C            input:
C            ------
C              triangulation (nv,nt,q,me,ng) and
C                problem data
C                (right hand side f of Laplace's equation and
C                Dirichlet boundary condition stored in array
C                phi0) read from file filnam through
C                subroutines readT and readF;
C              parameter:nvMax (maximum number of vertices),
C                ntMax (maximum number of triangles)
C                of the triangulation, nBandMax (maximum
C                  bandwidth), nMLMax=(nvMax-1)*nBandMax+nvMax
C                (maximum size of array al) and nMAMax=8*nvMax
C                (maximum size of array aa).
C
C            output:
C            -------
C              ff:array containing the values of the
C                approximate solution at each of the mesh
C                vertices, stored in a file (whose
C                name is given by the user) in view of a possible
C                graphical exploitation
C
C
C            arrays:
C            -------
C            mesh:
C              q(2,nvMax):coordinates of the vertices of the mesh,
```

```
C           me(3,ntMax):numbering of the vertices of the
C             triangles,
C           ng(nvMax):array indicating whether a
C             vertex is on the boundary (ng(i)≠0) or not (ng(i)=0).
C
C         data:
C           f(nvMax):Laplacian right hand side
C           phi0(nvMax):Dirichlet boundary conditions
C
C         other arrays:
C           aa(nMAMax): array storing matrix A in CSR format,
C           ia(nvMax+1): index where first non-zero
C             element of row i of A is actually stored in aa,
C           ja(nMAMax): column index j of stored Aij element;
C           al(nMLMax), il(nvMax+1),jl(nMLMax): same
C             meaning as three previous arrays, but for
C             the Choleski factor L of A;
C           area(ntMax):area of each triangle
C           ff(nvMax):on startup, right hand side, of the
C             linear system to solve, then solution of
C             this system.
C
C         scalars:
C         --------
C           nBand: matrix bandwidth
C           cte: value of the very large constant enabling
C             to satisfy the Dirichlet boundary condition
C           err: error in H¹₀ norm between the exact solution
C             phi0 and the approximate solution ff.
C
C         files:
C         ------
C         filnam:name of file for storing the mesh,
C           the model problem data (Laplacian right
C           hand side and Dirichlet boundary condition), or the
C           approximate solution, for instance, for
C           future graphical exploitation.
C
C
C         subroutines:
C         ------------
C           readT:reads the mesh
C           readF:reads problem data;
C           getBand:computes bandwidth of matrix A;
C           getArea:computes area of each triangle;
C           CgetMat:computes matrix A and stores it in
C             CSR format in array aa;
C           RHS:computes, and stores in array ff, the
C             system right hand side;
C           CDirCL:program to satisfy Dirichlet boundary
C             conditions
```

```
C            Sort:reorders elements of array aa so that
C              elements on a same line of matrix A are stored
C              in this array in increasing column index order;
C            CCholeski:Choleski factorization of matrix
C              A stored in CSR format;
C            CSolvSys:solution of linear system by forward
C              Choleski and backward solves; uses the
C              factor L of A stored in array aa in CSR format;
C            writeF:writes approximate solution.
C
C
C         functions:
C         ----------
C            errH10:computes H_0^1 error between
C              exact solution phi0 and approximate
C              solution ff.
C
C
C         library (subroutines and functions):
C         ------------------------------------
C            utilP1.for:file containing functions deter
C              (to compute areas), errH10 and subroutines
C              readT, readF, writeF and getBand;
C            RHS:file containing subroutines getArea
C              and RHS;
C            elem.for:file containing function elemMat;
C            CgetMat.for:file containing subroutines
C              CgetMat and CDirCL;
C            CCholes.for:file containing subroutines
C              CCholeski et CSolvSys;
C            sort.for:file containing subroutine Sort.
C
C
          PARAMETER (nvMax=1800, ntMax=3400, nBandMax=70)
          PARAMETER (nMLMax=(nvMax-1)*nBandMax+nvMax)
          PARAMETER (nMAMax=8*nvMax)
          DIMENSION q(2,nvMax),me(3,ntMax),ng(nvMax)
          DIMENSION area(ntMax)
          DIMENSION phi0(nvMax),f(nvMax),ff(nvMax)
          DIMENSION aa(nMAMax),ia(nvMax+1),ja(nMAMax)
          DIMENSION al(nMLMax),il(nvMax+1),jl(nMLMax)
          CHARACTER*20 nomfic
C
          DATA cte/1.0e14/
C
          write(*,*) 'name of file where mesh is stored'
          read(*,'(a)') filnam
          CALL readT(filnam,nv,nt,q,ng,me,ngt)
          write(*,*) 'name of file where f is stored'
          write(*,*) '(Laplacian right hand side)'
          read(*,'(a)') filnam
```

```
            CALL readF(filnam,nv,f)
            write(*,*) 'name of file where phi0 is stored'
            write(*,*) '(Dirichlet boundary conditions)'
            read(*,'(a)') filnam
            CALL readF(filnam,nv,phi0)
C
C
            CALL getBand(me,ng,nt,nv,nBand)
            CALL getArea(q,me,nt,nv,area)
            CALL CgetMat(q,me,ng,area,nt,nv,aa,ia,ja)
            CALL RHS(me,f,nv,nt,area,ff)
            CALL CDirCL(aa,ia,phi0,nv,ng,ff,cte)
C
            DO 1 i=1,nv
1           CALL Sort(ja(ia(i)),aa(ia(i)), ia(i+1)-ia(i))
C
            CALL CCholeski(nv,aa,ia,ja,al,il,jl,nBand)
            CALL CSolvSys(nv,ff,al,il,jl)
C
            err=errH10(q,me,ff,phi0,nv,nt)
            write(*,*) 'error=',err
            write(*,*) 'name of file for storing the '
            write(*,*) 'approximate solution'
            read(*,'(a)') filnam
            CALL writeF(filnam,nv,ff)
            PAUSE
            END
C
C
C ***********************************
C
C           LIBRARY
C
C ***********************************
C
            INCLUDE 'elem.for'
            INCLUDE 'CgetMat.for'
            INCLUDE 'CCholes.for'
            INCLUDE 'utilP1.for'
            INCLUDE 'RHS.for'
            INCLUDE 'sort.for'
C
C
C File: CgetMat.for
C
C ---------------------------------------------------
C    SUBROUTINE CgetMat(q,me,ng,area,nt,nv,aa,ia,ja)
C ---------------------------------------------------
C This subroutine computes and stores the elements
C of the lower triangular part of the linear system
C matrix A associated with the solution of the
```

```
C Laplacian with Dirichlet boundary conditions
C written in variational form, and discretised with
C P1 finite elements. It also computes the actual
C size nElemA of array aa where matrix A is stored
C in CSR format.
C
C Element Aij of this matrix is stored in
C one-dimensional array aa at index k; ja(k)
C holds the column index j of this element.
C Array ia is defined by: ia(i) is the index
C in array aa of the first element of the ith row
C of A actually stored in this array.
C If we agree to let ia(nv+1)=nElemA+1,
C element Aii of matrix A is stored in array aa
C at index ia(i+1)-1.
C At the beginning of the computation, array ia has
C another meaning: ia(i) is the number of vertices
C strict neighbours of index i point (i.e. not
C including the point itself).
C
C
C           input:
C           ------
C           triangulation (nv,nt,q,me,ng) and areas of
C             triangles stored in array area.
C
C           output:
C           ------
C           arrays aa,ia,ja; size nElemA of array aa
C
C
C           important remarks:
C           ------------------
C           - we assume the triangles are positively oriented
C           - array ng is used to detect the boundary
C             edges in the following way: an edge is on
C             the boundary if and only if both its ends
C             are boundary vertices.
C           One must check that the triangulation under
C             study satisfies those two conditions.
C
C
            DIMENSION q(2,nv),me(3,nt),ng(nv),area(nt)
            DIMENSION aa(*),ia(*),ja(*)
C
            DO 110 i=1,nv
110           ia(i)=0
            DO 111 k = 1,nt
              DO 111 il = 1,3
                ilp = mod(il,3)+1
                i=me(il,k)
```

```
                  ip=me(ilp,k)
                  IF (ip.LE.i) THEN
                    ia(i) = ia(i)+1
                  ELSEIF ((ng(i).NE.0).AND. (ng(ip).NE.0)) THEN
                    ia(ip) = ia(ip)+1
                  ENDIF
111           CONTINUE
C
C ia(i) is the number of strict neighbours
C of index i vertex.
C
              nElemA=0
              DO 211 i = 1,nv
                ineighb=ia(i)
                ia(i)= nElemA + 1
                nElemA = nElemA + ineighb + 1
211           CONTINUE
              ia(nv+1)=nElemA+1
C
C ia is constructed; there remains to construct aa and ja.
C
              DO 1 k = 1, nElemA
              aa(k)=0.
1             ja(k)=0
              DO 2 kt = 1,nt
                DO 2 il=1,3
                  i = me(il,kt)
                  DO 2 jl=1,3
                    j=me(jl,kt)
                    k=0
                    IF (i.EQ.j) THEN
                      k=ia(i+1)-1
                      ja(k)=i
                    ELSEIF (i.GT.j) THEN
                      DO 3 l= ia(i), ia(i+1)-2
C
C j cannot be elsewhere
C
C
                        IF (ja(l).EQ.0) GOTO 4
                        IF ( ja(l) .EQ. j )THEN
                          k=l
                          GOTO 4
                        ENDIF
3                     CONTINUE
              write(*,*)'error in CgetMat.for at point i,j=',i,j
4                     IF (k.EQ.0) THEN
                        k = l
                        ja(l) = j
                      ENDIF
                    ENDIF
```

```
                IF(k.NE.0)aa(k)=aa(k)+
     >            elemMat(kt,il,jl,area(kt),q,me)
2           CONTINUE
            END
C
C
C -----------------------------------------------------------
      SUBROUTINE CDirCL(aa,ia,phi0,nv,ng,ff,cte)
C -----------------------------------------------------------
C This subroutine recovers the right Dirichlet
C boundary conditions for the boundary vertices.
C
C For each boundary vertex with index i (ng(i)≠0),
C we simultaneously modify the ith diagonal element
C of matrix A, replacing it by a very large constant cte,
C and the ith component of the right hand side ff,
C replacing it by the product cte*phi0, where phi0
C is the Dirichlet boundary condition. In this way,
C the ith row of the linear system becomes, after
C simplifying by the constant cte
C   phi(i) + small terms = phi0(i),
C and the boundary condition is satisfied up to
C a term of order (1/cte).
C
C
C           input:
C           ------
C             triangulation: nv,ng(nv);
C             matrix aa(*) and array ia(*);
C             Dirichlet boundary condition stored in aray phi0
C               (N.B. this array is assumed to be filled
C               for all vertices).
C             right hand side ff(nv) of linear system;
C             very large constant cte.
C
C           output:
C           -------
C             modified matrix aa and right hand side ff.
C
C
C
            DIMENSION aa(*),ia(nv+1),phi0(nv),ng(nv),ff(nv)
            DO 1 i=1,nv
              IF(ng(i).NE.0)THEN
                aa(ia(i+1)-1)=cte
                ff(i)=cte*phi0(i)
                ENDIF
1           CONTINUE
            END
C
C
```

```
C File CCholes.for
C
C ----------------------------------------
            SUBROUTINE MSolvSys(n,z,al,il,jl)
C ----------------------------------------
C This subroutine computes the solution x of system
C Ax = z, where A is a positive definite symmetric
C matrix of size n*n factored by the Choleski algorithm
C i.e. A = LL^T, with L a lower triangular matrix with
C positive diagonal.
C Matrix L is stored in CSR format in array al, with
C auxiliary arrays il and jl.
C
C The computation is performed in two steps:
C - a forward solve, where we solve system Ly = z,
C - a backward solve, where we solve system L^Tx = y.
C It is not necessary to use two additional arrays to
C store y and x; at the beginning, array z(n) represents
C the system right hand side, then successively, during
C the computation, solutions y then x.
C
C
            DIMENSION z(*),al(*),jl(*),il(*)
C
            DO 1 i=1,n
              sum=z(i)
              DO 2 k=il(i), il(i+1)-2
                sum=sum-al(k)*z(jl(k))
2             CONTINUE
              z(i)=sum/al(il(i+1)-1)
1           CONTINUE
C
            DO 3 i=1,n
3             z(i) = z(i)/al(il(i+1)-1)
            DO 4 i1=1,n
              i=n+1-i1
              DO 4 k=il(i), il(i+1)-2
                jlk=jl(k)
                aljj=al(il(jlk+1)-1)
                z(jlk) = z(jlk) - al(k)*z(i)/aljj
4             CONTINUE
            END
C
C
C ------------------------------------------------------
C     SUBROUTINE CCholeski(n,aa,ia,ja,al,il,jl,nBand)
C ------------------------------------------------------
C This subroutine computes the Choleski factor L of a
C symmetric positive definite matrix A of size n*n, of
C which only the lower triangular part is stored in CSR
C format in array aa, with additional arrays ia and ja.
```

```
C The Choleski factor is stored in CSR format in array
C al, with auxiliary arrays il and jl, constructed first.
C The factorisation is then performed column by column
C beginning by the computation of the diagonal element.
C We assume the bandwidth nBand is known.
C
C N.B. This factorisation assumes the elements of
C array aa satisfy:
C jl(k+1) > jl(k), ∀k ∈ [il(i), il(i+1)-2],   ∀i
C
C
C
C           input:
C           -------
C           bandwidth nBand of matrix A(n,n), i.e.
C             nBand = max{|i-j| : i > j, A(i,j) ≠ 0};
C           arrays aa,ia and ja for CSR storage of
C             matrix A
C
C
C           output:
C           -------
C           arrays al,il and jl for CSR storage of the
C             Choleski factor L of A.
C
C
            DIMENSION aa(*),ia(*),ja(*)
            DIMENSION al(*),il(*),jl(*)
            k=0
            DO 41 i=1,n
              il(i)=k+1
              if(ja(ia(i)).GT.i)write(*,*)i,ia(i), ja(ia(i)),k
              DO 41 j=ja(ia(i)),i
                k=k+1
                jl(k)=j
41            CONTINUE
              il(n+1)=k+1
C
              DO 34 j=1,n
                aljj=aa(ia(j+1)-1)
                DO 33 k=il(j), il(j+1)-2
                  aljj=aljj-al(k)**2
33              CONTINUE
                aljj=sqrt(aljj)
                al(il(j+1)-1)=aljj
                iMax=min(j+nBand,n)
                DO 34 i=j+1,iMax
                  DO 32 k=ia(i),ia(i+1)-2
                    IF(ja(k).GE.j)GOTO 37
32                CONTINUE
37                IF(ja(k).NE.j) THEN
```

```
                   alij=0.
                 ELSE
                   alij=aa(k)
                 ENDIF
                   lk=0
                   ik=il(i)
                   jk=il(j)
                   maxik=il(i+1)-2
                   maxjk=il(j+1)-2
35                 IF(ik.LE.maxik) THEN
                     jlik=jl(ik)
                     IF(jlik.EQ.j) lk=ik
36                   IF(jk.LE.maxjk) THEN
                       jljk=jl(jk)
                       IF(jlik.GT.jljk)THEN
                         jk=jk+1
                         GOTO 36
                       ELSEIF(jlik.LT.jljk)THEN
                         ik=ik+1
                         GOTO 35
                       ELSE
                         alij=alij-al(ik)*al(jk)
                         ik=ik+1
                         jk=jk+1
                         GOTO 35
                       ENDIF
                     ENDIF
                   ENDIF
                   IF(lk.NE.0)al(lk)=alij/aljj
34             CONTINUE
               END
```

4 The Finite Element Method: Optimisation of the Method

Summary

In Chapters 2 and 3, we proposed a program for the solution, using triangular finite elements, of the model problem, that is the Laplacian with Dirichlet boundary conditions on a bounded domain Ω of $\mathbb{R}^2$. The iterative method used in Chapter 2 for minimising the energy, required little memory space, but was relatively costly in computer time; in contrast, the direct method proposed in the previous chapter for solving the variational problem was fast but greedy for memory space. We shall take advantage of these two experiences to propose a method for solving the variational problem via preconditioned conjugate gradient, that combines the advantages of the two previous methods. At the end of the chapter, we shall discuss the influence of the mesh on the quality of the final solution, and shall propose an 'automatic mesh refinement' method.

4.1 *PRECONDITIONED CONJUGATE GRADIENT*

The starting point of our study is the same as that in the previous chapter, that is the variational formulation

find $\varphi \in H^1_g(\Omega)$ such that

$$\forall w \in H^1_0(\Omega), \quad \int_\Omega \nabla\varphi(\vec{x}) \cdot \nabla w(\vec{x}) \mathrm{d}\vec{x} = \int_\Omega f(\vec{x}) w(\vec{x}) \mathrm{d}\vec{x}. \tag{4.1}$$

of the model problem

$$-\Delta\varphi(\vec{x}) = f(\vec{x}), \quad \forall \vec{x} \in \Omega, \tag{4.2}$$

$$\varphi(\vec{x}) = g(\vec{x}), \quad \forall \vec{x} \in \Gamma, \tag{4.3}$$

discretised by P^1 finite elements on a triangulation of the domain Ω. We denote by $q^j, j \in \{1, \ldots, n_v\}$ the vertices of this triangulation. The problem then amounts to the solution of a linear system that allows the determination of approximate values of the solution at each of the internal vertices of the triangular mesh.

By virtue of Proposition 3.1 and of Remark 3.3, this system can be written in an approximate fashion (i.e. up to a term of order $(1/C)$, where C is a very large constant),

$$A\Phi = F, \tag{4.4}$$

where A is the $n_v \times n_v$ symmetric and positive definite matrix whose entries are

$$A_{ij} = \int_{\Omega_h} \nabla w^i(\vec{x}) \cdot \nabla w^j(\vec{x}) \mathrm{d}\vec{x},\ i \neq j, \quad A_{ii} = C, \tag{4.5}$$

and where F is a vector of dimension n_v whose components are

$$F_j = \int_{\Omega_h} (f\, w^j)(\vec{x}) \mathrm{d}\vec{x}, \text{ if } q^j \notin \Gamma, \quad F_j = Cg(q^j), \text{ otherwise.} \tag{4.6}$$

The unknown of system (4.4) is the vector $\Phi = (\varphi_i)_{i \in \{1,\dots,n_v\}}$.

In the previous chapter, we solved this system with a direct method, by factorising the matrix A by the Choleski algorithm. We also took into account the particularly sparse band structure (i.e. within the band, an enormous number of zeros remain) of this matrix, to store it in a very 'compact' format , the Compressed Sparse Row format. Unfortunately, the Choleski factor does not inherit the same structure: it only has a band structure, and one cannot hope for anything better. Thus, this method, though fast, is costly in memory space.

The idea here is to not perform an actual Choleski factorisation, but an 'approximate' factorisation, i.e. we shall seek a matrix L^* 'close' to L, but with a sparse structure. By doing this, we cannot solve system (4.4) exactly, and we shall choose to solve it iteratively.

An example of an iterative method would be to write schematically (4.4) in the form

$$L^* L^{*T} \Phi = (L^* L^{*T} - A)\Phi + F, \tag{4.7}$$

then, starting from any initial data Φ^0, to define Φ^{m+1} as a function of Φ^m via the recurrence relation:

$$L^* L^{*T} (\Phi^{m+1} - \Phi^m) = -A\Phi^m + F.$$

This iterative method is convergent if the matrix $L^* L^{*T}$ is 'close' to the matrix A in the sense that the eigenvalues of $M = Id - (L^* L^{*T})^{-1} A$ must be of modulus strictly less than 1. In fact, if we denote by $\hat{\Phi}$ the solution of (4.7), then the error at iteration $m + 1$ can be written as a function of the error at the previous iteration $\epsilon^m = \Phi^m - \hat{\Phi}$ through the relation: $\epsilon^{m+1} = M\epsilon^m$.

Among all possible iterative methods, we choose the conjugate gradient method for its speed. The construction of L^* that we present is carried out via an 'incomplete' Choleski factorisation of A, and the conjugate gradient algorithm is preconditioned by the matrix L^*L^{*T}, which makes it especially efficient. We actually have several possible choices for the preconditioning matrix, knowing that the initial aim is to choose a relatively sparse matrix; but we shall see that the closer the preconditioning matrix is to the inverse of A, the faster the algorithm converges: there is thus a compromise to be found! Last, let us note that the methods presented in the previous two chapters are particular cases of the preconditioned conjugate gradient algorithm that follows, and which we shall abbreviate by PCG.

4.1.1 Incomplete Choleski Factorisation

The principle is simple: rather than computing the Choleski factor L of the matrix A, we shall build an approximation L^*. To do this, we could for instance decide to 'select' some indices, and decide that L^* coincides with L on this set of indices, and is zero outside. In other words, $L^*_{ij} = L_{ij}$ if $(i, j) \in I_0$, $L^*_{ij} = 0$ otherwise, where

$$I_0 \subset \{(i, j) : \ 1 \leq j \leq i \leq n_v\}. \tag{4.8}$$

This method is too costly, and we shall not use it, since it requires the knowledge of L.

We prefer to combine tightly the construction of L^* with that of L. Specifically, this means that we shall add a test in the Choleski algorithm, to determine whether a couple (i, j) belongs to the index set I_0: otherwise, we shall not construct the corresponding L_{ij} element. A natural choice for the set I_0 is given by the very structure of the matrix A, that is:

$$I_0 = \{(i, j) : \ \ A_{ij} \neq 0, \ \ j \leq i\}. \tag{4.9}$$

This condition means, in particular, that $L_{ij} = 0$ if points q^i and q^j are not neighbours (but this is not a necessary and sufficient condition). Indeed, since this criterion allowed us to select which elements of the matrix A we would actually store in the *aa* array, we shall choose for the index set I_0 the set:

$$I_0 = \{(i, j) : j \leq i, \, q^i \text{ and } q^j \text{ neighbours } \}. \tag{4.10}$$

Another possible choice is to decide that $L_{ij} = 0$ if points q^i and q^j are not 'neighbours by marriage', i.e. if there is no index k such that q^i and q^k on the one hand, q^k and q^j on the other hand are neighbours. In other words, we demand that L^* has the same sparsity structure as the matrix A^2.

These different choices give an L^*L^{*T} matrix close to the starting matrix A, but nothing ensures us that this new matrix is positive definite! This is a drawback of the method, that fortunately only rarely occurs in practice. Schematically, the algorithm for constructing L^* is as follows:

```
DO j=1,ns
```

$$L^*_{jj} = \sqrt{A_{jj} - \sum_{k<j} L^{*2}_{jk}}$$

```
DO m=1,ns-j
   IF (j+m,j) ∈ I0 THEN
```

$$L^*_{j+m,j} = \left(A_{j+m,j} - \sum_{k<j} L^*_{j+m,k} L^*_{j,k} \right) / L^*_{jj}$$

```
   ELSE
```

$$L^*_{j+m,j} = 0$$

```
   ENDIF
  end m loop
end j loop
```

One should note that nothing, in this computation, ensures that the quantity under the square root is positive, nor that the denominator L^*_{jj} is not zero. One must, in theory, add a test to make sure the quantity under the square root in the expression that defines L^*_{jj} is strictly positive.

4.1.2 COMPUTATION AND CSR STORAGE OF L^*

With the choice (4.10) that we made for the index set, L^* will have the same CSR storage as A, so that it is unnecessary to create additional index arrays *il* and *jl*. Only one additional array *al* is necessary. The construction is done as in the following program, where, for a change, we have performed the 'pseudo factorisation' line by line, ending with the computation of the diagonal element. Up to this modification, the program is similar to, but obviously much simpler than, that of the previous chapter.

```
C File ICholes.for
C ----------------------------------------
C SUBROUTINE ICholeski(nv,aa,ia,ja,al)
C ----------------------------------------
C This subroutine performs the incomplete Choleski
C factorisation of the matrix A stored in CSR format
C in array aa with auxiliary index arrays ia and ja.
C ia(k): index in array aa of the first element of
C   ith line of A stored in aa
C ja(k): column index j of element Aij stored in array
C   aa at index k
C The result is stored in array al, that uses the same
C auxiliary index arrays ia and ja as aa
C The computation of the incomplete factorisation is done
C line by line, ending with the computation of the diagonal
C element
C
C
```

```
      INTEGER ia(*), ja(*)
      DIMENSION aa(*),al(*)
C
      DO 30 i=1,nv
        DO 34 k=ia(i),ia(i+1)-2
          j=ja(k)
          alij=aa(k)
          ik=ia(i)
          jk=ia(j)
35      IF(ik.LE.(ia(i+1)-2)) THEN
          jlik=ja(ik)
36        IF(jk.LE.(ia(j+1)-2)) THEN
            jljk=ja(jk)
              IF(jlik.GT.jljk)THEN
                jk=jk+1
                GOTO 36
              ELSEIF(jlik.LT.jljk)THEN
                ik=ik+1
                GOTO 35
              ELSE
                alij=alij-al(ik)*al(jk)
                ik=ik+1
                jk=jk+1
                GOTO 35
              ENDIF
            ENDIF
          ENDIF
          al(k)=alij/al(ia(j+1)-1)
34      CONTINUE
        alii=aa(ia(i+1)-1)
        DO 33 k=ia(i), ia(i+1)-2
          alii=alii-al(k)**2
33      CONTINUE
          al(ia(i+1)-1)=sqrt(abs(alii))
30    CONTINUE
      END
```

4.1.3 Preconditioned Conjugate Gradient Algorithm for Solving Linear Systems

Let us briefly recall the PCG algorithm for solving linear systems. We shall first assume that the matrix A, of size $N \times N$, is symmetric and positive definite, then we shall give the changes the algorithm needs if A is only semi-positive definite.

First of all, the simple conjugate gradient algorithm, i.e without preconditioning (which we shall abbreviate by CG), for solving the linear system $Ax = b$ consists in applying the conjugate gradient algorithm, as shown in chapter 2 to the minimisation of the quadratic functional

$$E(x) = \tfrac{1}{2}x^T Ax - b^T x, \tag{4.11}$$

whose gradient at a point x is precisely $Ax - b$. We shall sometimes call 'non-linear conjugate gradient' the general conjugate gradient introduced in chapter 2 for computing the minimum of the not necessarily quadratic functional, of which algorithm CG is a particular case.

Having chosen an invertible $N \times N$ matrix U, the preconditioned conjugate gradient can be seen as applying the previous algorithm to the solution of the system $(U^{-1})^T A U^{-1} z = (U^{-1})^T b$ obtained after the change of variables $z = Ux$; the matrix of this new system is, from the hypotheses on A and U, symmetric and positive definite. Let us note $C = U^T U$; this is a symmetric and positive definite matrix as soon as U is invertible, so that it induces an inner product $\langle . , . \rangle_C$ on $\mathbb{R}^N \times \mathbb{R}^N$, as well as a norm $||.||_C$, given by the relations:

$$\langle a, b \rangle_C = a^T C b; \quad ||a||_C = (a^T C a)^{\frac{1}{2}} \tag{4.12}$$

This matrix C is called the 'preconditioning matrix' and, after a change of variables to 'eliminate' the matrix U (whose only use was to derive algorithm PCG from algorithm CG), the conjugate gradient algorithm preconditioned by this matrix C can be written as:

General PCG Algorithm

0 Initialisation:
choose the preconditioning matrix $C \in R^{N \times N}$ (it must be symmetric and positive definite), the accuracy ϵ for the computation, an initialisation x^0 for x (0 for instance), as well as the following initialisation for g^0 and h^0: $g^0 = h^0 = -C^{-1}(Ax^0 - b)$; let $n = 0$; last, choose (for safety's sake) a maximum *kMax* for the number of iterations;

1 Compute:

$$\rho^n = \frac{\langle g^n, h^n \rangle_C}{\langle h^n, C^{-1} A h^n \rangle_C} = \frac{\langle g^n, h^n \rangle_C}{h^{nT} A h^n} \tag{4.13}$$

$$x^{n+1} = x^n + \rho^n h^n \tag{4.14}$$

$$g^{n+1} = g^n - \rho^n C^{-1} A h^n \tag{4.15}$$

$$\gamma^n = \frac{||g^{n+1}||_C^2}{||g^n||_C^2} \tag{4.16}$$

$$h^{n+1} = g^{n+1} + \gamma^n h^n \tag{4.17}$$

2 Evaluate the stopping test: if $||g^{n+1}||_C^2 < \epsilon ||g^0||_C^2$, stop ($x^{n+1}$ is the approximate solution of the problem, for the given accuracy ϵ); otherwise increment n by 1, and go back to step **1**.

We shall show that this algorithm is 'well defined' (we see it might not be if the denominators of (4.13) and (4.16) ever became zero); in other words, we shall study its 'degeneration cases' and show they are favourable cases, that is cases where the algorithm has already converged before degenerating. More generally, we shall now recall some properties of the PCG algorithm.

4.1.3.1 Properties of the PCG algorithm

Let us first make some remarks on this algorithm, which we said was a particular case of the non-linear conjugate gradient of chapter 2. In particular, we note that: i) by induction, and because of the definition of $g^0 = -C^{-1}(Ax^0 - b)$, one deduces from relations (4.14) and (4.15) that

$$\forall n, \quad g^{n+1} = -C^{-1}(Ax^{n+1} - b); \tag{4.18}$$

in other words, $-Cg^{n+1}$ is the gradient of the functional E defined by (4.11) evaluated at point x^{n+1};

ii) modulo this remark, the scalar ρ defined by (4.13) is exactly the same as the one in chapter 2, and this number realises the minimum on $\mathbb{R}$ of the functional $(\rho \to E(x^n + \rho h^n))$.

The scalar γ is chosen so as to ensure orthogonality between two successive descent directions, as we shall show in Lemma 4.1, where we shall prove, in particular, that $\gamma^n = -(g^{n+1T}Ah^n)/(h^{nT}Ah^n)$. The computation of γ actually results from a Gram–Schmidt orthogonalisation procedure. Let us assume we have constructed a basis $\{h^i\}_1^N$ of $\mathbb{R}^N$ satisfying the following property: for any $(i, j), i \neq j$, $h^{i^T}Ah^j = 0$; then to solve the linear system $Ax = b$, it is enough to find the components x_j of x in this basis, and as they satisfy the N relationships $h^{i^T}A(\sum x_j h^j) = h^{i^T}b$, they are given by $x_j = h^{j^T}b/(h^{j^T}Ah^j)$.

We refer the reader to the Appendix for the proof of the following result.

Lemma 4.1 Assume that for any integer $k \leq n_0$ we have $g^k \neq 0, h^k \neq 0$, and let n be any integer less or equal to n_0. The scalar γ^n satisfies

$$\gamma^n = -\frac{g^{n+1T}Ah^n}{h^{nT}Ah^n}, \tag{4.19}$$

and we have the following orthogonality relations:

$$h^{jT}Ah^n = 0, \quad \forall j < n, \tag{4.20}$$

$$\langle g^n, g^j \rangle_C = 0, \quad \forall j < n, \tag{4.21}$$

$$\langle g^n, h^j \rangle_C = 0, \quad \forall j < n. \tag{4.22}$$

We also have:

$$\rho^n = \frac{\|g^n\|_C^2}{h^{nT}Ah^n} \tag{4.23}$$

We shall deduce from these properties that the algorithm is well defined and that it is convergent. More precisely, we have

Corollary 4.2 Algorithm PCG is well defined, and it converges in at most N iterations.

PROOF Let us first study the 'degeneration cases' for the algorithm. These can only occur if $g^n = 0$, or if $h^n = 0$ (we have assumed A was positive definite). In the first case, it is clear from (4.18) that this means that x^n is the solution of the system $Ax = b$, and by the way, this is the reason the stopping test of the program uses the norm of g^n. In the second case, let us assume $g^n \neq 0$ (otherwise we are back to the first case), and let us assume that n is the first index for which $h^n = 0$; then from (4.17), we have $g^n = -\gamma^{n-1} h^{n-1}$. If we take the inner product of this equation with $(h^{n-1})^T C$, we obtain, from (4.22) $-\gamma^{n-1} \|h^{n-1}\|_C^2 = 0$, which implies $h^{n-1} = 0$, since $\gamma^{n-1} \neq 0$, and thus leads to a contradiction. In other words, h^n can only become zero if g^n itself becomes zero, and in that case x^n is the sought solution. In summary, we have shown that the degeneration cases for this algorithm are 'favourable', as they can only occur after the algorithm has converged, and thus after the program has stopped.

The algorithm is thus well defined; let us now show that it is convergent. This results directly from the orthogonality relations of Lemma 4.1, since the vectors g^n, if they are not zero, are all orthogonal to one another, and thus form an independent set in the space $\mathbb{R}^N$. There can thus be at most N non-zero g^n vectors,
which shows that the algorithm converges in at most N iterations. □

We give an additional property of the conjugate gradient (here C is the identity matrix) expressed in the following proposition, whose proof has been deferred to Appendix 4.A:

Proposition 4.3 *Let us assume that all the $g^i, i \in \{1, .., n_0\}$, are non-zero, and let $n \leq n_0 + 1$. The approximate solution x^n obtained at iteration n of the conjugate gradient satisfies*

$$E(x^n) \leq E(x), \quad \forall x \in x^0 + K_n, \tag{4.24}$$

where K_n is the vector space spanned by the n first iterates, with respect to the matrix A, of the initial residual, i.e.

$$K_n = \text{span}\{g^o, Ag^o, A^2 g^o, \ldots A^{n-1} g^o\}, \tag{4.25}$$

and where E is the functional defined by (4.11). The space K_n also satisfies:

$$K_n = \text{span}\{g^o, g^1, \ldots g^{n-1}\} = \text{span}\{h^o, h^1, \ldots h^{n-1}\}. \tag{4.26}$$

Remark 4.4 The space K_n is called a *Krylov space*, and we shall have an opportunity to meet it again in chapter 6, in connection with the GMRES algorithm, for the solution of not necessarily symmetric problems. □

Remark 4.5 Proposition 4.3 shows the optimal nature of the conjugate gradient method, since, according to the previous remark, x^n achieves the minimum of the functional E on the space $x^0 + \text{span}\{h^o, h^1, \ldots h^{n-1}\}$, whereas in the beginning, the aim was only to minimise iteratively along each descent direction. □

If now the symmetric matrix A is only semi-definite positive ($detA = 0$), one can ask the following questions: can we still use this algorithm? If so, under which conditions, and what changes should we make to it? Last, is the algorithm still convergent? The answers are in the following proposition, whose proof is in Appendix 4.A.

Proposition 4.6 *If A is only semi-definite positive, and if b lies in the range of A, then if one initialises the PCG algorithm with an x_0 such that Cx^0 lies in the range of A, the algorithm converges towards the unique solution x' of the linear system $Ax = b$ satisfying* $Cx' \in \text{range}(A)$.

4.1.3.2 Choice of the preconditioning matrix

The PCG method for a system of size N converges in at most N iterations, but N is usually a very large number. The following proposition, whose proof can be found for instance in Lascaux and Theodor (1984), Axelsson and Barker (1984) or Golub and Meurant (1982), shows that convergence is fast if the preconditioner is appropriately chosen.

Proposition 4.7 *Let x^* be the solution of the linear system $Ax = b$ and let x^k be the solution at iteration k of the conjugate gradient method. If we denote, analogously to (4.12), by $||x^k - x^*||_A = ((x^k - x^*)^T A(x^k - x^*))^{\frac{1}{2}}$, the error at iteration k, we have:*

$$||x^k - x^*||_A^2 \leq 4\left(\frac{\sqrt{\mu(A)} - 1}{\sqrt{\mu(A)} + 1}\right)^{2k} ||x^0 - x^*||_A^2 ,$$

where we denote by $\mu(A)$ the condition number of the matrix A.

We recall that the condition number of a matrix is defined, with respect to the choice of a matrix norm $||.||$, by the relation $\mu(B) = ||B||\,||B||^{-1}$. It is a number that, provided the matrix norm is subordinate to a vector norm, is always larger or equal to 1, and is equal to 1 if B is the identity matrix. In the case of a symmetric positive definite matrix, it can be expressed as the ratio between the largest and the smallest eigenvalue of the matrix.

If we want to apply the result of this proposition to the preconditioned conjugate gradient as we have previously built it, it is enough to substitute $\mu((U^{-1})^T AU^{-1})$ for $\mu(A)$. We deduce from this that, if we choose $U = A^{1/2}$ *, then this means on the one hand that $C = U^T U = A$, and on the other hand that

* A being a symmetric positive definite matrix, it is diagonalisable, with an orthonormal basis of eigenvectors, i.e. we have $A = PDP^{-1}$, with P a regular matrix and D a diagonal matrix all of whose entries are positive; $A^{1/2}$ denotes by definition the matrix $PD^{1/2}P^{-1}$, where $D^{1/2}$ is the diagonal matrix whose diagonal elements are the square roots of the diagonal elements of D.

$(U^{-1})^T A U^{-1}$ is the identity matrix, which gives an optimal result for the error estimate. The previous proposition thus proves that in the limit case where $C = A$ (and taking $x^0 = 0$), the method converges in one iteration, which is understandable since we have actually inverted A! In the general case, the 'closer' to A the preconditioning matrix C, the better the convergence.

There is obviously a compromise to be found in the choice of C, since on the one hand we wish to have a matrix C as simple and as sparse as possible, so we can easily invert it and save memory space, and on the other hand the previous proposition leads us to choose a matrix C as close to A as possible, so as to accelerate the convergence of the algorithm. Experience shows that the following choices, listed in increasing order of complexity (and thus also of performance), are good choices:

$$C_{ij} = A_{ii}\delta_{ij}(\text{ diagonal of } A) \tag{4.27}$$

$$C_{ij} = A_{ij} \text{ if } |i-j| \leq 1 \text{ (3 diagonals of } A), \quad 0 \text{ otherwise;} \tag{4.28}$$

$$C = L^*(L^*)^T, L^* \text{ incomplete Choleski factor of } A. \tag{4.29}$$

$$C = LL^T, L = \sqrt{\frac{\omega}{2-\omega}}\left(\frac{D}{\omega} - E\right)D^{-1/2}, \text{ with the decomposition}$$
$$A = D - E - E^T \text{ where D is the diagonal part of A and} \tag{4.30}$$
$$-E \text{ its strictly lower triangular part}$$

The last choice is called 'SSOR preconditioning' (Symmetric Successive Over Relaxation), the relaxation parameter ω being chosen larger than 1. For this method, let us point out the following result, whose proof is to be found in Joly (1990), or Axelsson (1994):

Lemma 4.8 If the relaxation parameter ω lies in the interval $]0,2[$, the eigenvalues of $C^{-1}A$ are real and lie in the interval $]0,1[$. In particular, for $\omega = 1$, the maximum eigenvalue of $C^{-1}A$ is 1.

Remark 4.9 Let us recall that, in theory, it is necessary for the matrix C to be positive definite, which is not always easy to prove, even for the choices we recommended. However, in practice, this usually poses no problems. □

Remark 4.10 One can also prove 'super convergence' results [see Polak (1971), for instance]: the sequence $\{x^k\}$ converges faster than any geometric progression, but this result assumes that the number of iterations is large with respect to N. □

4.2 PROGRAMMING THE METHOD

The algorithm is programmed thanks to formulae (4.13)–(4-17): we first give a sketch of the main loop, then we shall make some comments.

Program sketch for (4.13)–(4.17)

$$b = Ah^n$$

$$\rho = \frac{(cg)^n \cdot h^n}{h^n \cdot b}$$

$$(cg)^{n+1} = (cg)^n - \rho b$$

compute $C^{-1}b$; store the result in b

$$x^{n+1} = x^n + \rho h^n$$
$$g^{n+1} = g^n - \rho b$$
$$(gcg)^{n+1} = g^{n+1} \cdot (cg)^{n+1}$$
$$\gamma = \frac{(gcg)^{n+1}}{(gcg)^n}$$
$$h^{n+1} = g^{n+1} + \gamma h^n$$

In some textbooks b is not introduced. But then the computation of Ah^n would occur twice, so this vector is stored, at each iteration, in a vector which we denote, for the moment, by ah (in practice, it is not necessary to use an additional array to store this vector, as the array holding the right hand side b at the beginning can be destroyed at the end of the initialisation step, and can then be used as an auxiliary storage array). On a different issue, it is clear that the matrix C will *never be inverted*, which means that writing $y = C^{-1}z$ is to be understood as meaning *solve a linear system*, namely the system $Cy = z$ (here again, we shall actually use the array b to store temporarily the value of y). For that reason, we shall use an additional array, denoted by cg to store the product of the matrix C by the vector g, which enables us to split the computation in (4.15) into two steps: a first step where we compute the new value of cg, and a second step where we solve a linear system ($Cy = ah$), so as to compute the new value of g. We must also store the values of the inner products of the vectors g and cg, so as to compute γ; we denote by gcg the scalar in which this result is stored. Memory storage thus essentially includes two matrices A and C and five vectors: the unknown x, g and the gradient cg, the descent direction h and, last, the right hand side b.

We now give the full subroutine implementing (4.13)–(4.17).

```
C File GradCP.for
C
C -----------------------------------------------------------
C SUBROUTINE GradC(n,aa,ia,ja,al,b,x,g,h,cg,kMax,eps)
C -----------------------------------------------------------
C This subroutine solves the linear system Ax = b by the
C Preconditioned Conjugate Gradient method, in at most
C kMax iterations.
C The stopping test is |Ax^k - b| < eps |Ax^0 - b|
C The matrix A of the linear system, of size n x n, is
C stored in Compressed Sparse Row (CSR) format in array aa;
C this storage format also needs auxiliary index arrays
C ia and ja, as explained in Chapter 3.
C The preconditioning matrix C is the incomplete Choleski
C factor of A (C = L*(L*)^T) and the matrix L* is stored
C in CSR format in array al, with the same index
C arrays ia and ja as A
C
```

```
C     input:
C     ------
C       system matrix A stored in CSR format
C         in arrays (aa,ia,ja);
C       incomplete Choleski factor of A stored
C         in arrays (al,ia,ja);
C       equation right hand side stored
C         in array b;
C       maximum number of iterations kMax,
C         computational accuracy eps;
C
C
C
C     N.B. the right hand side computation is only
C       used during the initialisation step, the
C       array b is then overwritten and used as an
C       auxiliary array for the computation, in
C       particular for storing the product of the
C       matrix A by h, then for storing the solution
C       y of the system Cy = Ah.
C
C
C     output:
C     -------
C       x: array containing the values of the
C         solution at each vertex of the mesh ;
C       gcg: scalar containing the square of the norm
C         of g (in practice this number should
C         decrease during the iterations);
C
C     other arrays:
C     -------------
C       g, cg: arrays containing respectively the
C         values of g^n and Cg^n;
C       h: array containing the descent direction h^n
C         at the very beginning of the program, this
C         array is used to store the product of A by x^0);
C
C
C
C     subroutines:
C     ------------
C       IAmul: performs the product of a matrix by
C         a vector, with the matrix stored in CSR
C         format,
C       ISolvSys: solution of the linear
C         system LL^T x = y by forward and backward substitution,
C         for a matrix L stored in CSR format;
C       These two subroutines are in file GCPLaplac.for
C
C
```

```
      DIMENSION x(*),b(*),g(*),h(*),cg(*)
      DIMENSION aa(*),ja(*),ia(*),al(*)
C
      DO 1 k=1, kMax
C
        IF (k.EQ.1) THEN
          CALL IAmul(n,aa,ia,ja,x,h)
          DO 2 i=1,n
            cg(i)=b(i)-h(i)
            g(i)=cg(i)
2         CONTINUE
          CALL ISolvSys(n,g,al,ia,ja)
          gcg=0.
          DO 3 i=1,n
            gcg=gcg + g(i)*cg(i)
            h(i)=g(i)
3         CONTINUE
        ENDIF
C
        CALL IAmul(n,aa,ia,ja,h,b)
        hAh=0.
        DO 10 i=1,n
          hAh=hAh+h(i)*b(i)
10      CONTINUE
        ro=gcg/hAh
        DO 11 i=1,n
          cg(i) = cg(i) -ro*b(i)
11      CONTINUE
        CALL ISolvSys(n,b,al,ia,ja)
        DO 12 i=1,n
          x(i)=x(i)+ro*h(i)
          g(i)=g(i)- ro*b(i)
12      CONTINUE
        gamma=gcg
        gcg=0
        DO 13 i=1,n
          gcg=gcg + g(i)*cg(i)
13      CONTINUE
C
C
        write(*,*) 'iteration:',k,'gcg=',gcg
        IF(k.EQ.1) eps=eps*gcg
        IF (gcg.LT.eps)  RETURN
        gamma=gcg/gamma
        DO 14 i=1,n
          h(i)=g(i)+gamma*h(i)
14      CONTINUE
C
C
1     CONTINUE
      END
```

The complete program for solving the model problem using this method can be found in Appendix 4.B, together with the subroutines used by the previous program.

4.3 APPLICATION TO THE LAPLACE EQUATION

4.3.1 Numerical Results

We go back to the example at the end of Section 2.6.3, and we program the method developed in this chapter with two different preconditioners: a diagonal preconditioning (abbreviated by 'PCG Diag.' in the table below), and a preconditioning by incomplete Choleski factorisation (abbreviated 'PCG. In. Chol.'.) The second method is usually faster unless the mesh is very coarse,which corresponds to a very small number of unknowns: the time spent in the factorisation becomes very large with respect to the time due to the iterations, as the conjugate gradient algorithm then converges very rapidly. The following table gives the computing time (CPU time), and the number of iterations (nb iter.) necessary to get the residual down to 10^{-6} times the initial residual. By comparing this table with that given in the previous chapter, we can compare the PCG method with the direct Choleski method.

	PCG Diag.		*PCG In. Chol.*	
nv	nb iter.	CPU time	nb iter.	CPU time
49	14	3.8"	7	4"
170	30	14"	10	14"
625	57	85"	21	74"
1741	115	412"	35	346"

4.3.2 Other Preconditioners

The search for an efficient preconditioner for the Laplacian is a difficult subject, that is still an active research topic at the present time [see Joly (1990) or Axelsson (1994)]. Let us give an example of a preconditioner coming from an incomplete Choleski factorisation of A, with the same sparsity structure as the matrix A^2. Intuitively, if we keep the elements L_{ij} of the Choleski factor L of A, corresponding to 'neighbours by marriage' vertices q^i and q^j (i.e. there exists k such that q^i and q^j on the one hand, and q^j and q^k on the other hand are neighbours), the new matrix $\hat{L}$ so defined is closer to L than the previous matrix L^*. Now, it is easy to see that if $(A^2)_{ij} \neq 0$, then vertices q^i and q^j are 'neighbours by marriage'.We shall thus determine all pairs $\{i, j\}$ for which $(A^2)_{ij} \neq 0$, then apply incomplete Choleski factorisation to this index set. The algorithm formally splits in the following way:

1. Build index arrays $ia2, ja2$ corresponding to the CSR storage of matrix A^2, obviously without storing this matrix;

2. Store A in CSR format in an array $aa2$ associated with the previous index arrays $ia2, ja2$; this means that this new array $aa2$ contains the usual values of array aa, as well as zeros where A_{ij} is zero.
3. Use subroutine ICholeski with these arrays $aa2, ia2, ja2$ to build the desired incomplete factor $\hat{L}$.

However, experience shows that, although using this new preconditioner improves the speed of convergence of the algorithm, computation time remains almost the same, as each iteration requires many more computations, since the new matrix has many more non-zero coefficients! An idea for improving the method is to suppress all coefficients of the matrix that are too small. This leads to a fourth step in the algorithm, defined by:

4. Replace coefficients $\hat{L}_{ij}$ by 0 as soon as they satisfy: $\hat{L}_{ij} < \alpha \min\{A_{ii}, A_{jj}\}$ and $A_{ij} = 0$.

In the authors' experience, this is the best preconditioner of all those listed for the Laplacian [Bespalov *et al* (1992)]. We leave programming this algorithm as an exercise to the reader [Lucquin and Pironneau (1997)].

4.4 AUTOMATIC MESH REFINEMENT

4.4.1 Generalities

It is clear that the performance of the method is tightly linked to the 'quality' of the mesh. The computation time is at least proportional to the number of vertices and the accuracy is linked to the product of the mesh size by the norm of the second derivatives of the solution (cf. Theorem 2.28). An idea to improve this accuracy, without significantly increasing the computation time, is only to'refine' (i.e. increase the number of vertices of) the mesh in the areas where the second derivatives of the solution are large.

4.4.1.1 Choosing a refinement criterion

As we usually do not know the solution to the problem, we shall try to obtain an estimate of its second derivatives; several possible solutions lead to that goal, and we shall propose one. Let us call φ_h a continuous, piecewise P^1 (i.e. P^1 on each triangle), approximation to the solution φ. Let us interpolate the function $\nabla\varphi_h$, which is a piecewise constant function, by a continuous (vector valued) function v_h, piecewise linear, whose value at each vertex of the mesh is the average of the gradient of φ_h on all triangles with point q^i as a vertex; more precisely:

$$v^i = v_h(q^i) = \left\{ \sum_{k \in K_i} \nabla\varphi_h|_{T_k} |T_k| \right\} \Big/ \left\{ \sum_{k \in K_i} |T_k| \right\}, \tag{4.31}$$

where K_i is the set of all indices of the triangles with point q^i as a vertex. Then, $v_h = \sum_{i=1}^{nv} v^i w^i$, the functions w^i being the hat functions introduced in chapter 2 (i.e. w^i is a continuous function on the whole computational domain, its value is 1 at vertex q^i and is 0 at all other vertices), and ∇v_h is an approximation to the second derivative matrix of φ.

Another method consists in refining the mesh where the residual error is the largest. Let φ solve the equation $\Delta\varphi + f = 0$ and φ_h solve the approximate variational problem (cf. equation (3.15)):

$$\text{find } \varphi_h \in H^1_{gh} \quad \text{such that}$$
$$\forall w_h \in H^1_{0h}, \quad \int_{\Omega_h} \nabla\varphi_h(\vec{x}) \cdot \nabla w_h(\vec{x}) \mathrm{d}\vec{x} = \int_{\Omega_h} f(\vec{x}) w_h(\vec{x}) \mathrm{d}\vec{x}.$$

Let

$$r_h(w_h) = \int_\Omega [w_h(\Delta\varphi_h + f)](\vec{x})\mathrm{d}\vec{x}; \tag{4.32}$$

we have:

$$\begin{aligned} r_h(w_h) &= \sum_k \int_{T_k} [\, f w_h - \nabla\varphi_h \nabla w_h](\vec{x})\mathrm{d}\vec{x} + \sum_k \int_{\partial T_k} \left[\frac{\partial\varphi_h}{\partial n} w_h\right](\vec{x})\mathrm{d}\gamma(\vec{x}) \\ &= \sum_k \int_{\partial T_k} \left[\frac{\partial\varphi_h}{\partial n} w_h\right](\vec{x})\mathrm{d}\gamma(\vec{x}). \end{aligned} \tag{4.33}$$

The number $r_h(w^i)$ is a measure of the residual error $\Delta\varphi_h + f$ taken at point q^i, whereas $r_h(w^{i_1}) + r_h(w^{i_2}) + r_h(w^{i_3})$ is the error relative to triangle $(q^{i_1}, q^{i_2}, q^{i_3})$.

Other criteria are possible. For instance, one could compute, in the neighbourhood of a point q^i, a P^2 approximation $\tilde{\varphi}_h$ to φ by solving

$$\int_{\sigma_i} [\nabla(\tilde{\varphi}_h - \varphi_h)\nabla w_h](\vec{x})\mathrm{d}\vec{x} = 0, \forall w_h \in H^2_{0h},$$

where σ_i is the area of the plane formed by all triangles having point q^i as a vertex.

4.4.1.2 Subdividing the triangles

Once a refinement criterion has been chosen, as well as the computational accuracy ϵ, we shall decide to subdivide into four triangles each triangle for which the local error (for the chosen criterion) is larger than ϵ, according to the picture in Figure 4.1. In this picture, the triangle created at the centre has for its edges the midpoints of the edges of the initial triangle. An important task is then to *make the new triangulation admissible*, as it may no longer be so after triangles have been subdivided. To do this, we must count, for each triangle T_k that has not been subdivided, the number *ntdiv*(k) of 'neighbouring' triangles (i.e. those that share a common edge with it) that have been subdivided; there are four possibilities:

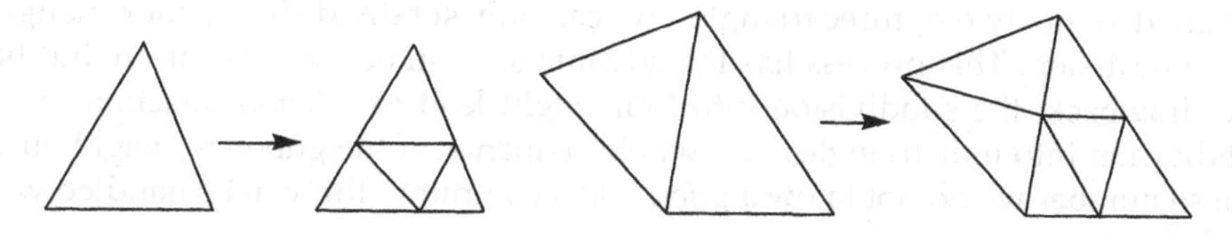

Figure 4.1 Subdivision of a triangle into four new triangles, and of a neighbouring triangle into two new triangles

ntdiv(k)=0: we do nothing;

ntdiv(k)=1: we subdivide triangle T_k into two triangles, as shown in Figure 4.1;

ntdiv(k)=2: we subdivide triangle T_k into three triangles, and this can be done in two ways (we choose the one for which the created triangles have the largest angles);

ntdiv(k)=3: we subdivide triangle T_k into four triangles, exactly as we did previously for its three neighbouring triangles.

The actual computation of this number *ntdiv*(k) presupposes that we have numbered the edges, as we explained in Section 3.3, then, we just have to go over the three edges of T_k, and check whether or not the other triangle having this edge in common with T_k has been subdivided into four triangles.

This process is reiterated as many times as is necessary, that is until the local error in each triangle becomes smaller than ϵ. The algorithm is the following:

Automatic mesh refinement algorithm

0. Choose ϵ and choose an error indicator r;

1. For $k = 1, \dots, n$, compute the local error $r|_{T_k}$; if this is larger than ϵ, subdivide triangle T_k into 4 triangles;

2. Make the triangulation admissible;

3. Go back to step **1** as many times as is necessary.

From a practical point of view, steps 1 and 2 can be done with 3 successive 'sweeps' over the mesh: in the first one, one marks triangles that will be subdivided into four, the second one marks those that must be subdivided (into two, three, or four) so as to make the mesh admissible, and in the last one the triangles are actually subdivided, that is, new triangles are created, and old ones are removed.

As it stands, this construction presents small problems from the standpoint of the regularity of the mesh. Indeed, let us imagine that the solution is very 'non-smooth' in a certain area; in the above process, the triangles neighbouring this area run the risk of being subdivided into two several times in a row, and always in the same way (i.e. the same angle always gets divided into two), so that some triangles could become very elongated very rapidly, which is unsuitable. One way we can remedy this difficulty is to decide that, if a triangle has already been subdivided into two or three triangles, we can only subdivide it into four triangles at the next step. This process has the advantage of smoothing the mesh, but has one drawback: the subdivision into four might lead to a 'chain reaction', i.e. a subdivision into four triangles of a whole sequence of 'neighbours' neighbours', whose number we do not know *a priori*. On a computer, this can be handled with the help of a tree.

Another way to 'smooth' the mesh is to transform it once in a while so as to make it Delaunay. We shall define this notion in Section 4.5.

Once the new mesh is created, we must rebuild the system matrix. This can be done *ab initio*. However, there exist methods that take into account the fact that the matrix has already been built on a coarser mesh: they are hierarchical bases methods, or multigrid methods, which we shall now present.

4.4.2 Hierarchical Bases

Let us introduce the method from a simple example [cf. Bank (1983)]. Let us imagine that a single new vertex, located on an edge, has to be added to the triangulation.Then only the two triangles sharing this edge must be subdivided, and they will each be subdivided into two new triangles, as shown on Figure 4.2.

The new linear system matrix differs little from the old one. Let us denote by $w^i, i = 1, ..., nv$, the basis 'hat' functions in the old triangulation, and $\tilde{w}^i$ the new 'hat' functions associated with the new mesh. We remark that the $w^i, i = 1, ..., nv$ functions still form an independent set, and that if we add the $\tilde{w}^m$ function associated with the newly created vertex, we obtain a basis for the discretisation space associated with the new mesh. Let us remark that this basis is different from the canonical 'hat functions' associated with the new mesh. The interest is that the new linear system to be solved can simply be written:

$$\begin{pmatrix} A & a \\ a^T & a_{mm} \end{pmatrix} \begin{pmatrix} \Phi \\ \tilde{\varphi}^m \end{pmatrix} = \begin{pmatrix} F \\ f^m \end{pmatrix}. \tag{4.34}$$

Let us explain the notations, and assume for simplicity's sake that the boundary conditions are homogeneous Dirichlet conditions ($\varphi|_\Gamma = 0$). The new matrix is nothing but the old one with an additional line and column, and we have:

$$\varphi_h(x) = \sum_{i \in I} \varphi_i w^i(x) + \tilde{\varphi}_m \tilde{w}^m(x), \tag{4.35}$$

$$A_{ij} = \int_{\Omega_h} (\nabla w^i \nabla w^j)(\vec{x}) \mathrm{d}\vec{x}, \; i, j \neq m,$$

$$a = (a_1, ..., a_{ns})^T, \; a_i = \int_{\Omega_h} (\nabla w^i \nabla \tilde{w}^m)(\vec{x}) \mathrm{d}\vec{x}, \tag{4.36}$$

$$a_{mm} = \int_{\Omega_h} (\nabla \tilde{w}^m \nabla \tilde{w}^m)(\vec{x}) \mathrm{d}\vec{x},$$

$$F = (F_1, ..., F_{ns})^T, \; F_j = \int_{\Omega_h} (f w^j)(\vec{x}) \mathrm{d}\vec{x}, \quad f^m = \int_{\Omega_h} (f \tilde{w}^m)(\vec{x}) \mathrm{d}\vec{x}. \tag{4.37}$$

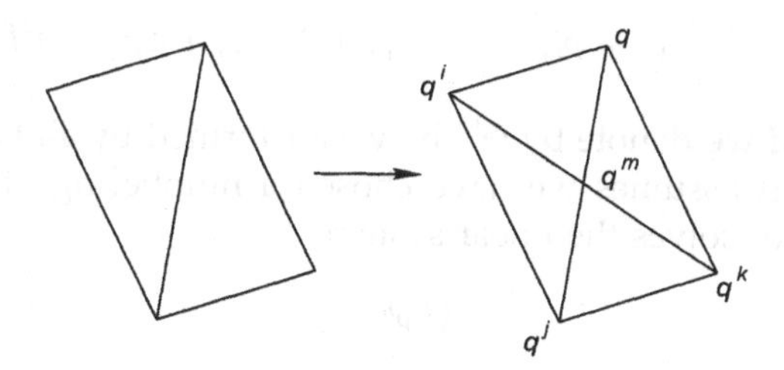

Figure 4.2 A simple illustrative hierarchical basis example

A programming sketch for this simple example is as follows:

1 Update the triangles by adding to the *me* array the indices for the four newly created small triangles; add to the coordinates array q those of the new vertex (whose index will be $nv+1$). Store the old numbers of vertices nv, and of triangles nt.

2 Update arrays aa, ia, ja. Here, CSR storage is particularly well suited to this task, since it is enough to add a line at the bottom of the lower part of the matrix. However, to compute the values of $aa(k)$, we need new information: the new vertex q^m is on an edge, and we have to know the indices of the triangles that share this edge. Thus, we must store the edges, a task we explained in chapter 3.

3 Complete the right hand side by computing f^m.

4 Update the incomplete Choleski factorisation. This task is also easy as it is only concerned with the last line of the matrix.

As far as the conjugate gradient method is concerned, we can initialise it with the old value of the approximate solution at the vertices of the old mesh, and for the new vertex, we can either assign it the value 0, or interpolate values at neighbouring vertices, provided they are old vertices.

In the general case, i.e. when there are several new vertices, the principle is exactly the same.

4.4.3 Multigrid Methods

If we apply this mesh refinement algorithm in a recursive manner, we obtain a whole mesh hierarchy imbricated within one another, with finer and finer meshes. In this particular case, it is possible to build a very efficient preconditioner for the conjugate gradient algorithm, from a multigrid method 'V-cycle'.

In order to explain the multigrid method, we shall use as an illustrative example the finite difference discretisation of the Laplace equation $-\Delta\varphi = f$ on a regular mesh in the plane, made up of elementary squares of side h (we refer to chapter 7 for details pertaining to the finite difference approximation of a partial differential equation). Let us denote by φ_{ij} the value of the approximate solution at vertex (ih, jh); when i and j vary, these values solve:

$$-(\varphi_{i-1,j} + \varphi_{i+1,j} + \varphi_{i,j+1} + \varphi_{i,j-1}) + 4\varphi_{ij} = h^2 f_{ij}. \tag{4.38}$$

In other words, if we denote by Φ^h the vector formed by all the unknowns φ_{ij} of this problem (this assumes we have chosen a numbering of the mesh vertices (ih, jh)), this vector solves the linear system

$$A^h \Phi^h = F^h, \tag{4.39}$$

where the matrix and the right hand side depend on the mesh size h. An iteration of the Gauss–Seidel relaxation method for solving this system can be written as:

$$\phi^n_{ij} = \tfrac{1}{4}(\phi^{n-1}_{i-1,j} + \phi^n_{i+1,j} + \phi^n_{i,j+1} + \phi^{n-1}_{i,j-1} + h^2 f_{ij}). \tag{4.40}$$

If we denote by $u^n = \phi^n - \varphi$ the error at iteration n, we have the recurrence relation:

$$u^n_{ij} = \tfrac{1}{4}(u^{n-1}_{i-1,j} + u^n_{i+1,j} + u^n_{i,j+1} + u^{n-1}_{i,j-1}). \tag{4.41}$$

This relation can be interpreted as a process for smoothing the error (the error at each point is replaced by its average over neighbouring points), and this explains why the error becomes smoother in the course of the Gauss–Seidel iterations (be careful: smoother does not mean smaller!).

Let us now assume that we have two grids, with one obtained from the other by a subdivision process analogous to the one presented above. We are provided with a fine mesh, of size h, and a coarser one, whose size is denoted by H ($h < H$). If we solve system (4.39) on each of the meshes, we can define r^h and r^H by

$$\left.\begin{array}{l} A^h\Phi^h = F^h \\ A^H\Phi^H = F^H \end{array}\right\} \Rightarrow \left\{\begin{array}{l} r^h \equiv A^h(\Phi^h - I_{H\to h}\Phi^H) = F^h - A^h I_{H\to h}\Phi^H \\ r^H \equiv A^H(\Phi^H - I_{h\to H}\Phi^h) = F^H - A^H I_{h\to H}\Phi^h, \end{array}\right. \tag{4.42}$$

where $I_{H\to h}$ is an interpolation operator that lets us construct, from a vector Φ^H defined on the coarse mesh, the values of ϕ^h at each of the vertices of the fine mesh. Conversely, the operator denoted by $I_{h\to H}$ is a restriction operator that selects values of vector Φ^h, so as to construct a new vector on the coarse mesh. Now r^h and r^H are sort of residual errors; let us change their definitions and take:

$$r^h \simeq I_{H\to h}[F^H - A^H\Phi^H], \quad r^H \simeq I_{h\to H}[F^h - A^h\Phi^h]. \tag{4.43}$$

Now consider the algorithm:

1 Choose an initialisation Φ^{0h} of Φ^h for the solution on the fine grid.

2 Perform a few iterations of the Gauss–Seidel method for solving $A^h\Phi^h = F^h$.

3 Compute the restriction of the residual error, $r^H = I_{h\to H}(F^h - A^h\Phi^h)$, on the coarse mesh.

4 Perform a few iterations of the Gauss–Seidel method for solving $A^H U^H = r^H$.

5 Compute an interpolation U^h of U^H by the relation: $U^h = I_{H\to h}U^H$.

6 Go back to step 3, substituting $U^h + \Phi^h$ for Φ^h.

Each iteration of this algorithm is called a V-cycle, because it starts with an error U^h defined on the fine mesh, continues with a computation of the error U^h on the coarse mesh, and ends on the fine mesh with a new error $[U^h]$. If there are several different grid levels, and not only two, step four can be solved by the same multigrid algorithm, and so on, until we reach the coarsest grid, where we can solve the linear system by a direct method.

If this method is only applied once, it can be seen as a linear mapping defining Φ^h from an initialisation Φ^{0h}. Let us denote be C^* the matrix of this mapping; we have: $\Phi^h = C^*\Phi^{0h}$. The matrix $C = (C^*)^{-1}$ can be used as a preconditioning matrix for the conjugate gradient algorithm, and the method is especially fast [Bank *et al* (1988)].

Remark 4.11 For programming the PCG algorithm, we need to know the matrix C explicitly. Let D be the diagonal matrix defined by $D_{ii} = A_{ii}$. Let L denote the lower triangular part of A, i.e. $L_{ij} = A_{ij}$, for $i > j$, and $L_{ij} = 0$ otherwise; likewise, $U = L^T$ is the upper triangular part of A, as A is symmetric. The Gauss–Seidel algorithm is based on the decomposition $A = (L + D) + U$, and the iteration matrix for this method is $-(L + D)^{-1}U$. We obtain $C = [-U^{-1}(L + D)]^n$, where n is the number of Gauss–Seidel iterations used in the above algorithm. In particular, computing $x \to Cx$ requires n 'backward solves', by virtue of the upper triangular nature of U. □

4.5 DELAUNAY TRIANGULATION

We shall describe a process allowing us to 'improve' a given triangulation, in the sense that we shall be able to maximise the lower bound for the angles of all triangles making up the triangulation. This triangulation, called a Delaunay triangulation, is defined by duality starting from the Voronoï polygon, which we shall now define. For further details we refer for instance to Risler (1991).

Let $(p^i)_{1\le i\le n}$ be a set of points in the plane, such that any four of them are never on a common circle. We have

Definition 4.12 The set of the V_i, $i \in \{1, ..., n\}$, defined below, is called the *Voronoï polygon* associated with the family of points p^i, $i \in \{1, ..., n\}$:

$$V_i = \{x \in \mathbb{R}^2, \text{ such that } |x - p^i| \le |x - p^j| \text{ for all } j \ne i\}, \tag{4.44}$$

where $|x - y|$ denotes the Euclidean distance between points x and y in the plane.

It is easy to see that the boundary of the V_i is made up from the medians of the segments $[p^i, p^j]$. Figure 4.3 shows the Voronoï polygon in case the set has either 3 or 6 points.

Definition 4.13 The *Delaunay triangulation* associated with the set of points p^i, $i \in \{1, ..., n\}$ is obtained by joining each point p^i to the various points p^j symmetric

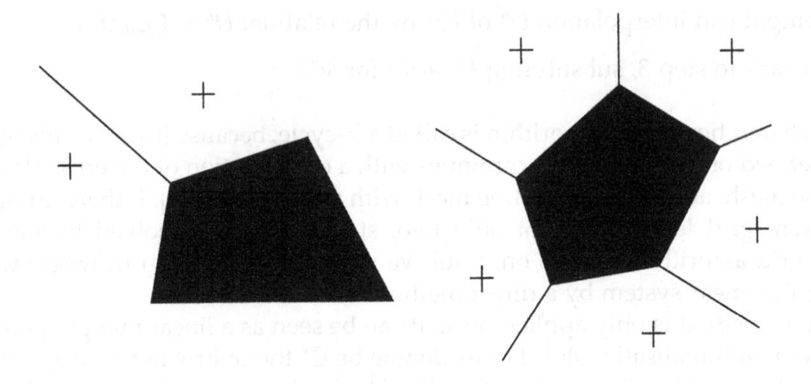

Figure 4.3 Examples of Voronoï polygons

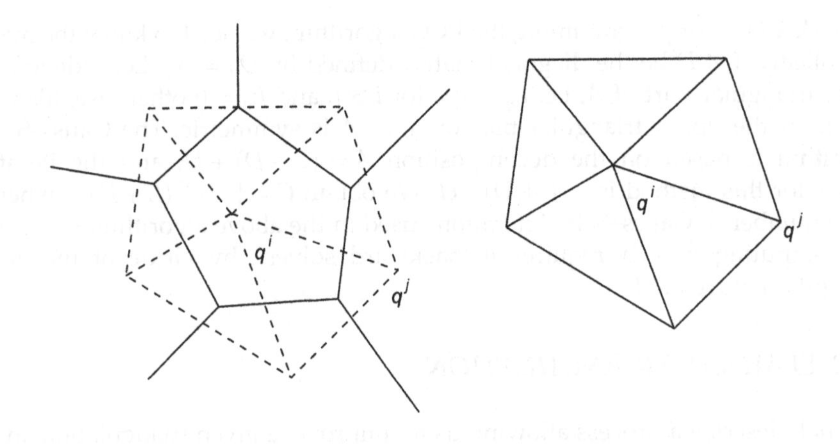

Figure 4.4 Voronoï polygon and Delaunay triangulation

of p^i with respect to the segments forming the boundary of V_i. This triangulation is obtained by 'duality' of the Voronoï polygon, i.e.:

(i) $[p^i, p^j]$ is an edge of the triangulation if and only if the intersection $V_i \cap V_j$ is not empty,

(ii) $p^i p^j p^k$ is a triangle if and only if $V_i \cap V_j \cap V_k$ is not empty.

Figure 4.4 shows the Delaunay triangulation and the Voronoï polygon in the neighbourhood of a point.

The Delaunay triangulation is unique, as is the Voronoï polygon, as soon as the family $p^i, i \in \{1, ..., n\}$, has no points on a common circle. Moreover, it satisfies the following property, whose proof can be found in Risler (1991).

Proposition 4.14 *The Delaunay triangulation $\mathcal{T}_D$ can be characterised by the fact that no circumcircle of a triangle in $\mathcal{T}_D$ contains a vertex p^i in its interior.*

From this geometric characterisation results the following corollary, whose proof is immediate, as can be seen in Figure 4.5.

Corollary 4.15 Let q^1, q^2, q^3 and q^4 be 4 non cocyclic points of the plane forming a convex quadrilateral; then

(i) there exist two triangulations of the convex hull of these four points: the Delaunay triangulation $\mathcal{T}_D$ and another triangulation $\mathcal{T}'$. They can be deduced from one another by 'swapping diagonals' in the quadrilateral $q^1q^2q^3q^4$.

(ii) the Delaunay triangulation is the one that maximises the lower bound for the triangle angles.

Property (ii) results from the fact that any angle in $\mathcal{T}_D$ is bigger than or equal to any angle in $\mathcal{T}'$ (cf. Figure 4.5). It also shows in which sense the Delaunay triangulation is a 'good' triangulation. However, let us remark that this triangulation does not minimize the maximum of the angles.

This corollary shows how to transform any triangulation so as to make it Delaunay. For each triangle T_k in the triangulation, whose vertices we denote by q^1, q^2 and q^3, we just have to go over its three edges, and for each one, to look for the fourth point q^4, i.e. the vertex not on this edge, but belonging to the triangle T_l located on the other side of the edge with respect to T_k. Then, we must check whether this point q^4 is in the interior of the circumcircle of triangle T_k. If that is the case, we modify the two triangles T_k et T_l by 'swapping diagonals' in quadrilateral $q^1q^2q^3q^4$ (to fix ideas, if the edge is segment $[q^1, q^3]$, triangles T_k and T_l are replaced by the triangles with vertices q^1, q^2, q^4 and q^4, q^2, q^3; we must of course take care to preserve the positive orientation when locally numbering the new triangles), then we redo the operation with that of the two newly formed triangles that has the smallest index. In the other case, we do nothing, as the triangulation $T_k \cup T_l$ is already Delaunay (which is precisely the case in Figure 4.5), and we go on to the next edge.

This process assumes we know the edge arrays defined in the previous chapter.

EXERCISE Carefully write and program the corresponding algorithm. □

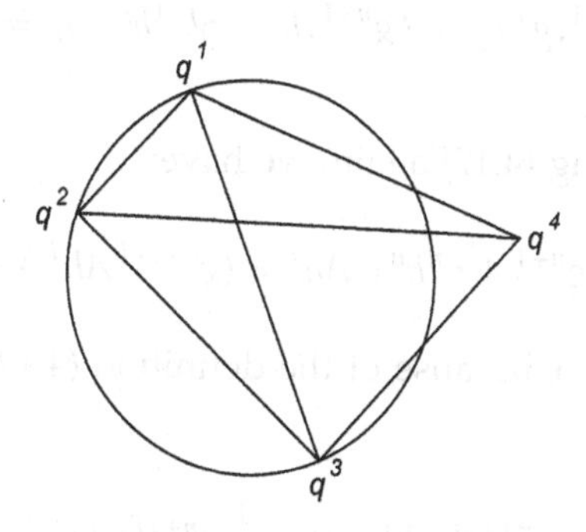

Figure 4.5 the Delaunay test

Remark 4.16 If we start this algorithm with the edge opposite the largest angle in the triangulation, then the process is convergent since the largest angle decreases strictly at each iteration, unless there are quadrilaterals whose vertices are cocyclic. In practice, we do not need to check this last point. □

APPENDIX 4.A

Proof of Lemma 4.1 and Propositions 4.3 and 4.6

PROOF OF LEMMA 4.1

Let us first assume the orthogonality relations (4.21)–(4.22) are satisfied, and let us prove the two other properties, starting with (4.23) which follows trivially from (4.17) and (4.22). We remark that for $n \leq n_0$, ρ^n cannot become zero, since by hypothesis, no g^n is zero. Let us now compute γ^n. We first have, from above

$$\gamma^n = \frac{\|g^{n+1}\|_C^2}{\langle g^n, h^n \rangle_C},$$

then

$$\gamma^n = -\frac{\langle g^{n+1} - g^n, g^{n+1} \rangle_C}{\langle g^{n+1} - g^n, h^n \rangle_C} = -\frac{g^{n+1T} A h^n}{h^{nT} A h^n},$$

the first equality follows from (4.21), (4.22) and the second one from (4.15), which proves (4.19).

The orthogonality relations (4.20)–(4.22) are proved by induction. Let n be less than n_0; let us assume these relations are true at all ranks $k \leq n$, and prove them at rank $n+1$.

(i) Let us take the inner product of (4.15) with h^j; we obtain

$$\langle g^{n+1}, h^j \rangle_C = \langle g^n, h^j \rangle_C - \rho^n \langle C^{-1} A h^n, h^j \rangle_C.$$

For $j = n$, the result is zero, by virtue of the definition (4.13) of ρ^n. For $j < n$, it is also zero, because of the induction hypothesis, by virtue of (4.22) and (4.20).

(ii) Let $j \leq n$; by (4.17) we have:

$$\langle g^{n+1}, g^j \rangle_C = \langle g^{n+1}, h^j - \gamma^{j-1} h^{j-1} \rangle_C = 0,$$

from point (i) above.

(iii) Eventually, by using (4.17) again, we have:

$$h^{n+1T} A h^j = (g^{n+1} + \gamma^n h^n)^T A h^j = (g^{n+1})^T A h^j + \gamma^n (h^n)^T A h^j$$

For $j = n$, this term is zero, because of the definition (4.19) of γ^n. Then, for $j < n$, we have

$$h^{n+1T} A h^j = (g^{n+1})^T A h^j + 0 = \frac{1}{\rho^j} g^{n+1T} C (g^{j+1} - g^j) = 0,$$

where we have used relation (4.15) for the second equality, the induction hypothesis on (4.20) for the first equality, and that on (4.21) for the last one.

We leave it to the reader to check that all the properties are true at rank $n = 1$. □

PROOF OF PROPOSITION 4.3

Let us start by proving (4.26) by induction, the equality between the various sets being trivially true (because of initialisation $g^0 = h^0$) for $n = 1$. Let us assume (4.26) is true at rank $n \le n_0$. Since by hypothesis $h^{n-1} \in K_n$ and $AK_n \subset K_{n+1}$, relation (4.15) shows that $g^n \in K_{n+1}$, then (4.17) shows that $h^n \in K_{n+1}$. Conversely, let us show that $A^n g^0 \in \{g^0, ..., g^n\}$. We have just shown that $g^n \in K_{n+1}$; thus,there exists scalars $\lambda_i, i \in \{0, ..., n\}$, such that

$$g^n = \sum_{i=0}^{i=n} \lambda_i A^i g^0, \tag{A.1}$$

and $\lambda_n \neq 0$, because $g^n \neq 0$ and from (4.22), g^n cannot belong to the space K_n, which coincides, by the induction hypothesis, with the space span$\{h^0, h^1, \dots h^{n-1}\}$. It follows from (A.1) that $A^n g^0 = \frac{g^n}{\lambda_n} + G$, where, by the induction hypothesis, G is a vector in the space generated by vectors $g^0, \dots, g^{n-1}$, which shows that $A^n g^0 \in$ span$\{g^0, ..., g^n\}$. Relation (4.17), which lets us express g^n as a function of h^n in the previous expression for $A^n g^0$, also shows that we also have $A^n g^0 \in$ span$\{h^0, ..., h^n\}$, which completes the proof of (4.26).

Let us note that hypothesis $n \le n_0 + 1$ shows that K_n is a vector space of dimension n, an orthogonal basis of which is formed by vectors $g^0, ..., g^{n-1}$.

Let us now prove (4.24). By construction, x^n is of the form $x^n = x^0 + \sum_{j=0}^{n-1} \rho^j h^j$, which means, because of (4.26), that $x^n \in x^0 + K_n$. Now, $x \in x^0 + K_n$ if and only if $x - x^n \in K_n$; thus, it is equivalent to show that

$$\forall y \in K_n, \quad E(x^n) \le E(x^n + y). \tag{A.2}$$

This inequality is equivalent to $(\nabla E(x^n))^T y = 0$, i.e. $(g^n)^T y = 0$, for all $y \in K_n$. But this last relation is actually satisfied, since, because of (4.21), it is true for all the elements $g^0, ..., g^{n-1}$ in a basis of K_n. This proves (A.2), and thus also (4.24). □

PROOF OF PROPOSITION 4.6

Let us first prove the following result:

$$z \in KerA \text{ and } Cz \in \text{range}(A) \quad \Rightarrow \quad z = 0. \tag{A.3}$$

If $Cz \in$ range(A), there exists y such that $Cz = Ay$. Since matrix C is invertible, we have $Az = AC^{-1}Ay$, so that the hypothesis $z \in KerA$ implies $0 = y^T Az = (Ay)^T C^{-1} Ay$, the last equality being due to the symmetry of A. As C^{-1} is a symmetric and positive definite matrix, we conclude that necessarily $Ay = 0$, i.e. $Cz = 0$. This is only possible if $z = 0$, because C is invertible, which proves the result.

Relation (A.3) allows us to say that if the PCG algorithm converges towards a solution x of the problem $Ax = b$, which moreover satisfies $Cx \in$ range(A), then this solution must be unique. There remains to show that this algorithm actually converges, that is that its 'degeneration cases' are, as previously, 'favourable' cases, and that it also converges towards an x such that $Cx \in \text{range}(A)$.

Let us start with this second property, whose proof by induction is immediate. Indeed, the hypothesis made on x^0 and on b ensures that $Cx^0, Cg^0, Ch^0 \in \text{range}(A)$. Now, relations (4.14), (4.15) and (4.17) show that

$$Cx^n, Cg^n, Ch^n \in \text{range}(A) \quad \Rightarrow \quad Cx^{n+1}, Cg^{n+1}, Ch^{n+1} \in \text{range}(A).$$

Let us now study the 'degeneration cases' for the algorithm. Case $g^n = 0$ is always favourable, since by virtue of (4.18), it means that x^n is the sought solution, and it is, moreover, the stopping test of the method. Let us now assume n is the first index for which $(h^n)^T Ah^n = 0$; the matrix A being symmetric and semi-positive definite, it is diagonalisable in an orthonormal basis of eigenvectors, its eigenvalues are positive or zero, and the above relation implies that $h^n \in KerA$. But we saw previously that $Ch^n \in \text{range}(A)$; from (A.3), we deduce that $h^n = 0$, and the proof is completed exactly as in the case where A is symmetric and positive definite (cf. Corollary 4.2).

APPENDIX 4.B

Fortran Program for Solving the Laplacian with Dirichlet Boundary Conditions by a Variational Formulation and Resolution of the Linear System by Conjugate Gradient Preconditioned by an Incomplete Choleski Factorisation of the System Matrix.

```
C File: GCPLaplac.for
C
      PROGRAM GCPLaplace
C
C This program solves the model problem, i.e. the
C Laplacian with Dirichlet boundary conditions:
C -ΔΦ = f,Φ given on the boundary;
C
C The problem is written in variational form, then
C discretised by P1 finite elements. The approximate
C solution at each vertex of the triangulation is
C determined by solving a linear system, which is
C solved by the Conjugate Gradient method preconditioned
C by C = L*(L*)T where (L*)T is the incomplete Choleski
C factor of A.
C Matrices are stored in CSR format.
C
C    input:
C    ------
C      triangulation (nv,nt,q,me,ng) and problem
C        data (right hand side f for Laplace's equation
```

```
C         and Dirichlet boundary condition stored
C         in array phi0) read in file filnam via subroutines
C         readT and readF;
C      parameter:nvMax (maximum number of vertices)
C         ntMax (maximum number of triangles) of the
C         triangulation, nMAMax=8*nvMax (maximum
C         size of array aa).
C
C    output:
C    -------
C      f:array containing the values of the
C         approximate solution at each of the vertices of the
C         mesh stored in a file (whose name is given by the
C         user) for a possible graphical exploitation
C         Remark: at the beginning, this array
C         contains the Laplacian right hand side.
C
C    arrays:
C    -------
C    mesh:
C      q(2,nvMax):coordinates of the mesh vertices
C      me(3,ntMax):numbering of the triangle vertices
C      ng(nvMax):array showing whether a vertex
C         is on the boundary (ng(i)≠0) or not (ng(i)=0).
C
C    data:
C      f(nvMax): at the beginning, Laplacian right hand side
C         (then values of the approximate solution),
C      phi0(nvMax): Dirichlet boundary conditions
C
C    arrays for the preconditioned conjugate gradient
C      cg(nvMax):array containing the negative of the
C         gradient of the functional to be minimised
C         (product of the preconditioning matrix by g),
C      g(nvMax):array containing g,
C      h(nvMax):array containing the descent direction
C
C
C    other arrays:
C      aa(nMAMax): array storing matrix A, in CSR format,
C      ia(nvMax+1): index in array aa of first
C         element of line i of matrix A actually stored in aa,
C      ja(nMAMax): index j of column of stored Aij element;
C      al(nMLMax), ia(nvMax+1),ja(nMLMax): same meaning
C         as the three previous arrays for incomplete
C         Choleski factor of A;
C      area(ntMax): area of each of the triangles;
C      ff(nvMax): right hand side of linear system
C         to be solved
C
C    scalars:
```

```
C    --------
C      cte: value of the very large constant for
C        satisfying the Dirichlet boundary
         condition
C      err: error in H_0^1 norm between exact solution phi0
C        and approximate solution ff.
C
C    files:
C    ------
C      filnam:name of file for model problem data
C        (source term of Laplacian and Dirichlet
C        boundary conditions)
C        or for the approximate solution, for
C        instance, in view of a graphical exploitation
C
C    subroutines:
C    ------------
C      readT:reads the mesh,
C      readF:reads problem data;
C      getArea:computes the area of each triangle;
C      MgetMat:computes and stores matrix A in
C        array a, in CSR format;
C      RHS:computes, and stores in array ff,
C        the system right hand side;
C      MDirCL:program satisfying Dirichlet
C        boundary conditions;
C      Sort:reorders elements of array aa so that
C        elements of a line of matrix A are stored in
         increasing order of their column indices;
C      ICholeski:incomplete Choleski factorisation of
C        matrix A stored in CSR format;
C      GradC:preconditioned conjugate
C        gradient program;
C      N.B.: this subroutine requires subroutine
C        IAmul and ISolvSys
C      writeF:writes the approximate solution
C
C
C    functions:
C    ----------
C      errH10:computes the H_0^1 error between exact
C        solution phi0 and approximate solution f.
C
C    library (subroutines and functions):
C    ------------------------------------
C      utilP1.for:file containing functions
C        deter (for area computations), errH10
C        and subroutines readT,readF, writeF
C        and getBand;
C      RHS.for:file containing subroutines
C        getArea et and RHS;
```

```
C      elem.for:file containing function
C        elemMat;
C      MgetMat.for:file containing subroutines
C        MgetMat et MDirCL;
C      ICholes.for:file containing subroutine
C        ICholeski;
C      GradCP.for:file containing subroutine
C        GradC;
C      sort.for:file containing subroutine Tri.
C
C
      PARAMETER (nvMax=1800, ntMax=3400, nMAMax=8*nvMax)
      DIMENSION q(2,nvMax),me(3,ntMax),ng(nvMax)
      DIMENSION area(ntMax)
      DIMENSION phi0(nvMax),f(nvMax),ff(nvMax)
      DIMENSION aa(nMAMax),ia(nvMax+1),ja(nMAMax)
      DIMENSION al(nMAMax)
      DIMENSION g(nvMax),cg(nvMax),h(nvMax)
      CHARACTER*20 filnam
C
      DATA cte/1.0e25/
      DATA eps/1.0e-12/
      DATA kMax/150/
C
      write(*,*) 'name of file storing the mesh'
      read(*,'(a)') filnam
      CALL readT(filnam,nv,nt,q,ng,me,ngt)
      write(*,*) 'name of file storing f'
      write(*,*) '(Laplacian right hand side)'
      read(*,'(a)') filnam
      CALL readF(filnam,nv,f)
      write(*,*) 'name of file storing phi0'
      write(*,*) '(Dirichlet boundary conditions)'
      read(*,'(a)') filnam
      CALL readF(filnam,nv,phi0)
C
C
      CALL getArea(q,me,nt,nv,area)
      CALL MgetMat(q,me,ng,area,nt,nv,aa,ia,ja)
      CALL RHS(me,f,nv,nt,area,ff)
      CALL MDirCL(aa,ia,phi0,nv,ng,ff,cte)
C
      DO 1 i=1,nv
1     CALL Sort(ja(ia(i)),aa(ia(i)), a(i+1)-ia(i))
C
      CALL ICholeski(nv,aa,ia,ja,al)
      DO 2 i=1,nv
        f(i)=0.
        if(ng(i).NE.0) f(i)=phi0(i)
2     CONTINUE
      CALL GradC(nv,aa,ia,ja,al,ff,f,g,h, cg,kMax,eps)
```

```
C
      err=errH10(q,me,f,phi0,nv,nt)
      write(*,*) 'error=',err
      write(*,*) 'name of file for storing the '
      write(*,*) 'approximate solution'
      read(*,'(a)') filnam
      CALL writeF(filnam,nv,f)
      PAUSE
      END
C
C
C -------------------------------------------------------
C     SUBROUTINE IAmul(nv,aa,ia,ja,x,b)
C -------------------------------------------------------
C This subroutine computes the product of matrix A, stored
C in Morse format in arrays (aa,ia,ja), by vector x;
C the result is stored in array b.
C The computation is performed by first computing the
C contribution of the lower triangular part of the matrix
C (that is stored in array aa), then that associated
C with the strictly upper triangular part.
C
C
      DIMENSION x(nv), b(nv)
      DIMENSION aa(*),ia(nv),ja(*)
      DO 1 i=1,nv
        sum=0.
        DO 2 k=ia(i),ia(i+1)-1
          sum=sum+aa(k)*x(ja(k))
 2      CONTINUE
        b(i)=sum
 1    CONTINUE
      DO 3 i=1,nv
        xi=x(i)
        DO 3 k=ia(i),ia(i+1)-2
          jak=ja(k)
          b(jak)=b(jak)+aa(k)*xi
 3    CONTINUE
      END
C
C -----------------------------------------------------------
C     SUBROUTINE ISolvSys(nv,z,al,il,jl)
C -----------------------------------------------------------
C This subroutine solves the linear system
C LL^Tz = r, where L is a lower triangular
C matrix of size nv x nv, stored in CSR
C format in arrays (al, il, jl). On input,
C array z contains the system right hand side;
C on output, it contains solution z.
C Resolution is in two steps: a forward solve,
C followed by a backward solve.
```

```
C
C
      DIMENSION z(*),al(*),il(*),jl(*)
C
      DO 1 i=1,nv
        sum=z(i)
        DO 2 k=il(i), il(i+1)-2
          sum=sum-al(k)*z(jl(k))
2       CONTINUE
        z(i)=sum/al(il(i+1)-1)
1     CONTINUE
      DO 3 i=1,nv
        z(i) = z(i)/al(il(i+1)-1)
3     CONTINUE
      DO 4 i1=1,nv
        i=nv+1-i1
        DO 4 k=il(i), il(i+1)-2
          jlk=jl(k)
          z(jlk) = z(jlk) - al(k)*z(i)/al (il(jlk+1)-1)
4       CONTINUE
      END
C
C ********************************
C
C LIBRARY
C
C ********************************
      INCLUDE 'elem.for'
      INCLUDE 'GradCP.for'
      INCLUDE 'MgetMat.for'
      INCLUDE 'ICholes.for'
      INCLUDE 'utilP1.for'
      INCLUDE 'RHS.for'
      INCLUDE 'sort.for'
```

5 Finite Element Method for General Elliptic Problems

Summary

In this chapter, we shall use the techniques introduced previously to study more general partial differential equations. We shall begin by considering several types of boundary conditions, then we shall study general second-order symmetric and elliptic operators. Last, we shall treat the case of systems. In this chapter, we shall also introduce approximations by finite elements of order higher than one.

5.1 NEUMANN OR ROBIN BOUNDARY CONDITIONS

5.1.1 Formulation of the Problem

Let Ω be a domain in the plane, with sufficiently smooth boundary Γ. We consider the following problem, in the unknown φ:

$$-\Delta\varphi = f \text{ in } \Omega, \tag{5.1}$$

$$a_1\varphi + \frac{\partial\varphi}{\partial n} = g \text{ on } \Gamma, \tag{5.2}$$

where a_1 is a given, positive, smooth function defined on Γ. If $a_1 = 0$, the boundary condition is a Neumann condition; otherwise, it is called a Robin, or sometimes Fourier, condition. From a theoretical point of view, those two cases are quite different, as we shall see. The function f is assumed square integrable on Ω, and g is at least square integrable on Γ.

Following the approach introduced in Chapter 2, we shall transform this problem into different equivalent forms: the variational formulation, or the minimisation of a certain energy functional. We first have the following result, which is the counterpart of Theorem 2.1:

Theorem 5.1 If $\varphi \in H^1(\Omega)$ and $\Delta\varphi \in L^2(\Omega)$, φ is a solution of (5.1)–(5.2) if and only if φ is a solution of the following variational problem:

find $\varphi \in H^1(\Omega)$ such that $\forall w \in H^1(\Omega)$,

$$\int_{\Omega}(\nabla\varphi\cdot\nabla w)(\vec{x})\mathrm{d}\vec{x}+\int_{\Gamma}(a_1\varphi w)(\vec{x})\mathrm{d}\gamma(\vec{x})=\int_{\Omega}(f\,w)(\vec{x})\mathrm{d}\vec{x}+\int_{\Gamma}(gw)(\vec{x})\mathrm{d}\gamma(\vec{x}),\quad(5.3)$$

where $\mathrm{d}\gamma$ denotes the measure on Γ.

PROOF In the proof, we only consider the case where φ is a smooth solution (i.e. $\varphi\in H^2(\Omega)$) of the original problem (5.1)–(5.2) (we refer to to Remark 2.3 for the generalisation to the case where φ is only in $H^1(\Omega)$ with $\Delta\varphi\in L^2(\Omega)$). We then prove, analogously to Theorem 2.1, that φ solves (5.3). Indeed, it suffices to multiply equation (5.1) by a function $w\in H^1(\Omega)$, to integrate on Ω, to use Green's formula, and to replace, in the boundary term, $(\partial\varphi/\partial n)$ by its value given by the boundary condition (5.2).

Conversely, let us assume that φ solves the variational problem (5.3), and in particular that (5.3) is satisfied for any function w in the space $D(\Omega)$ of C^∞ functions with compact support in Ω, whose dual is the space $D'(\Omega)$ of distributions on Ω. If $\varphi\in H^2(\Omega)$, equation (5.3) becomes, if we denote by $\langle .,.\rangle$ the duality bracket between these two spaces,

$$\langle -\Delta\varphi-f,w\rangle=0.$$

Then since $-\Delta\varphi-f$ is a function in $L^2(\Omega)$, it follows that this function is zero almost everywhere, i.e. that equation (5.1) is satisfied almost everywhere in Ω. To recover the boundary condition, we multiply equation (5.1) by $w\in H^1(\Omega)$, use Green's formula, and subtract the resulting equation from equation (5.3). We obtain

$$\forall w\in H^1(\Omega),\int_{\Gamma}\left(\frac{\partial\varphi}{\partial n}w\right)(\vec{x})\mathrm{d}\gamma(\vec{x})=\int_{\Gamma}(gw)(\vec{x})\mathrm{d}\gamma(\vec{x})-\int_{\Gamma}(a_1\varphi w)(\vec{x})\mathrm{d}\gamma(\vec{x}),$$

which implies (5.2), since the space $H^{1/2}(\Gamma)$ of traces on Γ of functions in $H^1(\Omega)$ is dense in $L^2(\Gamma)$. □

Because we assume that the function a_1 is positive, arguing exactly as in Theorem 2.2 enables us to prove that the variational problem (5.3) is equivalent to a minimisation problem, and we obtain

Theorem 5.2 Problem (5.3) is equivalent to the following minimisation problem:

$$\min_{\varphi\in H^1(\Omega)} E(\varphi),\quad(5.4)$$

where the functional E is defined by:

$$E(\varphi)=\frac{1}{2}\left[\int_{\Omega}|\nabla\varphi(\vec{x})|^2\mathrm{d}\vec{x}+\int_{\Gamma}(a_1\varphi^2)(\vec{x})\mathrm{d}\gamma(\vec{x})\right]\quad(5.5)$$
$$-\int_{\Omega}(f\varphi)(\vec{x})\,\mathrm{d}\vec{x}-\int_{\Gamma}(g\varphi)(\vec{x})\mathrm{d}\gamma(\vec{x}).$$

Remark 5.3 We state without proof the *existence results* for this type of problem. We refer to Dautray and Lions (1990) and Grisvard (1992) for a justification. There are several different cases.

(i) Let assume that the function a_1 is identically zero, which corresponds to a pure Neumann problem. Then problem (5.1)–(5.2) will only have a solution if the data f and g satisfy the following compatibility relation:

$$\int_\Omega f(\vec{x})\,\mathrm{d}\vec{x} + \int_\Gamma g(\vec{x})\mathrm{d}\gamma(\vec{x}) = 0. \tag{5.6}$$

To convince oneself of this fact, it suffices to write the variational formulation, taking as test function w the function identically equal to 1 on Ω.

Also, this solution will only be unique up to the addition of a constant, since it is clear that if φ solves (5.1)–(5.2), then $\varphi + k$ (k any real constant) still solves this problem.

(ii) However, if the function a_1 is bounded from below by a strictly positive constant, then problem (5.1)–(5.2) has one and only one solution, without any compatibility condition of the above type. It is clear that the proof rests on the coercivity of the functional E, which is lost in case (i) above.

Let us also recall *smoothness results* for the solution of this problem (when it exists). If we restrict ourselves to $f \in L^2(\Omega), g \in L^2(\Gamma)$, the solution $\varphi \in H^1(\Omega)$ only satisfies: $\Delta\varphi \in L^2(\Omega)$. However, if we assume more smoothness on g, i.e. $g \in H^{1/2}(\Gamma)$, then $\varphi \in H^2(\Omega)$. □

5.1.2 Discretisation of the Variational Formulation

We approximate the variational formulation (5.3) by P^1 finite elements, as in the previous chapters. Given a regular triangulation $(T_k)_{k\in\{1,\dots,nt\}}$ of the domain Ω, we define the space H_h^1 by

$$H_h^1 = \{\varphi_h : \Omega_h \to \mathbb{R}, \varphi_h \text{ continuous on } \Omega_h, \text{ and}$$
$$\forall k \in \{1,\dots,nt\}, \varphi_{h|T_k} \text{ linear function}\}, \tag{5.7}$$

then we solve (5.3) in this approximation space, which leads to the following approximate variational problem:

$$\begin{aligned} &\text{find } \varphi_h \in H_h^1 \quad \text{such that } \forall w_h \in H_h^1, \\ &\int_{\Omega_h} \nabla\varphi_h(\vec{x}) \cdot \nabla w_h(\vec{x})\mathrm{d}\vec{x} + \int_{\Gamma_h} (a_1\varphi_h w_h)(\vec{x})\mathrm{d}\gamma(\vec{x}) \\ &= \int_{\Omega_h} f(\vec{x})w_h(\vec{x})\mathrm{d}\vec{x} + \int_{\Gamma_h} (g w_h)(\vec{x})\mathrm{d}\gamma(\vec{x}), \end{aligned} \tag{5.8}$$

where $\Omega_h = \cup_{k=1}^{nt} T_k$ and $\Gamma_h = \partial\Omega_h$. We recall (cf. Propositions 2.14 and 2.17) that the space H_h^1 is a finite dimensional subspace of $H^1(\Omega)$, its dimension being the number nv of vertices q^i of the triangulation. A basis of this space is formed by the hat functions $w^i, i \in \{1, ..., nv\}$ $(w^i(q^j) = \delta_{ij})$, which leads us to seek the solution of (5.8) in the form of a linear combination of these functions:

$$\varphi_h(\vec{x}) = \sum_{i=1}^{nv} \varphi_i w^i(\vec{x}), \quad \forall \vec{x} \in \Omega. \tag{5.9}$$

If we let the test functions in (5.8) be any elements of this basis, problem (5.8) becomes equivalent to the solution of the linear system

$$A\Phi = F, \tag{5.10}$$

where Φ is the vector with entries $\varphi_i, i \in \{1, ..., nv\}$, with the matrix $A = ((A_{ij}))_{1 \le i,j \le nv}$ and the right hand side $F = (F_i)_{1 \le i \le nv}$ being defined by:

$$A_{ij} = \int_{\Omega_h} (\nabla w^i \cdot \nabla w^j)(\vec{x}) d\vec{x} + \int_{\Gamma_h} (a_1 w^i w^j)(\vec{x}) d\gamma(\vec{x}), \tag{5.11}$$

$$F_j = \int_{\Omega_h} (f\, w^j)(\vec{x}) d\vec{x} + \int_{\Gamma_h} (g w^j)(\vec{x}) d\gamma(\vec{x}). \tag{5.12}$$

Let us state the properties of the matrix A, in the two cases considered in Remark 5.3.

Proposition 5.4 *If the function a_1 is positive, then the matrix A is symmetric and positive semi-definite. Also*

(i) *if the function a_1 is identically zero, then the kernel of A is formed of constant functions;*

(ii) *if the function a_1 is bounded from below by a strictly positive constant, matrix A is positive definite.*

PROOF The proof is analogous to the one developed for Theorem 2.6. Let $z = (z_1, \ldots, z_{nv})$ be any vector in $\mathbb{R}^{nv}$; we have

$$z^T A z = \sum_{i,j=1}^{nv} z_i A_{ij} z_j = \sum_{i,j=1}^{nv} \int_{\Omega_h} \nabla w^i \nabla w^j z_i z_j + \int_{\Gamma_h} a_1 w^i w^j z_i z_j,$$

where, to simplify notation, we have suppressed the integration variables.

Let $z_h(\vec{x}) = \sum_{i=1}^{nv} z_i w^i(\vec{x})$; we have $z_h \in H_h^1$ and:

$$z^T A z = \int_{\Omega_h} |\nabla z_h|^2 + \int_{\Gamma_h} a_1 |z_h|^2 \ge 0,$$

which proves that, under the hypothesis $a_1 \ge 0$, the matrix A is positive semi-definite. Also, the above quantity can only become zero if ∇z_h and $(a_1 |z_h|)_{|\Gamma_h}$ are both zero almost everywhere, which implies that function z_h is constant on Ω_h. If a_1 is zero, we can say nothing more, and the kernel of A is formed by vectors z of the form $z = (k, \ldots, k)$, where k is an arbitrary real constant, which proves point (i). If a_1 is bounded from below by a strictly positive constant, the constant value of the function z_h is necessarily 0. In this case, it follows that the vector z is identically zero, which proves that A is positive definite. □

The discrete problem (5.8) thus satisfies the same properties (existence and uniqueness of a possible solution) as the continuous problem. In particular, system (5.8) will have one and only one solution in case (ii) of the previous proposition. In case (i), which is more difficult, it will only have a solution if the right hand side F is in the range of A, and it will then have infinitely many solutions, obtained from one another by adding an arbitrary constant. For the numerical study of this problem, i.e. the Laplacian, we must restrict to case (ii), which is well posed. In practice, if we must solve a Neumann problem, obviously assuming that the compatibility condition (5.6) is satisfied, we must add a term like $\epsilon\varphi$ (with ϵ small) to the Neumann condition, so as to reduce to a Robin boundary condition, even though we know the original problem actually has infinitely many solutions.

Let us now study how to modify the program as compared with the case of Dirichlet boundary conditions.

5.1.3 *Programming the Boundary Conditions*

The only modifications we must bring to the various solution programs proposed in the previous chapters are in the computation of the matrix A and the right hand side F, where additional terms will appear in contrast with the Dirichlet case.

We assume here, to simplify, that the functions g and a_1 are defined on all of $\bar{\Omega}$ and that their values are given at each vertex of the mesh (we shall also denote by g and a_1 the arrays containing these values), which enables us to approximate them by affine functions (respectively denoted by g_h and a_{1h}) on each triangle of the triangulation.

Let us begin by the boundary term occurring in the right hand side.

5.1.3.1 Computing the term $\int_{\Gamma_h}(gw^j)(\vec{x})\mathrm{d}\gamma(\vec{x})$

Let us expand g_h along the hat functions basis. We have

$$g_h(\vec{x}) = \sum_{i=1}^{nv} g(q^i)w^i(\vec{x})$$

so that

$$\int_{\Gamma_h}(gw^j)(\vec{x})\mathrm{d}\gamma(\vec{x}) = \sum_{i=1}^{nv} g(q^i)\int_{\Gamma_h}(w^iw^j)(\vec{x})\mathrm{d}\gamma(\vec{x}), \tag{5.13}$$

and thus it suffices to compute coefficients

$$b_{ij} = \int_{\Gamma_h}(w^iw^j)(\vec{x})\mathrm{d}\gamma(\vec{x}). \tag{5.14}$$

Since the support of the function w^i is the union of all triangles surrounding the point q^i, these coefficients are zero if one of the vertices q^i or q^j is not on the boundary Γ_h. Let us assume this is not the case and use Simpson's quadrature formula (formula (2.57)) to compute the integral in (5.14) (the computation is exact in that case). If $i \neq j$ and if $[q^i, q^j]$ is a boundary edge, we have $b_{ij} = |q^i - q^j|/6$, where $|q^i - q^j|$ denotes the length of the segment $[q^i, q^j]$. If $i = j$, we obtain $b_{ii} = (|q^i - q^{ipf}| + |q^i - q^{imf}|)/3$, where *ipf* and *imf* are the indices of the two boundary points surrounding the point q^i.

The integral over Γ_h is a sum of integrals over the boundary edges of the domain, and these can be found by a loop over all triangles, which enables us to integrate in a simple way this computation in subroutine RHS given in Chapter 3. The following lines of code are part of subroutine 'gRHS' given below; array 'ff' contains, for each vertex of the mesh with index i, the value of the right hand side F_i given by (5.12). The extract we give obviously deals only with the part where g occurs.

Let us note an important point: in the following program, *a boundary edge is characterised by the fact that both its ends are boundary vertices,* and we must make sure that the mesh satisfies this property.

```
      DO 3 k=1,nt
        DO 3 il=1,3
          i=me(il,k)
          IF (ng(i).NE.0) THEN
            ilp=mod(il,3)+1
            ilpp=mod(ilp,3)+1
            ip=me(ilp,k)
            ipp=me(ilpp,k)
            IF (ng(ip).NE.0) THEN
              dx=q(1,i)-q(1,ip)
              dy=q(2,i)-q(2,ip)
              dist=sqrt(dx*dx+dy*dy)
              ff(i)=ff(i)+(2*g(i)+g(ip))*dist/6.
            ENDIF
            IF (ng(ipp).NE.0) THEN
              dx=q(1,i)-q(1,ipp)
              dy=q(2,i)-q(2,ipp)
              dist=sqrt(dx*dx+dy*dy)
              ff(i)=ff(i)+(2*g(i)+g(ipp))*dist/6.
            ENDIF
          ENDIF
3       CONTINUE
```

5.1.3.2 Computing the term $\int_{\Gamma_h} (a_1 w^i w^j)(\vec{x}) d\gamma(\vec{x})$

As above, this boundary integral is a sum of integrals over the boundary edges of the domain, and these are all zero if one of the vertices q^i or q^j is not on the boundary Γ_h. Also, since we assume that the function a_{1h} is linear on each triangle, we can expand it along the hat functions basis in the following way: $a_{1h}(\vec{x}) = \sum_{l=1}^{nv} a_1(q^l) w^l(\vec{x})$, which leads us to set

$$c_{ijl} = \int_{\Gamma_h} (w^i w^j w^l)(\vec{x}) \mathrm{d}\gamma(\vec{x}). \tag{5.15}$$

These coefficients are computed exactly as above, since Simpson's formula is exact for degree 3 polynomials, and the restriction to an edge of the function $w^i w^j w^l$ is of degree 3. Still with the above notation, if $i \neq j$, c_{ijl} is non-zero for only two values of l, namely $l = i$ and $l = j$, and in both cases $c_{ijl} = |q^i - q^j|/12$. Now, if $i = j$, we have

$$c_{iil} = \frac{|q^i - q^l|}{12}, \text{for } l = ipf \text{ or } imf, \quad c_{iii} = \frac{|q^i - q^{ipf}|}{4} + \frac{|q^i - q^{imf}|}{4}.$$

This computation is integrated into the matrix coefficients computations as follows:

```
      i=me(il,k)
      j=me(jl,k)
      IF(ng(i).NE.0)
        ip=me(ilp,k)
        ipp=me(ilpp,k)
        IF (i.EQ.j) THEN
        IF(ng(ip).NE.0) THEN
          dx=q(1,i)-q(1,ip)
          dy=q(2,i)-q(2,ip)
          dist=sqrt(dx*dx+dy*dy)
          elem=elem+(3*al(i)+al(ip))*dist/12.
        ENDIF
          IF (ng(ipp).NE.0) THEN
            dx=q(1,i)-q(1,ipp)
            dy=q(2,i)-q(2,ipp)
            dist=sqrt(dx*dx+dy*dy)
            elem=elem+(3*al(i)+al(ipp))*dist/12.
          ENDIF
          ELSEIF(ng(j).NE.0) THEN
          dx=q(1,i)-q(1,j)
          dy=q(2,i)-q(2,j)
          dist=sqrt(dx*dx+dy*dy)
          elem=elem+(al(i)+al(j))*dist/12.
        ENDIF
      ENDIF
```

In the above, 'elem' denotes the elementary contribution, on triangle T_k, to matrix element A_{ij}; the lines of code will be included in the function 'geMat' that we shall give later, in the framework of more general elliptic operators. The indices il and jl are local indices of vertices in the triangle T_k; they are input to this funtion.

5.1.4 Mixed Boundary Conditions

It is possible to consider a more general framework for this kind of problem, involving both Dirichlet boundary conditions on one part of the boundary, and

Neumann or Robin boundary conditions on the rest of the boundary; this is what we call 'mixed boundary conditions'. More precisely, let $\{\Gamma_1, \Gamma_2\}$ be a partition of Γ, meaning that $\Gamma_1 \cup \Gamma_2 = \Gamma$ and $\Gamma_1 \cap \Gamma_2$ has empty interior; we seek φ solution of (5.1) and of

$$\varphi = \varphi_0 \text{ on } \Gamma_1, \quad a_1\varphi + \frac{\partial \varphi}{\partial n} = g \text{ on } \Gamma_2. \tag{5.16}$$

The variational formulation of this problem can be written:

$$\text{find } \varphi \in H^1_{\varphi_0,\Gamma_1}(\Omega), \quad \text{such that } \forall w \in H^1_{0,\Gamma_1}(\Omega), \tag{5.17}$$

$$\int_\Omega (\nabla\varphi \cdot \nabla w)(\vec{x})\mathrm{d}\vec{x} + \int_{\Gamma_2} (a_1\varphi w)(\vec{x})\mathrm{d}\gamma(\vec{x}) = \int_\Omega (f\,w)(\vec{x})\mathrm{d}\vec{x} + \int_{\Gamma_2} (gw)(\vec{x})\mathrm{d}\gamma(\vec{x}),$$

where we have set ($\psi = \varphi_0$ or 0):

$$H^1_{\psi,\Gamma_1}(\Omega) = \{w \in H^1(\Omega),\, w = \psi \text{ on } \Gamma_1\}. \tag{5.18}$$

Let us note that, from a theoretical point of view, it suffices that Γ_1 has non-zero measure for this variational problem to be well posed, whatever the function $a_1 \geq 0$.

From a practical point of view, it suffices to extend g and a_1 by 0 on Γ_1, so that we can reuse the previous computations of the right hand side and the matrix, then to use the, now well known, 'trick' of replacing, for each vertex $q^i \in \Gamma_1$, the diagonal element A_{ii} by a very large constant *cte*, and of simultaneously replacing F_i by the product $cte * \varphi_0(q^i)$.

We could also simply mix these two ideas by extending the array a_1 to Γ_1 by a very large constant *cte*, and simultaneously extending g by $cte * \varphi_0(q^i)$, so we can program the Robin boundary condition on all of Γ, and avoid modifying the matrix A. However, accuracy will not be as good with this method, so we shall retain the first method for the complete program we give in Appendix 5.A. To simplify the computations, we have used a conjugate gradient preconditioned by the diagonal.

The only point left to specify is: how do we distinguish between the points of Γ_1 and those of Γ_2? The answer could be found in array ng, which we could define in the following way:

— or a boundary vertex q^i, $ng(i) \neq 0$;
— for any vertex q^i located on Γ_1, $ng(i) > 0$;
— for any vertex q^i located on Γ_2, $ng(i) < 0$;

However, this method is not really desirable, because it mixes purely geometric properties linked to the mesh, represented in array *ng*, with properties depending only on the type of partial differential equation under consideration. We have thus decided to characterise array *ng* only by: the vertex q^i is an internal vertex if and only if $ng(i) = 0$. Points on the boundary Γ_1 will then be characterised by

$$\Gamma_1 = \{q^i \in \Gamma, \mathrm{phi0}(i) \neq 0\}, \tag{5.19}$$

if we denote by phi0 the array representing the Dirichlet data φ_0. *This assumes that the array phi0 contains no zero value*! What can we do if the function φ_0 in (5.16) becomes zero at some points q^i of Γ_1? In this case, we must modify the value of phi0(i), replacing it by a very small constant ϵ_0; this task must be done *beforehand, before calling the program given in Appendix 5.A.*, as this program just reads the file containing the values of the array phi0.

Using the above trick, we extend all boundary integrals to Γ. If this were not the case, we would have to characterise a boundary edge in Γ_2 by the fact that at least one of its ends is not on Γ_1. Let us note, on this subject, that Dirichlet boundary conditions are imposed pointwise (on each boundary vertex), whereas Neumann or Robin conditions are imposed edgewise (through the computation of an integral on the domain boundary).

5.2 GENERAL SYMMETRIC AND LINEAR ELLIPTIC EQUATIONS

5.2.1 Generalities

We can consider more general symmetric and linear elliptic operators in the left hand side of equation (5.1), of the type: $(\varphi \to a_0\varphi - \nabla.(M\nabla\varphi))$, where we have set (to simplify notation, we denote here by (x_1, x_2) the coordinates of the generic point $\vec{x}$ in the plane):

$$\nabla \cdot (M\nabla\varphi)(\vec{x}) = \sum_{i=1}^{2}\sum_{j=1}^{2} \frac{\partial}{\partial x_i}\left[M_{ij}\frac{\partial\varphi}{\partial x_j}\right](\vec{x}),$$

where M is a two by two symmetric square matrix that depends continuously on $\vec{x}$. We further assume that there is a strictly positive constant α such that

$$\forall\vec{x} = (x_1, x_2) \in \bar{\Omega}, \quad \sum_{i=1}^{2}\sum_{j=1}^{2} M_{ij}(\vec{x})x_i x_j \geq \alpha \sum_{i=1}^{3} x_i^2,$$

which shows in particular that the matrix M is positive definite. We assume that the function a_0 is sufficiently smooth on $\bar{\Omega}$.

Thus the problems we can now treat can be written as: find a function φ solution of the following boundary value problem:

$$-\nabla \cdot (M\nabla\varphi) + a_0\varphi = f \text{ in } \Omega, \tag{5.20}$$

$$\varphi = \varphi_0 \text{ on } \Gamma_1, \quad a_1\varphi + n^T M\nabla\varphi = g \text{ on } \Gamma_2, \tag{5.21}$$

where n denotes the normal to Γ oriented towards the exterior of Ω. Let us note that the boundary condition on Γ_2 has a new expression, which will find its justification in Green's formula, as we shall see in the next section where we shall write the variational formulation of the problem.

5.2.2 *Variational Formulation*

We have:

Theorem 5.5 The variational formulation of problem (5.20)–(5.21) is written: find $\varphi \in H^1_{\varphi_0,\Gamma_1}(\Omega)$ such that $\forall w \in H^1_{0,\Gamma_1}(\Omega)$, we have

$$\int_\Omega \left[(\nabla w)^T M \nabla\varphi + a_0 \varphi w\right](\vec{x})\mathrm{d}\vec{x} + \int_{\Gamma_2} (a_1\varphi w)(\vec{x})\mathrm{d}\gamma(\vec{x}) \tag{5.22}$$
$$= \int_\Omega (f\,w)(\vec{x})\mathrm{d}\vec{x} + \int_{\Gamma_2} (gw)(\vec{x})\mathrm{d}\gamma(\vec{x}),$$

where the function space $H^1_{\psi,\Gamma_1}(\Omega)$ is defined in (5.18). Furthermore, if $\varphi \in H^2(\Omega)$, φ solves the boundary value problem (5.20)–(5.21) if and only if it solves the above variational formulation.

PROOF Let us assume that φ is a solution in $H^2(\Omega)$ of problem (5.20)–(5.21); let us then multiply equation (5.20) by w, where w is any function in the space $H^1_{0,\Gamma_1}(\Omega)$, then let us integrate the resulting equation over Ω. Using Green's formula we have, suppressing the integration variables to simplify notation:

$$-\int_\Omega w\nabla.(M\nabla\varphi) = \int_\Omega (\nabla w)^T M\nabla\varphi - \int_{\Gamma_2} w(M\nabla\varphi)\cdot n,$$

so that we obtain:

$$\int_\Omega (\nabla w)^T M\nabla\varphi + a_0\varphi w - \int_{\Gamma_2} w(M\nabla\varphi)\cdot n = \int_\Omega f\,w, \tag{5.23}$$

which gives precisely (5.22), because φ satisfies the Robin condition.

Conversely, let us assume that $\varphi \in H^2(\Omega)$ is a solution of the variational problem (5.22). Then, if we consider the variational formulation with functions w in the space $D(\Omega)$ of C^∞ functions with compact support in Ω, we see that the partial differential equation (5.20) is satisfied in the sense of distributions, and also almost everywhere on Ω, since all terms in (5.20) are square integrable functions on Ω. Since the Dirichlet boundary condition is included in the definition of the function space, it is automatically satisfied by φ. It remains to verify the boundary condition on γ_2. To do this, let us proceed as in the beginning of this proof, by multiplying (5.20) by w (any function in the space $H^1_{0,\Gamma_1}(\Omega)$), integrating over Γ, and using Green's formula; we obtain (5.23), which by subtraction from (5.22) (recall that φ solves (5.22)) gives eventually:

$$\forall w \in H^1_{0,\Gamma_1}(\Omega), \quad \int_{\Gamma_2} a_1\varphi w + \int_{\Gamma_2} w(M\nabla\varphi)\cdot n = \int_{\Gamma_2} gw.$$

We conclude by using the density in the space $L^2(\Gamma_2)$ of the traces on Γ of functions in $H^1_{0,\Gamma_1}(\Omega)$, which gives us (5.21) (almost everywhere on Γ_2). □

EXERCISE Can you express this variational problem in terms of a minimisation problem? Specify the relevant functional and function space. Show, when it is possible to do so, the equivalence with problem (5.22). □

Remark 5.6 From a theoretical point of view, if a_0 is continuous and positive on $\bar{\Omega}$, and if Γ_1 has non-zero measure, the variational problem (5.22) has one and only one solution. This results from the ellipticity of the bilinear form that occurs on the left hand side of (5.22), which itself stems from the hypotheses on M and from the positivity of a_1. If, however, Γ_1 has zero measure, the conclusion remains identical if a_0 is bounded from below by a strictly positive constant on all of $\bar{\Omega}$, or if this is the case for a_1 on Γ. □

Remark 5.7 Most problems arising from the conservation principles of physics fall into the formalism of (5.21) if φ satisfies (5.20). If it does not as in this example

$$-\Delta\varphi = f \quad \text{in } \Omega$$

$$a_1\varphi + \frac{\partial\varphi}{\partial n} + b\frac{\partial\varphi}{\partial s} = g \quad \text{on } \Gamma$$

where s is the arc length coordinate on Γ, then the variational formulation becomes non-symmetric:

$$\int_\Omega \nabla\varphi\nabla w + \int_\Gamma a_1\varphi w + b\frac{\partial\varphi}{\partial s}w = \int_\Omega f\,w + \int_\Gamma g w$$ □

5.2.3 *Discretisation and Programming*

Discretisation is obtained as above by replacing the function space $H^1(\Omega)$ by the space H_h^1 defined in (5.7). The test functions space V_h is written as

$$V_h = \{\varphi_h \in H_h^1 \text{ and for any vertex } q^i \text{ of } \Gamma_1, \quad \varphi_h(q^i) = 0\}, \tag{5.24}$$

and the discrete variational formulation is as follows:
find φ_h such that $\varphi_h - \varphi_{0h} \in V_h$, and such that $\forall w_h \in V_h$,

$$\int_\Omega \left[(\nabla w_h)^T M \nabla\varphi_h + a_0\varphi_h w_h\right](\vec{x})\mathrm{d}\vec{x} + \int_{\Gamma_2}(a_1\varphi_h w_h)(\vec{x})\mathrm{d}\gamma(\vec{x}) \tag{5.25}$$
$$= \int_\Omega (f\,w_h)(\vec{x})\mathrm{d}\vec{x} + \int_{\Gamma_2}(g w_h)(\vec{x})\mathrm{d}\gamma(\vec{x}),$$

where φ_{0h} is the P^1 interpolant of the function φ_0 at each of the vertices of the mesh, i.e. (we have assumed for simplicity's sake that φ_0 is defined on $\bar{\Omega}$):

$$\varphi_{0h} = \sum_{i=1}^{nv} \varphi_0(q^i)w^i. \tag{5.26}$$

Proceeding as in Section (5.1.2), we expand the unknown function φ on the hat functions basis, and problem (5.25) becomes equivalent to the solution of a linear system

$$A\Phi = F, \tag{5.27}$$

the A_{ij} element of matrix A being now defined by

$$A_{ij} = \int_{\Omega_h} \left[(\nabla w^i)^T M \nabla w^j + a_0 w^i w^j \right] (\vec{x}) \mathrm{d}\vec{x} + \int_{\Gamma_{2h}} (a_1 w^i w^j)(\vec{x}) \mathrm{d}\gamma(\vec{x}), \tag{5.28}$$

where we have denoted by Γ_{2h} the part of the boundary Γ_h that links vertices q^i in Γ_2.

We have been deliberately vague with the size of the linear system (5.27), as well as with the unknown Φ, it being understood that in theory the number of degrees of freedom of this problem is equal to $nv - nf_1$, denoting by nv the total number of vertices of the triangulation, and by nf_1 the number of vertices located on Γ_1. Actually, following our usual approach, we shall 'enlarge' this system, solving it on *all the vertices* of the mesh. The unknown Φ is then a vector of size nv whose index i entry represents an approximation to the value of the solution of the continuous problem at the point q^i. The Dirichlet boundary conditions on Γ_1 are satisfied using either one of the methods set forth in Section 5.1.3. The coefficients A_{ij} of the matrix A are defined for all indices i and j in the set $\{1, ..., nv\}$ by formula (5.28), and the right hand side is computed at each vertex of the mesh via the expression:

$$F_j = \int_{\Omega_h} (f w^j)(\vec{x}) \mathrm{d}\vec{x} + \int_{\Gamma_{2h}} (g w^j)(\vec{x}) \mathrm{d}\gamma(\vec{x}). \tag{5.29}$$

Remark 5.8 Remarks pertaining to the 'positive definite' nature of this matrix are of the same kind as in Proposition 5.4 and are directly connected with the 'well posedness' of the continuous problem that was alluded to in Remark 5.6. We leave the details as an exercise for the reader.

It now remains to construct the matrix of this system. It is composed of two surface integrals and a line integral, whose computation has already been treated in section 5.1.3.2.

Let us start with the computation of the surface integral featuring the matrix M. We assume that this matrix is given at each vertex of the mesh, which enables us to approximate each of its coefficients M_{ij} by its P^1 interpolant denoted by M_{ijh}; we denote by M_h the matrix whose entries are M_{ijh}. The integral computation then proceeds along the same lines as is for the case of the Laplacian (for which $M = Id$), by locally computing the contribution from each triangle T_k of the triangulation. The hat functions are linear on T_k, so their gradients are constant there, and an exact integration using the quadrature formula (2.59) then gives (for simplicity's sake we denote the generic point in the plane by $\vec{x} = (x_1, x_2)$)

$$\int_{T_k}\left[(\nabla w^i)^T M_h \nabla w^j\right](\vec{x})\mathrm{d}\vec{x} = |T_k|\left[m_{11}\frac{\partial w^i}{\partial x_1}\frac{\partial w^j}{\partial x_1} + m_{12}\left(\frac{\partial w^i}{\partial x_1}\frac{\partial w^j}{\partial x_2} + \frac{\partial w^j}{\partial x_1}\frac{\partial w^i}{\partial x_2}\right) + m_{22}\frac{\partial w^i}{\partial x_2}\frac{\partial w^j}{\partial x_2}\right], \tag{5.30}$$

where

$$m_{ij} = \frac{1}{3}\sum_{l=1}^{3} M_{ij}(q^l(T_k)), \tag{5.31}$$

if we denote by $q^1(T_k)$, $q^2(T_k)$ and $q^3(T_k)$ the three vertices of the triangle T_k and by $|T_k|$ the area of this triangle. We notice that, for this computation, it is not necessary to store the values of the coefficients M_{ij} at all vertices of the mesh, but only the m_{ij} coefficients defined in (5.31) on all triangles of the triangulation (we choose to approximate, from the beginning, the matrix M by a constant matrix, not a linear one, on each triangle).

Let us recall the expression of the gradient of the function w^i on one of the triangles T_k surrounding the vertex q^i (the gradient would be zero on any other triangle!). This expression is obtained from formulae (2.42), (2.64) and (2.65). If we denote by *ip* the index of the 'next' vertex after vertex q^i in triangle T_k (i.e. along the orientation of the triangle, assumed positively oriented), and by *ipp* the index of the following vertex after *ip*, this gives

$$\frac{\partial w^i}{\partial x_1} = \frac{1}{2|T_k|}\left(q_2^{ip} - q_2^{ipp}\right) \quad \frac{\partial w^i}{\partial x_2} = \frac{1}{2|T_k|}\left(q_1^{ipp} - q_1^{ip}\right), \tag{5.32}$$

where (q_1^j, q_2^j) are the coordinates of point q^j.

As far as the last integral is concerned, we shall assume that function a_0 is known at each vertex of the mesh, which enables us to approximate this function by its P^1 interpolant; by doing this, and if we follow the strategy of Section 5.1.3, we are led to compute, on each triangle, the product of three linear functions. To get an exact result, this requires the use of quadrature formulae exact for P^3. In order to simplify programming, we shall be content with using the 'edge mid-points' formula, i.e. formula (2.60), which is exact for second degree polynomials. Then, even if this entails an error, we choose to approximate the function a_0 by its P^0 interpolant, which we denote by a_{0h}: a_{0h} is the piecewise constant function on each triangle, whose value on triangle T_k is one third the sum of the values of a_0 at each vertex of this triangle. Then we have:

$$\int_{\Omega_h}(a_{0h}w^i w^j)(\vec{x})\mathrm{d}\vec{x} = \sum_{k=1}^{nt}\bar{a}_0(T_k)d_{ij}(T_k), \quad \bar{a}_0(T_k) = \frac{1}{3}\sum_{l=1}^{3} a_0(q^l(T_k)), \tag{5.33}$$

with

$$d_{ij}(T_k) = \frac{|T_k|}{3}\sum_{m=1}^{3}(w^i w^j)(\bar{q}^m), \tag{5.34}$$

if we denote by $q^1(T_k)$, $q^2(T_k)$ and $q^3(T_k)$ the three vertices of triangle T_k, $|T_k|$ its area and $\bar{q}^1$, $\bar{q}^2$ and $\bar{q}^3$ the midpoints of the edges of T_k. There remains to evaluate these coefficients $d_{ij}(T_k)$; they are zero if one of the indices i or j is not that of a vertex of triangle T_k. Otherwise, we have $d_{ii}(T_k) = |T_k|/6$, and if indices i and j are different, $d_{ij}(T_k)$ is equal to $|T_k|/12$. In summary, if i and j are indices of vertices of the triangle T_k, we have, using the same notation as above:

$$\int_{T_k} (a_{0h} w^i w^j)(\vec{x}) d\vec{x} = \frac{a_0(q^i) + a_0(q^{ip}) + a_0(q^{ipp})}{3} \begin{cases} \frac{|T_k|}{12}, & \text{if } i \neq j, \\ \frac{|T_k|}{6}, & \text{otherwise.} \end{cases} \tag{5.35}$$

We can now program the method. In the general case, for each triangle we must store three coefficients m_{11}, m_{12} and m_{22} defined in (5.31), and for each vertex, we must store two coefficients a_0 and a_1.

So as to minimise the computations, we shall particularise certain operators used very frequently (diagonal operators, for instance). We shall distinguish between four types of operators $(\varphi \to -\nabla(M\nabla\varphi))$, of increasing generality, to which we associate four increasing values of a single parameter denoted by *iM*:

- the Laplacian (matrix M is the identity matrix, and no storage is required), in which case the flag *iM* is set equal to 0;
- the case of operators whose associated matrix M is, up to a multiplicative constant (depending on $\vec{x}$), the identity matrix, in which case it suffices to store the value of $m_{11} = m_{22}$ on each triangle; in that case the flag satisfies: $iM \leq 1$;
- the case of operators with a diagonal matrix both of whose coefficients have to be stored on each triangle (m_{11} and m_{22}), and the parameter *iM* is such that $iM \leq 2$;
- last, the general case, where we must store three coefficients for each triangle (m_{11}, m_{12} and m_{22}) and *iM* is assigned the value 3.

To save on notation, the values of these various coefficients m_{ij} are stored in a one-dimensional array M of size $iM * nt$. The first nt locations in this array store the m_{11} coefficients, the next nt locations store m_{22}, and the last locations store the off-diagonal coefficients m_{12}.

In the same spirit, we define a parameter *ia*0 enabling us to detect whether the a_0 coefficient is identically zero ($ia0 = 0$), or not ($ia0 = 1$), since the first case is particularly frequent.

In a similar way, we particularise some usual boundary conditions, through a parameter *ia*1 equal to 0 if function a_1 is identically zero (which corresponds to Neumann conditions), and to 1 otherwise. Let us note that, since the values of the various parameters *iM*, *ia*0 and *ia*1 are known by the user according to the problem under study, they will be prescribed (and not computed) at the beginning of the program.

The complete program for solving problem (5.27), from the variational formulation (5.25), using the conjugate gradient preconditioned by the diagonal of the matrix A can be found in Appendix 5.A. There, we specify the way data is input. So as not to change the conjugate gradient program as compared to the previous chapter, we have kept the array cg, even though it could be dispensed with here.

We now give the subroutines that differ most from those in the previous chapter, that is the new computations 'gRHS' of the right hand side and 'geMat' of the elementary contribution, on each triangle, to the matrix coefficient A_{ij}, as well subroutine as 'gDirBC' for recovering the right Dirichlet boundary condition on the part Γ_1 of the domain boundary.

```
C  File: gelem.for
C
C  ----------------------------------------------------
   SUBROUTINE gRHS(me,f,nv,nt,area,ff,ng,q)
C  ----------------------------------------------------
C  This subroutine computes the right hand side
C       ff(i) = ∫Ω fw^i + ∫Γ2 gw^i
C  of the linear system to be solved, and stores the
C  result in array ff.
C  This computation is in several steps:
C
C  - we first compute the second integral, because
C    on entry to this program, the array f contains
C    the values of g at each vertex of the mesh
C    (the array f can then be erased);
C
C  - we then read from a file, whose name is given by
C    the user at run time, the values of the right hand
C    side f, and store them in array f; reading is via
C    subroutine 'readF' from file utilP1.for;
C
C  - we then compute the first integral.
C
C
C        input:
C        ------
C        triangulation (nv,nt,q,me,ng,area),
C        array f containing the values of g at each
C           vertex of the mesh
C
C        output:
C        -------
C        array f contains the values of the right hand side
C           f of the PDE at each vertex of the mesh
C        array ff contains the right hand side of the linear
C           system to be solved.
C
C
      DIMENSION q(2,*),me(3,*),ng(*),area(*)
      DIMENSION f(*),ff(*)
      DIMENSION inext(4)
```

```
      CHARACTER*20 filnam
C
      DATA inext/2,3,1,2/
      DO 2 j=1,nv
        ff(j)=0.
2     CONTINUE
C
      DO 3 k = 1,nt
      DO 3 il=1,3
        i=me(il,k)
        IF (ng(i).NE.0) THEN
          ip=me(inext(il),k)
          ipp=me(inext(il+1),k)
          IF (ng(ip).NE.0) THEN
            dx=q(1,i)-q(1,ip)
            dy=q(2,i)-q(2,ip)
            dist=sqrt(dx*dx+dy*dy)
            ff(i)=ff(i)+(2*f(i)+f(ip))*dist/6.
          ENDIF
          IF (ng(ipp).NE.0) THEN
            dx=q(1,i)-q(1,ipp)
            dy=q(2,i)-q(2,ipp)
            dist=sqrt(dx*dx+dy*dy)
            ff(i)=ff(i)+(2*f(i)+f(ipp))*dist/6.
          ENDIF
        ENDIF
3     CONTINUE
C
C
C
      write(*,*) 'name of file where f is stored'
      write(*,*) '(PDE right hand side)'
      read(*,'(a)') filnam
      CALL readF(filnam,nv,f)
C
      DO 4 k = 1,nt
        areak=area(k)
        DO 4 il=1,3
          i=me(il,k)
          ip=me(inext(il),k)
          ipp=me(inext(il+1),k)
          ff(i)=ff(i)+(2*f(i)+f(ip)+f(ipp))*areak/12.
4     CONTINUE
      END
C
C
C ------------------------------------------------------
      FUNCTION geMat(k,il,jl,areak,q,me,ng,nv,nt,AM,
     > iM,a0,ia0,a1,ia1)
C ------------------------------------------------------
C This function computes the elementary contribution on
```

```
C  the index k triangle to the A_ij coefficient of
C  the matrix of the linear system to be solved, in the
C  case of general symmetric second-order elliptic
C  operator
C       a0φ - ∇.(M∇φ),
C  with Robin boundary conditions on the boundary Γ2
C       a1φ + n^T M∇φ = g.
C
C  The computation is in 3 steps:
C  - computation of ∫Ω(∇w)^T M∇φ,
C  - computation of ∫Ω a0φw,
C  - computation of ∫Γ a1φw.
C
C  Each of these computations takes into account the
C  possible particular structure of the problem.
C  We recall that:
C  - iM = 0 if M = Id,
C  - iM ≤ 1 if M = M11 Id,
C  - iM ≤ 2 if M is diagonal
C  - iM = 3 in the other cases;
C
C  We also have: ia0 = 0 if a0 is the zero function,
C  and ia0 = 1 otherwise; the same goes for ia1 and a1.
C  Array AM here contains average values m on each triangle
C  of matrix M (i.e. one third of the sum of the values
C  of this matrix at each vertex of the triangle). It is
C  filled in the following way:
C  - the first nt indices store m11,
C  - the next nt indices store m22,
C  - eventually, the last nt indices store the off-diagonal
C  coefficients m12.
C
C
C  N.B.  This subroutine assumes the triangles are
C        positively oriented.
C
C

       DIMENSION q(2,*),me(3,*),ng(*),inext(4)
       DIMENSION AM(*),a0(*),a1(*)
C
       DATA inext/2,3,1,2/
C
C  step 1: matrix M in the PDE
C
       nkt=k-nt
       IF (iM.GT.0) THEN
         nkt=nkt+nt
         AM11=AM(nkt)
       ELSE
         AM11=1.
       ENDIF
```

```
      IF (iM.GT.1) THEN
        nkt=nkt+nt
        AM22=AM(nkt)
      ELSE
        AM22=AM11
      ENDIF
      i=me(il,k)
      ip=me(inext(il),k)
      ipp=me(inext(il+1),k)
      j=me(jl,k)
      jp=me(inext(jl),k)
      jpp=me(inext(jl+1),k)
      gradWi1= q(2,ip)-q(2,ipp)
      gradWi2= q(1,ipp)-q(1,ip)
      gradWj1= q(2,jp)-q(2,jpp)
      gradWj2= q(1,jpp)-q(1,jp)
      elem=AM11*gradWi1*gradWj1+AM22*gradWi2*gradWj2
C
      IF (iM.EQ.3) THEN
        nkt=nkt+nt
        elem=elem+AM(nkt)*(gradWi1*gradWj2+gradWi2* gradWj1)
      ENDIF
      elem=elem/(4*areak)
C
C  step 2: constant term a0 of the PDE
C
      IF (ia0.EQ.1) THEN
        a0k=(a0(i)+a0(ip)+a0(ipp))/3.
        IF (i.EQ.j) THEN
          elem=elem+a0k*areak/6.
        ELSE
          elem=elem+a0k*areak/12.
        ENDIF
      ENDIF
C
C  step 3: term a1 of the boundary condition
C
      IF (ia1.EQ.1) THEN
        IF (ng(i).NE.0) THEN
          IF (i.EQ.j) THEN
            IF (ng(ip).NE.0) THEN
              dx=q(1,i)-q(1,ip)
              dy=q(2,i)-q(2,ip)
              dist=sqrt(dx*dx+dy*dy)
              elem=elem+(3*a1(i)+a1(ip))*dist/12.
            ENDIF
            IF (ng(ipp).NE.0) THEN
              dx=q(1,i)-q(1,ipp)
              dy=q(2,i)-q(2,ipp)
              dist=sqrt(dx*dx+dy*dy)
              elem=elem+(3*a1(i)+a1(ipp))*dist/12.
```

```
            ENDIF
          ELSEIF (ng(j).NE.0) THEN
            dx=q(1,i)-q(1,j)
            dy=q(2,i)-q(2,j)
            dist=sqrt(dx*dx+dy*dy)
            elem=elem+(a1(i)+a1(j))*dist/12.
          ENDIF
        ENDIF
      ENDIF
      geMat=elem
      END
C
C --------------------------------------------------------
C  SUBROUTINE gDirBC(aa,ia,ff,phi0,ng,nv,cte)
C --------------------------------------------------------
C  This subroutine recovers the Dirichlet boundary
C  condition on the part Γ1 of the domain boundary
C  corresponding to vertices q^i that do
C  not satisfy phi0(i)=0.
C
C  At each of these points, we simultaneously modify
C  - the diagonal element aa by replacing it by a very
C     large constant cte,
C  - ff(i) by replacing it by the product phi0(i)*cte.
C
      DIMENSION aa(*),ia(*),phi0(*),ng(*),ff(*)
C
      DO 1 i = 1,nv
        IF ((ng(i).NE.0).AND.(phi0(i).NE.0)) THEN
          aa(ia(i+1)-1)=cte
          ff(i)=phi0(i)*cte
        ENDIF
1     CONTINUE
      END
```

5.2.4 *An Example*

As an example of a general elliptic problem, we set about to compute the temperature φ of a fluid at rest between two coaxial cylinders with radii R_1 and R_2, with $R_2 < R_1$. The temperature distribution inside the inner cylinder has such a small conductivity that we may take a zero flux condition on the interface between the two cylinders. The outer cylinder has a thermal conductivity μ, and we assume that its surface is held at a temperature T_0. The problem is thus to find φ such that

$$\nabla.(\mu\nabla\varphi) = 0 \quad \text{in} \quad \Omega, \tag{5.36}$$

$$\mu\frac{\partial\varphi}{\partial n} = 0 \text{ on } \Gamma_2, \quad \varphi = T_0 \text{ on } \Gamma_1,$$

where, in cylindrical coordinates (r, θ, z), z being the abscissa along the cylinder axis, and (r, θ) polar coordinates in the plane orthogonal to the axis,

$$\Omega = \{(r,\theta,z) : R_2 < r < R_1,\ 0 \le \theta \le 2\pi,\ z \in \mathbb{R}\},$$
$$\Gamma_1 = \{(r,\theta,z) \in \bar{\Omega},\ r = R_1\},$$
$$\Gamma_2 = \{(r,\theta,z) \in \bar{\Omega},\ r = R_2\}.$$

Since the problem has an axisymmetric geometry, we shall reformulate it in cylindrical coordinates. The interest lies in the fact that the unknown φ being independent of the polar angle, this change of variables reduces the problem to a two dimensional problem, which is much less costly. Let us then denote by Ω' the planar domain that is the intersection of Ω with the plane $\theta = 0$, and by Γ_1' and Γ_2', respectively, the intersections of this plane with Γ_1 and Γ_2; if μ is constant, we must solve the new problem:

$$\frac{1}{r}\frac{\partial}{\partial r}\left(r\frac{\partial\varphi}{\partial r}\right) + \frac{\partial^2\varphi}{\partial z^2} = 0 \quad \text{in } \Omega', \tag{5.37}$$
$$\frac{\partial\varphi}{\partial n} = 0 \text{ in } \Gamma_2', \quad \varphi = T_0 \text{ in } \Gamma_1'.$$

After multiplying by r and changing notation ($r \to x_1$, $z \to x_2$), we see that (5.37) has the form (5.20)–(5.21) with: $a_0 = a_1 = M_{12} = M_{21} = 0$, $M_{11} = M_{22} = x_1$, $g = 0$ and $\varphi_0 = T_0$:

$$\frac{\partial}{\partial x_1}\left(x_1\frac{\partial\varphi}{\partial x_1}\right) + \frac{\partial}{\partial x_2}\left(x_1\frac{\partial\varphi}{\partial x_2}\right) = 0 \quad \text{in } \Omega'$$

5.3 SECOND ORDER FINITE ELEMENTS

The finite elements discretisation of a boundary value problem of the above kind, can be carried out with piecewise polynomials of any arbitrary degree. This is interesting because the higher the degree of these polynomials, the better the method accuracy. Without going into the theory, let us recall that if the polynomials are of degree 2, the error in the norm $H^1(\Omega)$ as estimated in Theorem 2.28 of Chapter 2 becomes of order h^2 (instead of being of order h), at least for polygonal domains Ω. We refer the reader to the abundant bibliography for this type of error estimates, as our goal here is rather on describing how to implement the method.

5.3.1 Discrete Problem

To keeps things simple, let us go back to the model problem, i.e. the Laplacian with Dirichlet boundary conditions

$$-\Delta\varphi(\vec{x}) = f(\vec{x}), \quad \forall\vec{x} \in \Omega, \tag{5.38}$$
$$\varphi(\vec{x}) = g(\vec{x}), \quad \forall\vec{x} \in \Gamma, \tag{5.39}$$

written in variational form:
find $\varphi \in H^1_g(\Omega)$ such that

$$\forall w \in H_0^1(\Omega), \quad \int_\Omega \nabla\varphi(\vec{x}) \cdot \nabla w(\vec{x}) \mathrm{d}\vec{x} = \int_\Omega f(\vec{x}) w(\vec{x}) \mathrm{d}\vec{x}. \tag{5.40}$$

Let us give ourselves a regular triangulation of Ω and let us define $\Omega_h = \cup_{k=1}^{nt} T_k$; let us denote by $q^i, i \in \{1, ..., nv\}$ the vertices of this triangulation. Let us then define the discrete space V_h by:

$$V_h = \{v_h \in C^0(\Omega_h), \forall k \in \{1, ..., nt\}, \quad v_h|_{T_k} \in P^2\}, \tag{5.41}$$

where P^2 denotes the set of all degree 2 polynomials with respect to both variable x_1, x_2. We have the following result, which is the counterpart of Propositions 2.12 and 2.15, and for which we give a short proof.

Proposition 5.9 *Functions in V_h are entirely determined by their values at all of the vertices $q^i, i \in \{1, ..., nv\}$ as well as at the midpoints $q^{ij} = (q^i + q^j)/2$ of all the ne edges $[q^i, q^j]$ of the mesh. These points are called the 'nodes' of the mesh, and the dimension of the space V_h is equal to the total number $nv + ne$ of all these nodes. A basis of V_h is formed of functions in V_h equal to 1 at one node and 0 at all the other nodes.*

PROOF Let us first show that the restriction of a function v_h in V_h to an arbitrary triangle T of the triangulation is entirely determined by its values at the three vertices $q^{i_1}, q^{i_2}, q^{i_3}$ as well as at the three midpoints of the edges of this triangle. The restriction of v_h to T is a polynomial of degree 2, so there are six coefficients a, b, c, d, e and f such that:

$$\forall \vec{x} = (x_1, x_2) \in T, \; v_h(\vec{x}) = a + bx_1 + cx_2 + dx_1^2 + ex_2^2 + fx_1x_2.$$

If we express this relation by taking as particular values of $\vec{x}$ the three vertices $q^{i_j}, j \in \{1, 2, 3\}$, then the midpoints of the three edges of T, we obtain a system of six equations in six unknowns; but this system is invertible, since one can easily show (by working in a local coordinate system in the triangle) that the associated homogeneous system only has the trivial $(0, 0, 0, 0, 0, 0)$ solution, which proves the first point of the proposition.

As in Proposition 2.14, it is easy to show that the functions built that way, triangle by triangle, are continuous at the interface between two triangles, and so they are globally continuous on all of Ω, which proves that the dimension of V_h is equal to $ne + nv$, since this number represents the 'degrees of freedom' of functions in V_h.

There are $ne + nv$ functions in V_h equal to 1 at one node and 0 at all other nodes; to prove that they form a basis of V_h, it is thus sufficient to prove that they form an independent set, which is trivially shown as in proposition 2.17 of Chapter 2. □

EXERCISE Up to now, we have only considered triangular finite elements, because they allow for treating domains Ω of arbitrary geometry. For certain kinds of domain, it could be interesting to use rectangular finite elements. Let us assume our domain Ω is covered by nr rectangles R_k whose sides are parallel to the coordinate axes. We denote by Q^1 the space of polynomials of degree at most 1 in each of the variables x_1, x_2 (this space has dimension four, and a basis is $1, x_1, x_2, x_1x_2$). We then consider the following approximation space:

$$V_h = \{v_h \in C^0(\Omega_h), \forall k \in \{1, ..., nr\}, \quad v_h|_{R_k} \in Q^1\}.$$

Show that functions in V_h are entirely determined by their value at each vertex of all the nr rectangles R_k; deduce the dimension of V_h. Construct a basis of this space. □

To simplify notation, we shall denote by w^i ($i \in \{1, \ldots, nv\}$) the basis function associated with vertex q^i of the mesh and by w^{nv+i} ($i \in \{1, \ldots, ne\}$), the basis function associated with the midpoint of the edge with index i of the mesh (this assumes we have numbered the edges: cf. chapter 3 in this book). By analogy, the midpoint of the ith edge will also be denoted by q^i, with here $i \in \{nv+1, \ldots, nv+ne\}$. With this notation, the basis functions of V_h are simply defined by

$$\forall (i,j) \in \{1, \ldots, nv+ne\}^2, \quad w^i(q^j) = \delta_{ij}, \tag{5.42}$$

and among the functions defined above, those associated with the internal nodes of the mesh form a basis of the space

$$V_{0h} = \{v_h \in V_h, \quad v_h|_{\Gamma_h} = 0\}. \tag{5.43}$$

The discrete problem associated with the variational problem (5.40) is obtained, as usual, by solving (5.40) in the space V_{gh} of functions in V_h that coincide with g at each boundary node of the mesh, the 'tests' functions w_h running over all elements in the basis of V_{0h}, and the approximate solution φ_h of this new problem is determined from its components in the above basis.

Following our usual trick, we shall actually expand φ_h over *all nodes of the mesh*, by writing:

$$\varphi_h(\vec{x}) = \sum_{i=1}^{nv+ne} \varphi_i w^i(\vec{x}), \quad \forall \vec{x} \in \Omega. \tag{5.44}$$

Then we solve the linear system that enables us to determine the components of φ_h on all nodes of the mesh, or

$$A\Phi = F, \tag{5.45}$$

where Φ is the vector with entries $\varphi_i, i \in \{1, ..., nv+ne\}$, the matrix $A = ((A_{ij}))_{1 \leq i,j \leq nv+ne}$ and the right hand side $F = (F_i)_{1 \leq i \leq nv+ne}$ being defined by:

$$A_{ij} = \int_{\Omega_h} (\nabla w^i \cdot \nabla w^j)(\vec{x}) d\vec{x}, \tag{5.46}$$

$$F_i = \int_{\Omega_h} (f\, w^i)(\vec{x}) d\vec{x}, \tag{5.47}$$

with the usual modifications if q^i is a boundary point:

$$A_{ii} = cte \quad \text{and } F_i = cte * g(q^i), \quad \text{if } q^i \in \Gamma, \tag{5.48}$$

cte being a 'very large constant'. There now remains to compute the coefficients of this new matrix A in the P^2 approximation, as well as the right hand side.

5.3.2 *Construction of the System Matrix and Right Hand Side*

The coefficients A_{ij} are computed locally on each triangle, i.e.

$$A_{ij} = \sum_{k=1}^{nt} A_{ij}(T_k), \quad A_{ij}(T_k) = \int_{T_k} (\nabla w^i \cdot \nabla w^j)(\vec{x})d\vec{x}, \tag{5.49}$$

by noting that the elementary contribution $A_{ij}(T_k)$ is zero if neither q^i or q^j is a node of triangle T_k.

To simplify notation in this section, let us denote by q^i, $i \in \{1,2,3\}$, the vertices of triangle T_k (we assume that this triangle is positively oriented), and by $q^{\{i,ip\}}$ the midpoint of edge $[q^i, q^{ip}]$, where ip denotes the index of the point following the point with index i in the local numbering of the vertices of the triangle T_k ($ip = i$ mod $3+1$). By analogy, the basis functions associated with those nodes will be respectively denoted by w^i and $w^{\{i,ip\}}$, $i \in \{1,2,3\}$; they are defined by:

$$w^i \in V_h, \quad w^i(q^j) = \delta_{ij}, \quad w^i(q^{\{j,jp\}}) = 0, \quad \forall j \in \{1,2,3\}, \tag{5.50}$$

$$w^{\{i,ip\}} \in V_h, \quad w^{\{i,ip\}}(q^{\{j,jp\}}) = \delta_{ij}, \quad w^{\{i,ip\}}(q^j) = 0, \; \forall j \in \{1,2,3\}. \tag{5.51}$$

Computing the matrix coefficients then reduces locally to computing the following coefficients

$$a_{ij}(T_k) = \int_{T_k} (\nabla w^i \cdot \nabla w^j)(\vec{x})d\vec{x}, \quad (i,j) \in I^2, \tag{5.52}$$

which is, taking into account the symmetries of A, a total of 21 coefficients to compute. Without going into the details of the actual computation of these coefficients, [cf. Lucquin and Pironneau (1997)], we shall give the method, which is based on using the barycentric coordinates in triangle T_k.

Let us recall (cf. lemma 2.29) that each point $\vec{x}$ in triangle T_k can be written as the center of mass of the vertices $q^i, i \in \{1,2,3\}$ of T_k, i.e.:

$$\vec{x} = \sum_{i=1,2,3} \lambda_i(\vec{x})q^i, \quad 1 = \sum_{i=1,2,3} \lambda_i(\vec{x}). \tag{5.53}$$

The coefficients $\lambda_i(\vec{x}), i \in \{1,2,3\}$ are the barycentric coordinates of point $\vec{x} = (x_1, x_2)$; their expression as a function of the Cartesian coordinates (x_1, x_2) is obtained by solving the above linear system, which has a unique solution since its determinant is twice the area of the triangle. More precisely, we obtain for $i \in \{1,2,3\}$

$$\lambda_i = \frac{1}{2|T_k|}\left[x_1\left(q_2^{ip} - q_2^{ipp}\right) + x_2\left(q_1^{ipp} - q_2^{ip}\right)\right], \tag{5.54}$$

where ipp denotes the index of the point following the point with index ip along the orientation of triangle T_k. We note that the barycentric coordinates are affine functions of the Cartesian coordinates; their gradient is thus constant on T_k and is given by the relations:

$$\frac{\partial \lambda^i}{\partial x_1} = \frac{1}{2|T_k|}\left(q_2^{ip} - q_2^{ipp}\right), \quad \frac{\partial \lambda^i}{\partial x_2} = \frac{1}{2|T_k|}\left(q_1^{ipp} - q_1^{ip}\right), \tag{5.55}$$

where (q_1^j, q_2^j) are the coordinates of vertex q^j.

The interest of these barycentric coordinates lies in the fact that certain particular lines in the triangle have a very simple equation: the line going through point q^i and parallel to the opposite side $[q^{ip}, q^{ipp}]$ has $\lambda_i = 1$ as an equation, the one going through the midpoints $q^{\{i,ip\}}$ and $q^{\{ipp,i\}}$ of the two edges surrounding vertex q^i has $\lambda_i = 1/2$ as an equation, and last the line (q^{ip}, q^{ipp}) has $\lambda_i = 0$ as an equation. Because of that, the basis functions can be written as

$$w^i = \lambda_i(2\lambda_i - 1), \quad w^{\{i,ip\}} = 4\lambda_i\lambda_{ip}, \quad i \in \{1,2,3\}, \tag{5.56}$$

and their gradients are easily computed thanks to (5.55). To complete the computation of the $a_{ij}(T_k)$ coefficients, it then suffices to use the following result whose proof is left as an exercise to the reader [Lucquin and Pironneau (1997)]:

$$\int_{T_l} \lambda_i^m \lambda_j^n \lambda_k^p(\vec{x})d\vec{x} = \frac{2|T_l|\, m!n!p!}{(2+m+n+p)!}, \quad i \neq j \neq k \neq i. \tag{5.57}$$

Computing the right hand side follows the same strategy. On each triangle T_l, since the function f is assumed to be linear, we can write it in the form $f = \sum_{i=1}^{3} f_i\lambda_i$, then, using (5.57), we obtain

$$\int_T (\lambda_i w^j)(\vec{x})d\vec{x} = \frac{|T|}{30} \text{ if } i = j, \quad = \frac{-|T|}{60} \text{ otherwise;} \tag{5.58}$$
$$\int_T (\lambda_i w^{\{j,jp\}})(\vec{x})d\vec{x} = \frac{2|T|}{15} \text{ if } i = j \text{ or } i = jp, \quad = \frac{|T|}{15} \text{ otherwise}$$

It follows that:

$$\int_T (f w^j)(\vec{x})d\vec{x} = \frac{|T|}{60}\left[2f_j - f_{jp} - f_{jpp}\right] \tag{5.59}$$
$$\int_T (f w^{\{j,jp\}})(\vec{x})d\vec{x} = \frac{|T|}{15}\left[2f_j + 2f_{jp} + f_{jpp}\right].$$

From a programming point of view, there is but one minor modifications (item 3) to bring to the programs from the previous chapters; we must:

1 complete the array q by numbering the midpoints of the edges after the vertices of the mesh, as alluded to above, i.e. between indices $nv+1$ and $nv+ne$ (we refer to Chapter 3 for numbering the edges);

2 create a subroutine for computing the elementary contributions $a_{ij}(T_k)$ on each triangle T_k;

3 complete the array A for storing the matrix in CSR format and modify the computation of its elements so as to take into account the new degrees of freedom;

4 modify the subroutine for computing the right hand side, according to formulae (5.59).

Remark 5.10 Now that we have numbered the edges, it is possible to construct the linear system matrix by a loop over the edges rather than looping over the triangles. This remark may be useful for computing on a machine with vector processors, because loops are then easier to vectorise or parallelise. We leave the subsequent modifications to subroutine 'CgetMat' as an exercise. □

In this section, we have developed a computing technique coherent with the one used for the approximation by P^1 finite elements, i.e. we have computed in each triangle the elementary contribution to the linear system matrix. These computations depend on two very different factors that are: the particular geometry of the actual triangle (which governs, among other things, the computation of the barycentric coordinates as functions of the Cartesian coordinates), and the elementary operations on the basis functions (derivation, multiplication, integration) that we can envision to carry out once and for all on a 'model' triangle, which would reduce the amount of computation. This is precisely what we shall explain in the next section, choosing as model triangle the right-angled isoceles triangle of unit length on both axes: this triangle is traditionally called the 'reference triangle' and denoted by $\widehat{T}$. This will also allow true $O(h^2)$ accuracy when the boundary of Ω is curved by using curved elements.

5.3.3 The Reference Triangle Technique

Let us denote by $\widehat{q^1}, \widehat{q^2}, \widehat{q^3}$ the vertices of the right-angled isoceles triangle $\widehat{T}$ in the plane, as shown in Figure 5.1, and by q^1, q^2, q^3 the vertices of the triangle T_k on which we wish to carry out the computations. There is an affine transformation $\widehat{F}$ such that the image of each vertex $\widehat{q^i}$, $i \in \{1,2,3\}$ is the vertex q^i of T_k. This mapping sends the triangle $\widehat{T}$ into the triangle T_k and is defined analytically by:

$$\forall \widehat{x} = (\widehat{x_1}, \widehat{x_2}) \in \widehat{T}, \quad x = \widehat{F}(\widehat{x}) = \widehat{x_1}(q^2 - q^1) + \widehat{x_2}(q^3 - q^1) + q^1. \tag{5.60}$$

To any function ϕ defined on T_k we can associate a function $\widehat{\phi}$ defined on $\widehat{T}$ by the relation $\widehat{\phi}(\widehat{x}) = \phi(x)$. Expression (5.60) enables us to express the gradient of ϕ as a function of that of $\widehat{\phi}$ and we have:

$$\nabla\phi(x) = \left\{ \left[\widehat{J}(\widehat{x})\right]^T \right\}^{-1} \widehat{\nabla}\,\widehat{\phi}(\widehat{x}), \tag{5.61}$$

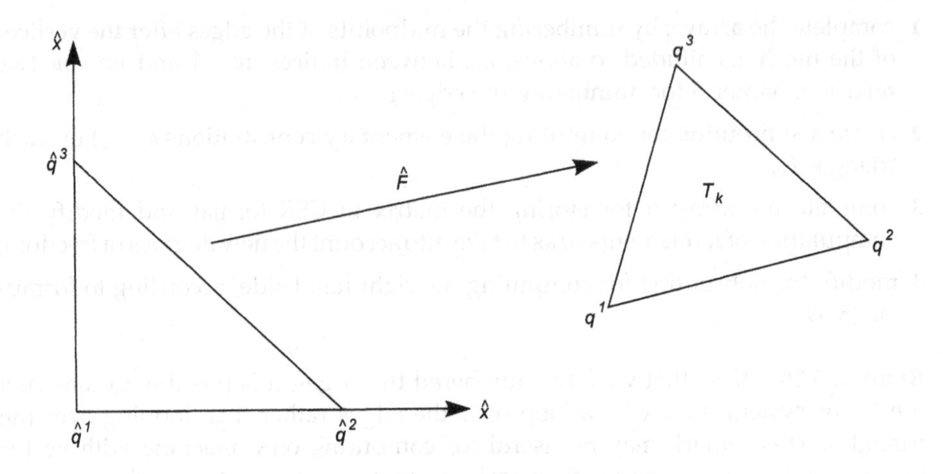

Figure 5.1 Reference triangle

where $\widehat{J}(\widehat{x})$ is the Jacobian matrix of the change of variable $(x \to \widehat{x})$; it is independent of $\widehat{x}$ and is defined by:

$$\widehat{J} = \begin{pmatrix} (q^2 - q^1)_1 & (q^3 - q^1)_1 \\ (q^2 - q^1)_2 & (q^3 - q^1)_2 \end{pmatrix}. \tag{5.62}$$

If we make this change of variables in (5.52) and notice that the Jacobian is equal to twice the area of T_k, we obtain:

$$a_{ij}(T_k) = \frac{1}{2|T_k|} \int_{\widehat{T}} \widehat{\nabla} \widehat{w^i}^T(\widehat{x}) \, B \, \widehat{\nabla} \, \widehat{w^j}(\widehat{x}) \mathrm{d}\widehat{x}, \quad (i,j) \in I^2, \tag{5.63}$$

where B is the matrix with entries

$$B_{11} = \|q^3 - q^1\|^2, B_{22} = \|q^2 - q^1\|^2, B_{12} = B_{21} = -(q^2 - q^1) \cdot (q^3 - q^1). \tag{5.64}$$

Let us set for $l, m \in \{1, 2, 3\}$:

$$b_{lm} = \frac{1}{2|T_k|} B_{lm} \tag{5.65}$$

$$\widehat{a}_{ij,lm} = \int_{\widehat{T}} \frac{\partial \widehat{w^i}}{\partial \widehat{x^l}}(\widehat{x}) \, \frac{\partial \widehat{w^j}}{\partial \widehat{x^m}}(\widehat{x}) \mathrm{d}\widehat{x};$$

we have the expression

$$a_{ij}(T_k) = \sum_{l=1}^{2} \sum_{m=1}^{2} \widehat{a}_{ij,lm} \, b_{lm}, \tag{5.66}$$

where we have separated the terms b_{kl} that are purely linked to the geometry of the triangle T_k and the terms $\widehat{a}_{ij,kl}$ that do not depend on it; the latter coefficients are computed in the reference triangle, and given their small number (78 if we take symmetries into account), we can store them once and for all.

To estimate those coeficients $\widehat{a}_{ij,kl}$, we have several possibilities. We could, as before, explicitely compute the basis functions in triangle $\widehat{T}$

$$\begin{aligned}
&\widehat{w^1}(\widehat{x}) = (1-\widehat{x}_1-\widehat{x}_2)(1-2\widehat{x}_1-2\widehat{x}_2),\ \widehat{w^2}(\widehat{x}) = \widehat{x}_1(2\widehat{x}_1-1),\\
&\widehat{w^3}(\widehat{x}) = \widehat{x}_2(2\widehat{x}_2-1),\ \widehat{w^{\{2,3\}}}(\widehat{x}) = 4\widehat{x}_1\widehat{x}_2,\\
&\widehat{w^{\{3,1\}}}(\widehat{x}) = 4\widehat{x}_2(1-\widehat{x}_1-\widehat{x}_2),\ \widehat{w^{\{1,2\}}}(\widehat{x}) = 4\widehat{x}_1(1-\widehat{x}_1-\widehat{x}_2),
\end{aligned} \tag{5.67}$$

then differentiate them, multiply them out and eventually integrate the products, taking into account the following relationship, which is similar to (5.57), and is easy to prove:

$$\int_{\widehat{T}} \widehat{x}_1{}^i\, \widehat{x}_2{}^j \,d\widehat{x}_1\, d\widehat{x}_2 = \frac{i!\,j!}{(2+i+j)!}, \quad i \neq j. \tag{5.68}$$

We could also, more simply, store in an array all the coefficients of the basis functions determined in (5.67), then create two 'symbolic' computing subroutines, one that computes the derivative with respect to the index k variable of a polynomial of degree 2, the other taking the product of two polynomials of degree 1; it then suffices to use (5.68) for $i,j \leq 2$ to deduce the final expression of $\widehat{a}_{ij,kl}$.

Remark 5.11 Until now we have used a preconditioned conjugate gradient method, to solve the linear system (3.8), with matrix A stored in CSR format. We could very well envision, to save on memory, to use the same algorithm, but without storing the matrix A. In this P^2 approximation, it would suffice to store the coefficients $a_{ij,kl}$, *independently of the mesh.* We could also envision, so as not to recompute the b_{kl} coefficients at each conjugate gradient step, to store, for each triangle, the 21 coefficients $a_{ij}(T_k)$.

Remark 5.12 The technique of the reference triangle, as well as the direct computation of the previous Section, can obviously be extended to the case of more general partial differential operators, as those we have studied at the beginning of this chapter.

Curved isoparametric elements The main interest of the reference triangle technique is that it can be adapted to the case of curved finite elements, which enables us to treat problems in domains whose boundary is not necessarily polygonal.

Let us assume we have defined a finite element in the reference triangle $\widehat{T}$, i.e. [cf. Ciarlet (1978)] a triple $(\widehat{T}, \widehat{P}, \widehat{\Sigma})$, where $\widehat{P}$ is a vector space of real valued functions defined on $\widehat{T}$ with finite dimension N (usually a space of polynomial functions) and where $\widehat{\Sigma}$ is a set of N linearly independent linear forms ϕ_i defined on $\widehat{P}$ (these forms are written $\widehat{\phi}_i(p) = p(a_i)$, the points a_i being the 'nodes' of the triangle; these forms represent the 'degrees of freedom' of the finite element). Let us furthermore assume that this finite element is 'unisolvent', meaning that, given N scalars α_i, there is one and only one element $\widehat{p} \in \widehat{P}$, such

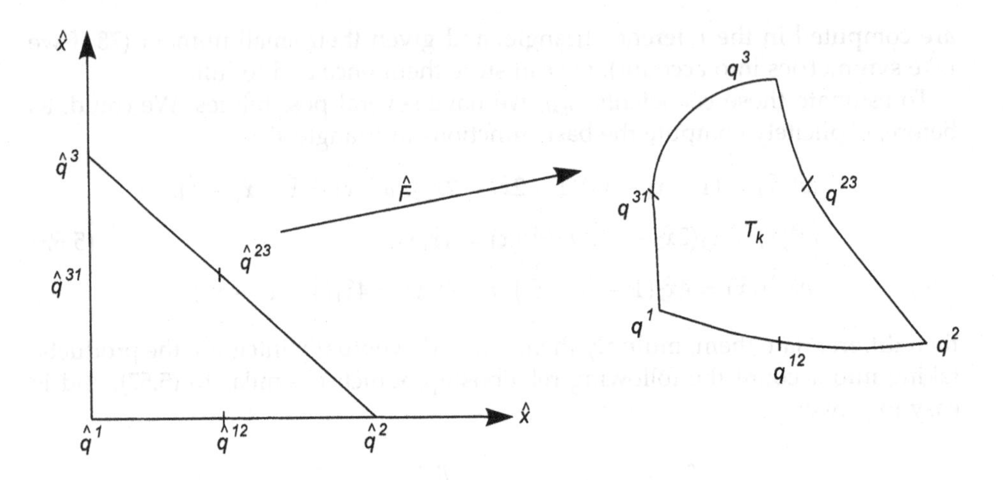

Figure 5.2 Isoparametric P^2 finite element

that $\forall i \in \{1, ..., N\}$, $\widehat{\phi}_i(\widehat{p}) = \alpha_i$. Let $\widehat{F}$ be a mapping defined on $\widehat{T}$ all of whose components are in $\widehat{P}$ and let us set $T = \widehat{f}(\widehat{T})$. Then, if $\widehat{F}$ is a bijection, the finite element (T, P, Σ) defined by

$$P = \{\widehat{p} \circ (\widehat{F})^{-1}, \widehat{p} \in \widehat{P}\}, \quad a_i = \widehat{F}(\widehat{a}_i), \ \Sigma = \{\phi_i, \phi_i(p) = p(a_i), i \in \{1, 2, 3\}\}$$

is unisolvent. It is called an *isoparametric finite element* and we say it is *isoparametrically equivalent to the reference finite element* $(\widehat{T}, \widehat{P}, \widehat{\Sigma})$ (let us note that this relation is usually not symmetric, contrary to the case seen above). Let us note that in practice, the mapping $\widehat{F}$ is actually defined from the data of the node a_i.

To fix ideas, given a 'curved' triangle with vertices q^1, q^2, q^3 (cf. Figure 5.2), we can define the P^2 isoparametric finite element as the 'image' by the mapping $\widehat{F}$ of the P^2 finite element defined on the reference triangle $\widehat{T}$ by:

$$\widehat{P} = P^2(\widehat{T}), \quad \widehat{\Sigma} = \{\widehat{q^1}, \widehat{q^2}, \widehat{q^3}, \widehat{q^{12}}, \widehat{q^{23}}, \widehat{q^{31}}\},$$

where $\widehat{q^1}, \widehat{q^2}, \widehat{q^3}$ are the vertices of $\widehat{T}$ and $\widehat{q^{12}}, \widehat{q^{23}}, \widehat{q^{31}}$ the midpoints of the edges of this triangle. The 'nodes' of triangle T are the images by the mapping $\widehat{F}$ of the six above nodes; in the general case, this mapping is thus defined as a function of the barycentric coordinates $\lambda_i(\widehat{x}), i \in \{1, 2, 3\}$ in the reference triangle by the relation

$$\widehat{F} \in (P^2(\widehat{T}))^2, \quad \widehat{F}(\widehat{x}) = \sum_{i=1}^{3} \lambda_i(\widehat{x})(2\lambda_i(\widehat{x}) - 1)q^i + \sum_{i=1}^{3} 4\lambda_i(\widehat{x})\lambda_{ip}(\widehat{x})q^{i,ip},$$

if we denote, as usual, by *ip* the index of the point 'following' the point with index i (i.e. in the local orientation of the triangle, which is assumed positively oriented).

Once this correspondence $\widehat{F}$ is established, the computations are carried out as above, in the reference triangle.

5.4 SECOND ORDER SYSTEMS

We shall study two examples of second order system: that of the Lamé equations, which occur in structural mechanics, and that of the Stokes equations for the study of weakly viscous incompressible fluid flows.

5.4.1 Lamé's Equations

The equations governing the displacements of a two dimensional structure subjected to an external force field $\vec{f}$ are written as

$$-\lambda\Delta\vec{u} - \mu\nabla(\nabla\cdot\vec{u}) = \vec{f}, \tag{5.69}$$

with the notation $\nabla\cdot\vec{u} = div(\vec{u}) = \frac{\partial u_1}{\partial x_1} + \frac{\partial u_2}{\partial x_2}$, which leads to the system of partial differential equations

$$-\lambda\left[\frac{\partial^2 u_i}{\partial x_1^2} + \frac{\partial^2 u_i}{\partial x_2^2}\right] - \mu\frac{\partial}{\partial x_i}\left[\frac{\partial u_1}{\partial x_1} + \frac{\partial u_2}{\partial x_2}\right] = f_i, \quad i = 1,2, \tag{5.70}$$

where λ and μ are two material characteristic constants. Various boundary conditions can be considered for this type of problem; we consider the case of homogeneous Dirichlet boundary conditions

$$\vec{u} = 0 \text{ on } \Gamma, \tag{5.71}$$

which corresponds physically to clamped structures. The variational formulation of this problem is written:

$$\text{find } \vec{u} = (u_1, u_2) \in [H_0^1(\Omega)]^2, \text{ such that for any } \vec{v} \in [H_0^1(\Omega)]^2 \text{ we have :}$$
$$\int_\Omega [\lambda\nabla\vec{u} : \nabla\vec{v} + \mu(\nabla\cdot\vec{u})(\nabla\cdot\vec{v})](\vec{x})d\vec{x} = \int_\Omega (f\cdot\vec{v})(\vec{x})d\vec{x}, \tag{5.72}$$

if we denote by $\nabla\vec{u} : \nabla\vec{v}$ the contracted product of the two matrices $\nabla\vec{u}$ and $\nabla\vec{v}$, that is $\sum_{i=1,2} \nabla u_i \cdot \nabla v_i$.

After we have triangulated the domain Ω, the discretisation by P^1 finite elements consists in approximating the function space $[H_0^1(\Omega)]^2$ by the space V_{0h} of functions defined and continuous on the computational domain Ω_h, with values in $\mathbb{R}^2$, that are zero on the boundary $\partial\Omega$, and linear on each triangle. A basis of this space is formed by all the functions $(w^i, 0)$ and $(0, w^i)$, where w^i is the usual one dimensional 'hat' function associated with the internal vertex of the mesh with index i. We seek the approximate solution in the form of a vector $\vec{u}_h$ with components $u_{hj}, j = 1,2$ defined by

$$u_{hj}(x) = \sum_{i=1}^{nv} u_{ji}w^i(x), \quad j = 1,2, \quad u_{ji} = 0 \text{ if } q^i \in \Gamma_h, \tag{5.73}$$

and we insert this expression in the approximate variational formulation that is then written as:

$$\int_{\Omega_h} [\lambda \nabla \vec{u}_h : \nabla \vec{v}_h + \mu(\nabla \cdot \vec{u}_h)(\nabla \cdot \vec{v}_h)](\vec{x})\mathrm{d}\vec{x} = \int_{\Omega_h} (f \cdot \vec{v}_h)(\vec{x})\mathrm{d}\vec{x}, \quad \forall \vec{v}_h \in V_{0h}. \tag{5.74}$$

Then we express (5.74) for the elements $\vec{v}_h$ of the basis of V_{0h}, and we obtain the following linear system

$$AU = F, \quad \text{with} \quad A = \begin{pmatrix} A_{11} & A_{12} \\ A_{12}^T & A_{22} \end{pmatrix}, \quad U = \begin{pmatrix} u_1 \\ u_2 \end{pmatrix}, \tag{5.75}$$

and for all $(i,j) \in \{1, ..., nv\}^2$:

$$(A_{kk})_{ij} = \int_{\Omega_h} \left[\lambda \nabla w^i \cdot \nabla w^j + \mu \left(\frac{\partial w^i}{\partial x_k} \frac{\partial w^j}{\partial x_k} \right) \right] (\vec{x})\mathrm{d}\vec{x}, \quad k \in \{1,2\}, \tag{5.76}$$

$$(A_{kl})_{ij} = \int_{\Omega_h} \left[\mu \frac{\partial w^i}{\partial x_k} \frac{\partial w^j}{\partial x_l} \right] (\vec{x})\mathrm{d}\vec{x}, \quad (k,l) \in \{1,2\}^2, k \neq l.$$

The main change at the programming level lies on the one hand in the fact that for each triangle, we must now compute three elementary contributions for each pair of indices $\{i,j\}$, and no longer one, and on the other hand in the different organisation of the data, and the numbering of the unknowns.

5.4.2 The Stokes Problem

The velocity $\vec{u}$ and pressure p of a Newtonian viscous and incompressible fluid are given, at low Reynolds number, by the equations

$$-\Delta \vec{u} + \nabla p = 0 \quad \text{and} \quad \nabla \cdot \vec{u} = 0 \quad \text{in} \quad \Omega, \quad \vec{u}|_\Gamma = \vec{u}_0, \tag{5.77}$$

where $\vec{u}_0$ is the velocity of the domain boundary, or the velocity at infinity if the domain Ω is not bounded. In two dimensions, this problem is thus written as:

$$-\Delta u_i + \frac{\partial p}{\partial x_i} = 0 \quad \text{and} \quad \sum_{i=1}^{2} \frac{\partial u_i}{\partial x_i} = 0 \quad \text{in} \quad \Omega, \quad u_i|_\Gamma = u_{0i}, \quad i = 1,2. \tag{5.78}$$

This problem has been the topic of numerous theoretical and numerical studies [cf. Girault and Raviart (1986) for example]; briefly, this problem admits the following variational formulation:

$$\text{find } \vec{u} \in [H^1(\Omega)]^2, \quad \vec{u}|_\Gamma = \vec{u}_0 \quad \text{such that :}$$

$$\int_\Omega [\nabla \vec{u} : \nabla \vec{v}](\vec{x})\mathrm{d}\vec{x} + \int_\Omega [\vec{v} \cdot \nabla p](\vec{x})\mathrm{d}\vec{x} = 0, \quad \forall \vec{v} \in [H_0^1(\Omega)]^2, \tag{5.79}$$

$$\int_\Omega [q \nabla \cdot \vec{u}](\vec{x})\mathrm{d}\vec{x} = 0, \quad \forall q \in L^2(\Omega).$$

The discrete problem consists in finding a function $\vec{u}_h$ that coincides with $\vec{u}_0$ at each boundary node of the mesh, and that solves:

$$\int_{\Omega_h} [\nabla \vec{u}_h : \nabla \vec{v}_h](\vec{x}) d\vec{x} + \int_{\Omega_h} [\vec{v}_h \cdot \nabla p_h](\vec{x}) d\vec{x} = 0, \quad \forall \vec{v}_h \in V_{0h}, \tag{5.80}$$
$$\int_{\Omega_h} [q_h \nabla \cdot \vec{u}_h](\vec{x}) d\vec{x} = 0, \quad \forall q_h \in Q_h.$$

Let us denote by $\mathcal{T}_{h/3}$ the triangulation obtained from triangulation $\mathcal{T}_h$ by dividing each of its triangle in three, which can be done by joining the three vertices of a triangle to its center of mass. Then, if we define the discrete spaces by

$$\begin{aligned} V_h &= \{v_h : \Omega_h \to \mathbb{R}^2, \forall i \in \{1,2\}, v_{hi} \text{ is continuous on } \Omega_h \\ &\qquad \text{and linear on each triangle of } \mathcal{T}_{h/3}\}, \\ V_{0h} &= \{v_h \in V_h, v_h|_\Gamma = 0\}, \\ Q_h &= \{q_h : \Omega_h \to \mathbb{R}, \text{continuous on } \Omega_h \text{ and linear on each triangle of } \mathcal{T}_h\}, \end{aligned} \tag{5.81}$$

problem (5.80) is well posed, meaning that u_h exists and is unique, whereas p_h is unique up to addition of a constant.
A basis of the space Q_h is made up of the 'hat' functions of the mesh $\mathcal{T}_h$, and completing this basis by the 'hat' functions associated with the centre of mass of the triangles in $\mathcal{T}_h$ gives a basis of V_h.

APPENDIX 5.A

The Fortran Program for the Solution of a Partial Differential Equation, with Very General Boundary Condition, by Conjugate Gradient Preconditioned by the Diagonal.

```
C  File: gPDE.for
C
            PROGRAM gPDE
C
C  This program solves the following general second order
C  elliptic problem (Γ=Γ1∪Γ2):
C        a0φ - ∇.(M∇φ) = f in Ω,
C        φ = φ0 on Γ1,
C        a1φ + n^T M∇φ = g on Γ2,
C  written in variational form, then discretised by P1
C  finite elements. The approximate solution at each vertex
C  of the triangulation solves the linear system AΦ = F.
C  This system is solved by a conjugate gradient method
C  preconditioned by the diagonal of A. The matrix A
C  of the system is stored in CSR format.
C
C
```

```
C input:
C ------
C     triangulation (nv,nt,q,me,ng) read in file
C       filnam through subroutine readT;
C     problem data read in file filnam, whose
C       name is given by the user every time
C       the program is run, thanks to subroutine
C       readF; the data are read in the following
C       order:
C     - Dirichlet boundary conditions read in
C         subroutine readDat and stored (if they
C         actually exist) in array phi0; with this
C         array is associated a flag parameter
C         iDir equal to 0 if there are no Dirichlet
C         conditions and to 1 otherwise;
C     - matrix M and function a0 of the PDE
C         read in subroutine readDat, as well as
C         functions a1 and g of the Robin
C         boundary conditions; array AM contains
C         the average on each triangle of the entries
C         of matrix M, array a0 contains the values
C         at each vertex of the mesh of function a0,
C         and the same goes with array a1 associated
C         with function a1, as well as array f where
C         function g is stored.
C         To the 3 arrays AM, a0, a1 are associated
C         flag parameters respectively denoted by
C         iM, ia0, ia1 (cf. subroutine readDat for
C         details);
C     - right hand side f of the PDE read in subroutine
C         gRHS (after having computed the boundary
C         integral associated with function g).
C     parameters:nvMax (maximum number f vertices),
C       ntMax (maximum number of triangles) of the
C       triangulation, nMAMax=8*nvMax (maximum size
C       of array aa).
C
C output:
C -------
C     f:array containing the value of the approximate
C       solution at each vertex of the mesh stored in
C       file (whose name is given by the user) in view
C       of a possible graphical exploitation.
C     Remark: this array is used all along the program
C       as an auxiliary array (before it contains the
C       solution).
C
C arrays:
C -------
C mesh:
C     q(2,nvMax):coordinates of the mesh vertices,
```

```
C       me(3,ntMax):numbering of the triangles vertices
C       ng(nvMax):array showing whether a vertex is on the
C          boundary (ng(i)≠0) or is internal (ng(i)=0).
C
C
C  data:
C       f(nvMax): at the beginning, coefficients of matrix
C          M, then function g, last PDE right hand
C          side f, (at the end: values of the
C          approximate solution);
C       phi0(nvMax): Dirichlet boundary condition;
C          N.B.: this array phi0 characterizes vertices
C             q^i in Γ1 by the relation
C             phi0(i) ≠ 0; this means that
C             before this program is called, it is
C             necessary to modify the values in this
C             array at Dirichlet points q^i where
C             we would have: φ0(q^i)=0
C             (replace them by a very small value).
C       AM(3*ntMax): average of the coefficients of matrix
C             M on each triangle; the first nt indices
C             store the coefficients with index (1,1), the
C             next nt indices store the coefficients with
C             index (2,2), and the last nt indices store the
C             coefficients with index (1,2), according to
C             the possible values of the flag parameter iM
C             (cf. subroutine readDat);
C       a0(nvMax),a1(nvMax): values of functions a0 and a1
C             at each vertex of the mesh, with flag parameters
C             ia0 and ia1 (array a0 is then used as an
C             auxiliary array in the conjugate gradient).
C
C       arrays for the preconditioned conjugate gradient:
C          cg(nvMax):array containing minus the gradient of
C             the functional to be minimised (product of the
C             preconditioning matrix C by g),
C       g(nvMax):array containing g,
C       h(nvMax):array containing the descent direction,
C       a0(nvMax): array containing the inverse of the
C          diagonal elements of the matrix.
C       N.B. in the case of diagonal preconditioning as
C          chosen here, we could have dispensed with array
C          cg; we have kept it so as not to modify to much
C          the conjugate gradient subroutine with respect
C          to the previous chapter.
C
C
C  other arrays:
C       aa(nMAMax): array for storing matrix A in
C          CSR format,
C       ia(nvMax+1): index in array aa of the first
```

```
C        element of the ith line of A actually stored in aa
C      ja(nMAMax): index of column j of stored Aij
C        element; area(ntMax):area of each triangle,
C      ff(nvMax):right hand side of linear system to be
C        solved;
C
C  scalars:
C  --------
C      cte: value of very large constant that helps
C        satisfying the Dirichlet boundary condition,
C      err: error in norm H_0^1 between the exact solution
C        phi0 and approximate solution ff.
C
C  files:
C  ------
C      filnam:name of file where the mesh, the problem
C        data or the approximate solution (for a
C        graphical exploitation) are stored.
C
C  subroutines:
C  ------------
C      readT:reads the mesh
C      getArea:computes the area of each triangle;
C      readDat: reads problem data: matrix M of the
C        partial differential operator,functions a1 and g
C        of the Robin boundary condition;
C      gDirBC:program to satisfy the Dirichlet boundary
C        conditions;
C      gRHS: computes, and stores in array ff, the
C        system right hand side;
C      ggetMat: computes and stores matrix A in CSR
C        format in array aa;
C      Sort:reorders the elements of array aa so that
C        elements of a line of matrix A are stored in
C        aa in increasing column index order;
C      gGradC: preconditioned conjugate gradient;
C      writeF:writes the approximate solution.
C
C
C  functions:
C  ----------
C      errH10:computes the error in norm H_0^1 between
C      the exact solution phi0 and approximate solution f.
C
C  library (subroutines and functions):
C  ----------------------------------
C      utilP1.for:file containing functions deter
C        (to compute areas), errH10 and subroutines
C        readT, readF and writeF;
C      RHS:file containing subroutine getArea
C      gelem.for:file containing function geMat and
```

```
C        subroutines gDirBC and gRHS;
C       sort.for:file containing subroutine Sort;
C
         PARAMETER (nvMax=1800, ntMax=3400,nMAMax=8*nvMax)
         DIMENSION q(2,nvMax),me(3,ntMax),ng(nvMax)
         DIMENSION area(ntMax)
         DIMENSION AM(3*ntMax),a0(nvMax),a1(nvMax)
         DIMENSION phi0(nvMax),f(nvMax),ff(nvMax)
         DIMENSION aa(nMAMax),ia(nvMax+1),ja(nMAMax)
         DIMENSION g(nvMax),cg(nvMax),h(nvMax)

         CHARACTER*20 filnam
C
         DATA cte/1.0e25/
         DATA eps/1.0e-12/
         DATA kMax/150/
C
         write(*,*) 'name of file where mesh is stored'
         read(*,'(a)') filnam
         CALL readT(filnam,nv,nt,q,ng,me,ngt)
         CALL getArea(q,me,nt,nv,area)
         CALL readDat(q,me,ng,nv,nt,AM,iM,a0,ia0,a1,ia1,f
     >          ,phi0,iDir)
         CALL gRHS(me,f,nv,nt,area,ff,ng,q)
         CALL ggetMat(q,me,ng,area,nv,nt,AM,iM,a0,ia0,a1,
     >          ia1,aa,ia,ja)
         IF(iDir.EQ.1) THEN
         CALL gDirBC(aa,ia,phi0,ng,nv,cte)
         ENDIF
C
C
         DO 1 i=1,nv
1        CALL Sort(ja(ia(i)),aa(ia(i)),ia(i+1)-ia(i))
C
         DO 2 i=1,nv
           f(i)=0.
           if(ng(i).NE.0) f(i)=phi0(i)
2        CONTINUE
         CALL gGradC(nv,aa,ia,ja,a0,ff,f,g,h,cg,kMax,eps)
C
         IF(iDir.EQ.1) THEN
           err=errH10(q,me,f,phi0,nv,nt)
           write(*,*) 'error=',err
         ENDIF
C
         write(*,*) 'name of file for storing the '
         write(*,*) 'approximate solution'
         read(*,'(a)') filnam
         CALL writeF(filnam,nv,ff)
         PAUSE
         END
```

```
C ----------------------------------------------------
      SUBROUTINE radDat(q,me,ng,nv,nt,AM,iM,a0,ia0,a1,
   >                    ia1,f,phi0,iDir)
C ----------------------------------------------------
C
C This subroutine reads (in files whose names are
C given by the user at run-time) and stores some of the
C problem data, in he following order:
C
C - it asks the user whether there are Dirichlet
C   boundary conditions (iDir=1) or not (iDir=0);
C   in the first case, it reads in a file, whose name is
C   given by the user at run time, the values of phi0 and
C   stores them in array phi0.
C   N.B. We recall that it is necessary for the remainder
C        of the program that this array contains no
C        zero value.
C
C - it asks the user whether the PDE has a constant term
C   a0 (ia0=1) or not (ia0=0); in the former case,
C   it reads this function and stores its values at each
C   vertex of the mesh in array a0;
C
C - it asks the user whether the coefficient a1
C   in the Robin boundary condition is zero (ia1=0)
C   or not (ia1=1); in the latter case, it reads this
C   function and stores its values at each vertex of the
C   mesh in array a1, then zeroes out the values in this
C   array that correspond to points in the boundary Γ1.
C
C - it asks the user what is the type of the PDE under
C   study:
C * if it is the Laplacian (M = Id), the flag parameter
C   iM equal to 0 and we do nothing;
C * if the matrix M = M11Id, but different from the
C   identity matrix, iM is set to1, we read the values
C   of function M11 at each vertex of the mesh,
C   then we compute the average of this function over
C   each triangle (i.e. one third of the sum of the
C   values at each vertex of the triangle); the result
C   is stored in the first nt indices of array AM;
C * if the matrix is diagonal, we read the values of
C   function M22 at each vertex of the mesh and we
C   store the average value over each triangle in array
C   AM after the previous ones; if furthermore the matrix
C   is not of the previous type, parameter iM
C   is set to 2;
C * in the general case, we also read the values of M12
C   and we again store the average value over each
C   triangle in aray AM between indices 2*nt+1 and 3*nt;
C   furthermore, if the matrix is not of the previous
```

```
C     type, parameter iM is set to 3.
C
C   - eventually it reads and stores in array f the values of
C     function g that occurs in the Robin boundary condition
C     and extends it to 0 for points in the boundary Γ1.
C
C
        PARAMETER (nvMax=1800, ntMax=3400,nMAMax=8*nvMax)
        DIMENSION q(2,nvMax),me(3,ntMax),ng(nvMax)
        DIMENSION AM(3*ntMax),a0(nvMax),a1(nvMax),phi0 (nvMax)
        DIMENSION f(nvMax)
        CHARACTER*20 filnam
C
        write(*,*) 'are there Dirichlet conditions?'
        write(*,*) '(answer: 1=yes, 0=no)'
        read(*,*) iDir
        IF (iDir.EQ.0) THEN
          write(*,*) 'name of file where phi0 is stored'
          write(*,*) '(Dirichlet boundary conditions)'
          read(*,'(a)') filnam
          CALL readF(filnam,nv,phi0)
        ENDIF
C
        write(*,*) 'is there a constant term in the PDE?'
        write(*,*) '(answer: 1=yes, 0=no)'
        read(*,*) ia0
        IF (ia0.EQ.0) GO TO 1
        write(*,*) 'name of file where a0 is stored'
        write(*,*) '(constant coefficient in the PDE)'
        read(*,'(a)') filnam
        CALL readF(filnam,nv,a0)
C
C
1       write(*,*) 'is there a coeff a1 in the BC?'
        write(*,*) '(answer: 1=yes, 0=no)'
        read(*,*) ia1
        IF (ia1.EQ.0) GO TO 2
        write(*,*) 'name of file where a1 is stored'
        write(*,*) '(coefficient a1 in the Robin condition)'
        read(*,'(a)') filnam
        CALL readF(filnam,nv,a1)
C
        IF(iDir.EQ.1) THEN
          DO 5 i=1,nv
            IF((ng(i).NE.0).AND.(phi0(i).NE.0)) THEN
              a1(i)=0.
            ENDIF
5         CONTINUE
        ENDIF
C
2       write(*,*) 'description of matrix M of the PDE'
```

```
      write(*,*) 'enter the value of parameter iM'
      write(*,*) 'remainder: iM=0 for M=Id (Laplacian)'
      write(*,*) 'iM=1 for M=M11*Id and M11 different ''from 1'
      write(*,*) 'iM=2 for M diagonal and not of the ''above type'
      write(*,*) 'iM=3 otherwise'
      read(*,*) iM
C
      nkt=0
C
      IF (IM.EQ.0) GO TO 100
        write(*,*) 'name of file where M11 is stored'
        write(*,*) 'nom du fichier de stockage de M11'
        read(*,'(a)') filnam
        CALL readF(filnam,nv,f)
        DO 10 k=1,nt
          kt=nkt+k
          som=0.
          DO 11 il=1,3
            i=me(il,k)
            som=som+f(i)
11        CONTINUE
          AM(kt)=som/3.
10      CONTINUE
      nkt=nkt+nt
C
      IF (IM.EQ.1) GO TO 100
        write(*,*) 'name of file where M22 is stored'
        read(*,'(a)') filnam
        CALL readF(filnam,nv,f)
        DO 20 k=1,nt
          kt=nkt+k
          som=0.
          DO 21 il=1,3
            i=me(il,k)
            som=som+f(i)
21        CONTINUE
          AM(kt)=som/3.
20      CONTINUE
      nkt=nkt+nt
C
      IF (IM.EQ.2) GO TO 100
        write(*,*) 'name of file where M12 is stored'
        read(*,'(a)') filnam
        CALL readF(filnam,nv,f)
        DO 30 k=1,nt
          kt=nkt+k
          som=0.
          DO 31 il=1,3
            i=me(il,k)
            som=som+f(i)
31        CONTINUE
```

```
          AM(kt)=som/3.
30      CONTINUE
      nkt=nkt+nt
C
100   CONTINUE
C
C
      write(*,*) 'name of file where g is stored'
      write(*,*) '(Robin boundary condition)'
      read(*,'(a)') filnam
      CALL readF(filnam,nv,f)
C
      IF(iDir.EQ.1) THEN
        DO 40 i=1,nv
          IF((ng(i).NE.0).AND.(phi0(i).NE.0)) THEN
            f(i)=0.
          ENDIF
40      CONTINUE
      ENDIF
C
C
      END
C
C
C ------------------------------------------------------
          SUBROUTINE ggetMat(q,me,ng,area,nv,nt,
     >                  AM,iM,a0,ia0,a1,ia1,aa,ia,ja)
C ------------------------------------------------------
C
C This constructs the lower triangular part of the matrix
C of the linear system to be solved and stores the result in
C array aa in CSR format, with auxiliary index arrays
C ia and ja: if element Aij is stored in array aa
C at index k, then ja(k)=j. ia(i) is the index in array
C aa of the first coefficient of the ith line of A
C actually stored in this array.
C
C If we agree to set ia(nv+1)=nElemA+1,
C the element Aii of matrix A is store in
C array aa at index ia(i+1)-1.
C At the beginning of the computing, array ia
C has another meaning: ia(i) is the number of
C vertices that are strict neighbours of the point
C with index i (i.e. except the point itself).
C
C     input:
C     ------
C     triangulation (nv,nt,q,me,ng) and areas of
C       the triangles stored in array area.
C     problem data (AM, iM, a0, ia0, a1, ia1)
C
```

```
C       output:
C       -------
C         arrays aa,ia,ja; size nElemA of array aa
C
C
C       important remarks:
C       ------------------
C       - we assume the triangles are positively oriented
C       - array ng is used to detect boundary edges in the
C         following way: an edge is on the boundary if and
C         only if its two extremities are boundary vertices.
C
C       We must thus check that the triangulation under
C       study satisfies those two conditions.
C
C
        DIMENSION q(2,nv),me(3,nt),ng(nv),area(nt)
        DIMENSION aa(*),ia(*),ja(*)
C
        DO 110 i=1,nv
 110      ia(i)=0
        DO 111 k=1,nt
          DO 111 il=1,3
            ilp=mod(il,3)+1
            i=me(il,k)
            ip=me(ilp,k)
            IF (ip.LE.i) THEN
              ia(i)=ia(i)+1
            ELSEIF ((ng(i).NE.0).AND.(ng(ip).NE.0)) THEN
              ia(ip)=ia(ip)+1
            ENDIF
 111      CONTINUE
C
C  ia(i) is the number of vertices that are strictly
C  neighbours of the index i vertex.
C
        nElemA=0
        DO 211 i=1,nv
          ineighb=ia(i)
          ia(i)=nElemA+1
          nElemA=nElemA+ineighb+1
 211    CONTINUE
        ia(nv+1)=nElemA+1
C
C  ia is constructed,; there remains to construct aa and ja.
C
        DO 1 k=1, nElemA
        aa(k)=0.
 1      ja(k)=0
        DO 2 kt=1,nt
          DO 2 il=1,3
```

```
          i=me(il,kt)
          DO 2 jl=1,3
            j=me(jl,kt)
            k=0
            IF(i.EQ.j) THEN
              k=ia(i+1)-1
              ja(k)=i
            ELSEIF (i.GT.j) THEN
              DO 3 l=ia(i), ia(i+1)-2
C
C  j cannot be anywhere else
C
            IF (ja(l).EQ.0) GOTO 4
            IF ( ja(l) .EQ. j )THEN
              k=l
              GOTO 4
            ENDIF
3         CONTINUE
          write(*,*)'error in MgetMat.for at point
   >                     i,j=',i,j
4         IF (k.EQ.0) THEN
            k=l
            ja(l)=j
          ENDIF
          ENDIF
          IF(k.NE.0)aa(k)=aa(k)+
   >        geMat(kt,il,jl,area(kt),q,me,ng,nv,nt,
   >          AM,iM,a0,ia0,a1,ia1)
2     CONTINUE
      END
C
C
C ---------------------------------------------------
          SUBROUTINE gGradC(n,aa,ia,ja,a0,b,x,g,h,cg,
    >                       kMax,eps)
C ---------------------------------------------------
C This subroutine solves the linear system Ax = b by the
C preconditioned conjugate gradient, with at most kMax
C iterations. The stopping test is |Ax^k - b| < eps|Ax^0 - b|.
C The linear system matrix A, of size n x n is
C stored in CSR format in array aa; this storage also
C requires auxiliary index arrays ia and ja.
C The preconditioning matrix C is the diagonal of A
C (the inverses of the diagonal elements are stored in array a0)
C
C     input:
C     ------
C       system matrix A stored in CSR format in arrays
C         (aa, ia, ja);
C       equation right hand side stored in array b;
C       maximum number of iterations kMax, computation
C         accuracy eps;
```

```
C      N.B.: since the right hand side computation is only
C            used in the initialisation step, array b is
C            subsequently overwritten and used as auxiliary
C            array, in particular to store the product of
C            matrix A by h, then to store the solution y
C            of system Cy = Ah.
C
C      output:
C      -------
C        x: array containing the values of the solution
C           at each vertex of the mesh;
C        gcg: scalar containing the norm square of g
C           (in practice this number should decrease along
C           the iterations).
C
C      other arrays:
C      -------------
C        g, cg: arrays containing the values of g^n and
C           of Cg^n respectively;
C        h: array containing the descent direction h^n
C           (at the very beginning of the program,
C           this array is used to store the product of
C           A by x^0);
C
C
C      subroutines:
C      ------------
C        IAmul: carries out the product of a matrix
C           by a vector, the matrix being stored in
C           CSR format, This subroutine is in file
C           GCPLaplac.for
C
C
       DIMENSION x(*),b(*),g(*),h(*),cg(*)
       DIMENSION aa(*),ja(*),ia(*),a0(*)
C
       DO 20 i=1,n
       a0(i)=1./aa(ia(i+1)-1)
20     CONTINUE
C
       DO 1 k=1, kMax
C
         IF (k.EQ.1) THEN
           CALL IAmul(n,aa,ia,ja,x,h)
           DO 2 i=1,n
             cg(i)=b(i)-h(i)
             g(i)=cg(i)*a0(i)
2          CONTINUE
           gcg=0.
           DO 3 i=1,n
             gcg=gcg+g(i)*cg(i)
```

```
            h(i)=g(i)
3         CONTINUE
        ENDIF
C
        CALL IAmul(n,aa,ia,ja,h,b)
        hAh=0.
        DO 10 i=1,n
          hAh=hAh+h(i)*b(i)
10      CONTINUE
        ro=gcg/hAh
        DO 11 i=1,n
          cg(i)=cg(i)-ro*b(i)
          b(i)=b(i)*a0(i)
11      CONTINUE
        DO 12 i=1,n
          x(i)=x(i)+ro*h(i)
          g(i)=g(i)-ro*b(i)
12      CONTINUE
        gamma=gcg
        gcg=0
        DO 13 i=1,n
          gcg=gcg+g(i)*cg(i)
13      CONTINUE
C
        write(*,*) 'iteration:',k,'gcg=',gcg
        IF(k.EQ.1) eps=eps*gcg
        IF (gcg.LT.eps) RETURN
        gamma=gcg/gamma
        DO 14 i=1,n
          h(i)=g(i)+gamma*h(i)
14      CONTINUE
C
1     CONTINUE
      END
C
C
C -----------------------------------------------------
                SUBROUTINE IAmul(nv,aa,ia,ja,x,b)
C -----------------------------------------------------
C This subroutine computes the product of matrix
C A, stored in CSR format in arrays (aa,ia,ja) by
C vector x; the result is in array b.
C
C
      DIMENSION x(nv), b(nv)
      DIMENSION aa(*),ia(nv),ja(*)
      DO 1 i=1,nv
        sum=0.
        DO 2 k=ia(i),ia(i+1)-1
          sum=sum+aa(k)*x(ja(k))
2       CONTINUE
```

```
        b(i)=sum
1     CONTINUE
      DO 3 i=1,nv
        xi=x(i)
        DO 3 k=ia(i),ia(i+1)-2
          jak=ja(k)
          b(jak)=b(jak)+aa(k)*xi
3     CONTINUE
      END
C
C
C ***************************************************
C
C LIBRARY
C
C ***************************************************
C
C      INCLUDE 'gelem.for'
       INCLUDE 'utilP1.for'
       INCLUDE 'RHS.for'
       INCLUDE 'sort.for'
C
C
```

6 Non-symmetric or Non-linear Partial Differential Equations

Summary

All partial differential equations are not symmetric: this is for example the case of the advection–diffusion equations in fluid mechanics. In this chapter we shall develop two different numerical methods to solve this kind of problems: *LU* factorisation and the GMRES algorithm. Last, we shall show on an example how the GMRES method can be generalised to the case of non-linear partial differential equations.

6.1 *SECOND ORDER NON-SYMMETRIC PROBLEMS*

6.1.1 *Statement of the Problem*

Let us consider the following problem: find φ solving

$$-\nabla\cdot(M\nabla\varphi)+u\cdot\nabla\varphi+a_0\varphi=f \text{ in } \Omega, \tag{6.1}$$

$$n^TM\nabla\varphi+a_1\varphi=g \text{ on } \Gamma, \tag{6.2}$$

where M is a matrix all of whose coefficients are L^∞ functions on Ω, $u=(u_1,u_2)\in(L^\infty(\Omega))^2$, $a_0\in L^\infty(\Omega)$ is a scalar function, and the source term f is supposed square integrable on Ω; last, a_1 and g are functions defined on the boundary Γ of Ω that satisfy: $a_1\in L^\infty(\Gamma)$ and $g\in L^2(\Gamma)$. The main difference with the previous chapter is that the matrix M is not necessarily symmetric, and the term

$$u\cdot\nabla\varphi=u_1\frac{\partial\varphi}{\partial x_1}+u_2\frac{\partial\varphi}{\partial x_2} \tag{6.3}$$

would make the operator non-symmetric anyway. In practice, u is a convection velocity, and φ could represent, for instance, the fluid temperature. If the fluid moves in a porous medium, M is the diffusion tensor. The source term f models a heat source in Ω, as does g albeit on Γ, with possible heat loss through radiation if $a_1 \neq 0$.

6.1.2 Variational Formulation

By arguing exactly as in Theorem 5.1 and 5.5 in the previous chapter, we obtain

Theorem 6.1 The variational formulation of problem (6.1)–(6.2) can be written as

$$\text{find } \varphi \in H^1(\Omega), \quad \text{such that } \forall w \in H^1(\Omega), \quad \text{we have}$$

$$\int_\Omega \left[(\nabla w)^T M \nabla \varphi + (u \cdot \nabla \varphi)\, w + a_0 \varphi w \right](\vec{x}) \mathrm{d}\vec{x} \int_\Gamma (a_1 \varphi w)(\vec{x}) \mathrm{d}\gamma(\vec{x}) \tag{6.4}$$
$$= \int_\Omega (fw)(\vec{x}) \mathrm{d}\vec{x} + \int_\Gamma (gw)(\vec{x}) \mathrm{d}\gamma(\vec{x}).$$

Also, if $\varphi \in H^2(\Omega)$, φ is a solution of the boundary value problem (6.1)–(6.2) if and only if it is a solution of the above variational formulation.

Remark 6.2 Here, we can no longer express the variational problem in terms of a minimisation problem, because of the non-symmetry of the bilinear form. □

Remark 6.3 From a theoretical point of view, it suffices that the following four hypotheses be fulfilled for the variational problem to have a unique solution:

– there exists a strictly positive constant α such that

$$\forall \vec{x} = (x_1, x_2) \in \bar{\Omega}, \quad \sum_{i=1}^{2} \sum_{j=1}^{2} M_{ij}(\vec{x}) x_i x_j \geq \alpha \sum_{i=1}^{3} x_i^2 \ ; \tag{6.5}$$

– the function a_0 is such that

$$a_0 - \tfrac{1}{2} \nabla \cdot u \geq C_0 \geq 0 \ ; \tag{6.6}$$

– the function a_1 is such that

$$a_1 + \tfrac{1}{2} u \cdot n \geq C_1 \geq 0 \ ; \tag{6.7}$$

– last, either one of the above two constants C_0 or C_1 is strictly positive.

Indeed, these conditions ensure the ellipticity of the bilinear form, as is shown in the following estimate, where, to simplify notation, we have suppressed the integration variables:

$$\int_\Omega \left[(\nabla \varphi)^T M \nabla \varphi + (u \cdot \nabla \varphi)\, \varphi + a_0 \varphi^2 \right] + \int_\Gamma a_1 \varphi^2$$
$$\geq \alpha \|\nabla \varphi\|_0^2 + \int_\Omega \left(a_0 - \tfrac{1}{2} \nabla \cdot u\right) \varphi^2 + \int_\Gamma \left(a_1 + \tfrac{1}{2} u \cdot n\right) \varphi^2 \tag{6.8}$$

To bound the second term (i.e. the one featuring the velocity field u) from below, we have made use of the identity $\nabla \cdot (u\varphi) = \nabla \cdot u\varphi + u \cdot \nabla\varphi$, and two successive applications of Green's formula:

$$\int_\Omega \nabla \cdot (u\varphi)\,\varphi = -\int_\Omega u\varphi \cdot \nabla\varphi + \int_\Gamma u \cdot n\,\varphi^2$$
$$= \frac{1}{2}\int_\Omega (\nabla \cdot u)\,\varphi^2 + \frac{1}{2}\int_\Gamma u \cdot n\,\varphi^2.$$

EXERCISE Apply the same argument when (6.2) is replaced by $n^T M \nabla\varphi + a_1\varphi + b\frac{\partial\varphi}{\partial s} = g$ on Γ. □

We shall now describe the P^1 finite element approximation of the variational formulation (6.4).

6.1.3 *Discretisation of the Variational Problem*

We assume that the domain Ω has been provided with a regular triangulation $(T_k)_{k\in\{1,\ldots,nt\}}$, and we denote by $q^i, i \in \{1,\ldots,nv\}$ the nv vertices. We define the discrete space H_h^1 by

$$H_h^1 = \{\varphi_h : \Omega_h \to \mathbb{R}, \varphi_h \text{ continuous on } \Omega_h, \text{ and}$$
$$\forall k \in \{1,\ldots,nt\}, \varphi_{h|T_k} \text{ is a linear function}\},$$

and we solve problem (6.4) in this space, which leads to the following approximate variational problem

find $\varphi_h \in H_h^1$ such that $\forall w_h \in H_h^1$,

$$\int_{\Omega_h} \left[(\nabla w_h)^T M \nabla\varphi_h + (u \cdot \nabla\varphi_h)\, w_h + a_0\varphi_h w_h\right](\vec{x})\mathrm{d}\vec{x} + \int_{\Gamma_h} (a_1\varphi_h w_h)(\vec{x})\mathrm{d}\gamma(\vec{x})$$
$$= \int_{\Omega_h} (f w_h)(\vec{x})\mathrm{d}\vec{x} + \int_{\Gamma_h} (g w_h)(\vec{x})\mathrm{d}\gamma(\vec{x}), \tag{6.9}$$

where $\Omega_h = \cup_{k=1}^{nt} T_k$ and $\Gamma_h = \partial\Omega_h$. According to Propositions 2.14 and 2.17, the space H_h^1 is a finite dimensional vector subspace of $H^1(\Omega)$. A basis of this space is formed by the hat functions $w^i, i \in \{1,\ldots,nv\}$ $(w^i(q^j) = \delta_{ij})$, which enables us to look for the solution of (6.9) in the form of a linear combination of these functions:

$$\varphi_h(\vec{x}) = \sum_{i=1}^{nv} \varphi_i w^i(\vec{x}), \quad \forall \vec{x} \in \Omega. \tag{6.10}$$

The variational problem (6.9) then reduces to the solution of the linear system

$$A\Phi = F, \tag{6.11}$$

where Φ is the vector with components $\varphi_i, i \in \{1,...,nv\}$, the matrix $A = ((A_{ij}))_{1\leq i,j\leq nv}$ and the right hand side $F = (F_i)_{1\leq i\leq nv}$ being defined by:

$$A_{ij} = \int_{\Omega_h} \left[(\nabla w^i)^T M \nabla w^j + (u \cdot \nabla w^j) w^i + a_0 w^i w^j\right](\vec{x}) \mathrm{d}\vec{x} \qquad (6.12)$$

$$+ \int_{\Gamma_h} (a_1 w^i w^j)(\vec{x}) \mathrm{d}\gamma(\vec{x}),$$

$$F_j = \int_{\Omega_h} (f w^j)(\vec{x}) \mathrm{d}\vec{x} + \int_{\Gamma_h} (g w^j)(\vec{x}) \mathrm{d}\gamma(\vec{x}). \qquad (6.13)$$

Remark 6.4 As in the previous chapter, remarks pertaining to the 'invertibility' of this system (6.11) are directly linked with the hypotheses ensuring the ellipticity of the continuous problem: under the conditions stated in Remark 6.3, the matrix A satisfies the following 'coercivity' property:

$$\exists C_A > 0, \quad \forall \Phi \in \mathbb{R}^{nv}, \quad \Phi^T A \Phi \geq C_A |\Phi|^2, \qquad (6.14)$$

which proves it is invertible.

What is new here is that, since the matrix A is no longer symmetric, we can no longer use the solution methods of the previous chapters (Choleski, conjugate gradient...) to solve the linear system (6.11). Its 'coerciveness' alluded to above will nevertheless enable us to use other methods.

By way of a direct method, we can use the LU factorisation of matrix A; we briefly recall its principle in the next section. We shall then dwell some more on an iterative method, close to the conjugate gradient algorithm: *the GMRES algorithm.*

6.1.4 *LU* Factorisation of Matrix *A*

The LU factorisation, or its close relative Gaussian elimination, [Ciarlet (1989), Schatzman (1991), Strang (1986)] consists of factorising matrix A as a product of two triangular matrices

$$A = LU, \qquad (6.15)$$

one of them, L being lower triangular, the other, U being upper triangular. This decomposition is furthermore unique if we impose, for instance, to the diagonal elements of U to all be equal to 1.

It is also easy to check that this factorisation, (as does the Choleski factorisation) preserves the 'band' structure of the matrix: if matrix A has at most $2\mu + 1$ non-zero diagonals centred around the main diagonal, L and U will have at most $\mu + 1$ non-zero diagonals. This remark lets us consider 'reduced' storage formats for these matrices: for example band storage (only one array is needed to store first A then L and U), or CSR storage (since the CSR structure of L and U are *a priori* not the same as that of A, several arrays are needed).

The construction of the matrices L and U is done in an identical way to the Choleski factorisation, by exploiting the definition

$$A_{ij} = \sum_{k \leq min\{i,j\}} L_{ik} U_{kj}, \tag{6.16}$$

which gives the following program sketch:

Algorithm 1 (Gauss)

$$i = 1, \ldots, nv:$$

$$L_{ii} = A_{ii} - \sum_{k<i} L_{ik}\, U_{ki}, \tag{6.17}$$

$$j = i+1, \ldots, n_v:$$

$$L_{ji} = A_{ji} - \sum_{k<i} L_{jk}\, U_{ki} \tag{6.18}$$

$$U_{ij} = (A_{ij} - \sum_{k<i} L_{ik}\, U_{kj}/L_{ii}). \tag{6.19}$$

EXERCISE We leave it to the reader to program this algorithm in Fortran, assuming the matrix A is stored in band format [Lucquin and Pironneau (1997)]. The C program is given in Appendix 6.B. □

This method requires a number of operations of the order of nv^3, and above all requires memory storage at least of the order of $nv^{\frac{3}{2}}$, as long as we assume that the vertices have been numbered in a sufficiently astute way so that the bandwidth of the matrix A is in $O(\sqrt{nv})$. As we did in the symmetric case, we shall now propose a more effective iterative method.

6.2 THE GMRES ALGORITHM

6.2.1 Introduction

Let us consider in general the problem of solving the linear system

$$Ax = b, \tag{6.20}$$

where A is a symmetric real square matrix of order N, invertible but not necessarily symmetric. For such a matrix, it is possible for the conjugate gradient algorithm to fail, as the following example shows:

EXAMPLE Let us consider the case

$$A = \begin{pmatrix} 0 & 1 \\ -1 & 0 \end{pmatrix}, \quad b = \begin{pmatrix} 1 \\ -1 \end{pmatrix}, \tag{6.21}$$

and initialise the conjugate gradient by $x^0 \neq (-1,1)^T$. Then, going back to the notation of Section 4.1.3, we have $h^0 \neq 0$, and yet $(h^0)^T A h^0 = 0$, which implies $\rho^0 = \infty$! □

In Section 4.1.3, we have proven some properties of the conjugate gradient algorithm that we shall briefly recall. First, the principle for solving (6.20) consists of minimising the functional J defined over $\mathbb{R}^N$ by (we denote by (x,y) the Euclidean inner product in $\mathbb{R}^N$):

$$J(x) = \tfrac{1}{2}(Ax,x) - (b,x). \tag{6.22}$$

Also, the solution x^n obtained at step n of this algorithm achieves the minimum of the functional J over the affine space $x^0 + K_n$, where K_n is the Krylov space defined by

$$\begin{aligned} K_n &= \text{Span}\{g^0, Ag^0, ..., A^{n-1}g^0\} \\ &= \text{Span}\{g^0, g^1, ..., g^{n-1}\} = \text{Span}\{h^0, h^1, ..., h^{n-1}\}, \end{aligned} \tag{6.23}$$

where g^i denotes the residual at step i of the algorithm $(g^i = b - Ax^i)$, $h^0 = g^0, h^1, ..., h^{n-1}$ are the successive descent directions, and $\text{Span}\{u_1, ..., u_n\}$ denotes the vector space generated by the n vectors u_i. Let us recall that if the conjugate gradient algorithm has not converged after the $(n-1)$st step, the space K_n has dimension n, with an orthonormal basis given by the vectors $g^0, ..., g^{n-1}$.

Motivated by this observation, the aim of the GMRES ('Generalised Minimum Residual') algorithm will be to *minimise the Euclidean norm of the residual* $(\|b - Ax^n\|)$ in the space $x^0 + K_n$. Before we explain its inner working and show its usefulness, we shall describe a first algorithm, conceptually simpler, also based on the construction of an orthonormal basis of the Krylov subspace.

6.2.2 *Arnoldi's Method*

Let us denote by x^0 an arbitrary vector in $\mathbb{R}^N$, and by $r^0 = b - Ax^0$ the initial residual, and assuming it is not zero (otherwise problem (6.20) is already solved), set $v_1 = r^0/\|r^0\|$; the vector v_1 is the first basis vector in the space K_n. Let us now assume that the Krylov subspace $K_n = \text{Span}\{v^1, Av^1, ..., A^{n-1}v^1\}$ has dimension n and let us construct a basis, denoted by $(v^1, ..., v^n)$, by a Gram - Schmidt type algorithm; this basis is defined iteratively in the following way:

Algorithm 2 (Arnoldi)

$j = 1, ..., n:$

$$h_{ij} = (Av^j, v^i), \quad 1 \leq i \leq j, \tag{6.24}$$

$$\widehat{v}^{j+1} = Av^j - \sum_{i=1}^{j} h_{ij} v^i, \tag{6.25}$$

$$h_{(j+1)j} = \|\widehat{v}^{j+1}\|, \tag{6.26}$$

$$v^{j+1} = \frac{\widehat{v}^{j+1}}{h_{(j+1)j}}. \tag{6.27}$$

If none of the $h_{(j+1)j}$ coefficients defined above is zero, it is easy to show by induction the following result:

Lemma 6.5 The vectors $(v^1, ..., v^n)$ form an orthonormal basis of the Krylov subspace K_n, and furthermore:

$$(Av^j, v^i) = \begin{cases} h_{ij}, \text{if } i \leq j+1, \\ 0, \text{if } i \geq j+2. \end{cases} \tag{6.28}$$

We denote by H^n the matrix with coefficients h_{ij}, for $1 \leq i, j \leq n$. According to this lemma, H^n is a Hessenberg matrix, i.e. all of its coefficients located below the diagonal immediately below the main diagonal are zero.

Once this basis is constructed, the method then consists in looking for the approximate solution x^n at step n of the algorithm in the space $x^0 + K_n$, which amounts to writing x^n in the form $x^n = x^0 + z^n, z^n \in K_n$. We then choose the element z^n so that the residual $r^n = b - Ax^n = r^0 - Az^n$ is orthogonal to K_n. If the matrix H^n is invertible such a construction is possible, and more precisely we have:

Proposition 6.6 *Denote by V^n the matrix of size $N \times n$ whose ith column is the vector v^i, $i \in \{1, ..., n\}$ (i.e. $V^n_{ij} = v^i_j$). Then, the matrix H^n is invertible, there exists a unique element z^n in K_n such that $r^n = r^0 - Az^n$ ($r^0 = b - Ax^0$) is orthogonal to K_n, and this element is defined by:*

$$z^n = \sum_{j=1}^{n} y_j v^j = V^n y^n, \text{with} \quad y^n = \|r^0\| \, (H^n)^{-1} \, e^n_1, \tag{6.29}$$

where we have denoted by $e^n_1 = (1, 0, ..., 0)$ the first basis vector in $\mathbb{R}^n$. Furthermore, the residual at step n can be written as;

$$r^n = -y_n \; h_{(n+1)n} \; v^{n+1}. \tag{6.30}$$

PROOF That z^n lies in the space K_n is expressed by the existence of n scalars y_j such that $z^n = \sum_{j=1}^n y_j v^j$, and this equality has a matrix interpretation in the form: $z^n = V^n y^n$, if we denote by y^n the vector in $\mathbb{R}^n$ with components $(y_1, ..., y_n)^T$. Let us now express that the residual $r^n = r^0 - Az^n$ is orthogonal to all basis vectors v^i of K_n. We obtain:

$$\begin{aligned} \forall i \in \{1, ..., n\}, 0 = (r^n, v^i) &= (r^0, v^i) - \sum_{j=1}^{n} y_j (Av^j, v^i) \\ &= \|r^0\| \delta_{1i} - \sum_{j=1}^{n} y_j h_{ij}, \end{aligned}$$

or in other words $||r^0||e_1^n = H^n y^n$, which proves (6.29).

The second point stems first from the fact that $r^n \in AK_n \subset K_{n+1}$, $K_{n+1} = \text{Span}\{v^1, ..., v^{n+1}\}$, and that because of the way z^n is constructed, r^n is orthogonal to the first n vectors v^i, and this proves it is actually proportional to the vector v^{n+1}. The proportionality constant is then computed by writing

$$(r^n, v^{n+1}) = (r^0, v^{n+1}) - \sum_{j=1}^{n} y_j (Av^j, v^{n+1}) = 0 - y_n h_{(n+1)n},$$

the last equality being due to (6.28), which ends the proof. □

The algorithm is then as follows:

Algorithm 3

1 initialisation: choose x^0, compute the initial residual $r^0 = b - Ax^0$; if $r^0 = 0$, the sought solution is x^0 and stop, otherwise, set $v^1 = r^0/||r^0||$ and go to step **2**;

2 iterations over $j \geq 1$:

$$n = j$$

$$h_{ij} = (Av^j, v^i), \quad 1 \leq i \leq j, \tag{6.24}$$

$$\widehat{v}^{j+1} = Av^j - \sum_{i=1}^{j} h_{ij} v^i, \tag{6.25}$$

$$h_{(j+1)j} = ||\widehat{v}^{j+1}||, \tag{6.26}$$

if $h_{(j+1)j} = 0$, go to **3** otherwise

$$v^{j+1} = \frac{\widehat{v}^{j+1}}{h_{(j+1)j}}. \tag{6.27}$$

3 compute the approximate solution:

$$x^n = x^0 + V^n y^n, \text{with } y^n = ||r^0||(H^n)^{-1} e_1^n. \tag{6.31}$$

Relation (6.30) shows that, *if the matrix H_n is invertible*, the algorithm is well defined, in the sense that it can 'degenerate', (i.e. $h_{(n+1)n} = 0$), only if $r_n = 0$, and this means that the algorithm has converged. Also, this algorithm converges in at most $n = N$ iterations, since there cannot be more than N vectors v^i, because they form an independent set in a space of dimension N.

A drawback of this method is that it requires the storage of all the vectors v^i, for $1 \leq i \leq n$, and furthermore the operation count might be very large if n is large. To alleviate this drawback, we can decide to limit the number of iterations on j to a fixed value m, even if it means restarting the whole process with $x^0 = x^m$ as many times as necessary i.e. until the residual becomes sufficiently small.

Another major drawback of this method lies in the fact that the matrices H^n may not be invertible, as the following example shows:

EXAMPLE Let us consider again the matrix A and the right hand side b in (6.21), and initialise the process with $x^0 = 0$. The first basis vector is $v^1 = (\sqrt{2}/2)\, b$, which gives $h_{11} = 0$, so H^1 is singular. However, $H^2 = -A$ is invertible. □

Thus, we shall now define a new algorithm, very close to the one above, but designed to avoid this problem.

6.2.3 *The GMRES Method*

6.2.3.1 Algorithm description

The starting idea is the same as in the previous algorithm, namely to look for the approximate solution at step n in the form $x^n = x^0 + z^n, z^n \in K_n$, but this time z^n will be chosen in so as to *minimise the residual* over K_n, i.e. in such a way that

$$||r^0 - Az^n|| \leq ||r^0 - Az||, \quad \forall z \in K_n, \tag{6.32}$$

where $||.||$ denotes the Euclidean norm in $\mathbb{R}^N$.

After n iterations of the Arnoldi algorithm, we have at our disposal a basis $(v^1, ..., v^{n+1})$ of the Krylov subspace K_{n+1} and a matrix of size $(n+1) \times n$, denoted by $\bar{H}^n$, whose non-zero elements are the coefficients h_{ij} generated by the algorithm. More precisely, this rectangular matrix is made of the matrix H^n bordered by an $n+1$st row, whose only non-zero element is, according to (6.28), the last coefficient $h_{(n+1)N}$:

$$\bar{H}^n = \begin{pmatrix} H^n \\ 0...0 \; h_{(n+1)n} \end{pmatrix}, \quad H^n = \begin{bmatrix} h_{11} & \dots & & & h_{1n} \\ h_{21} & h_{22} & \dots & & h_{2n} \\ 0 & h_{32} & h_{33} & \dots & h_{3n} \\ & \dots & \dots & & \dots \\ 0 & \dots & 0 & h_{(n-1)n} & h_{nn} \end{bmatrix}. \tag{6.33}$$

This matrix, which is Hessenberg, satisfies the relation

$$AV^n = V^{n+1}\bar{H}^n, \tag{6.34}$$

which is easily proven by comparing the columns of the two matrices.

Thanks to this remark, we can write the residual $r^n = b - Ax^n$ in the form: $r^n = r^0 - AV^n y^n = r^0 - V^{n+1}\bar{H}^n y^n$, or if we denote by e_k^{n+1} the kth basis vector $\mathbb{R}^{n+1}$ $\left((e_k^{n+1})_l = \delta_{kl}\right)$:

$$r^n = V^{n+1}\left(||r^0||e_1^{n+1} - \bar{H}^n y^n\right). \tag{6.35}$$

Since the matrix V^{n+1} is orthogonal, it preserves the Euclidean norm, and it follows that

$$||r^n|| = || \, ||r^0||e_1^{n+1} - \bar{H}^n y^n \, ||. \tag{6.36}$$

The GMRES algorithm is then the same as Algorithm 3 above, except for step 3, which is replaced by:

3′

$$x^n = x^0 + V^n y^n, \tag{6.37}$$
$$\text{where } y^n \in \mathbb{R}^n \text{ satisfies : } \forall y \in \mathbb{R}^n, J^n(y^n) \leq J^n(y), \tag{6.38}$$
$$\text{with } J^n(y) = || \, ||r^0||e_1^{n+1} - \bar{H}^n y \, ||. \tag{6.39}$$

Before we prove the properties of this algorithm, we shall describe the practical solution of step 3′.

6.2.3.2 Implementation of (6.38)–(6.39)

We first note that the minimum of the functional J^n over K_n is *a priori non-zero,* as the identity $||r^0||e_1^{n+1} = \bar{H}^n z$ leads to a system of $n+1$ equations in n unknowns.

The idea for solving this minimisation problem consists in constructing a 'QR' factorisation of the matrix $\bar{H}^n$, that is decomposing it in the form of the product of a square orthogonal matrix $Q^n \in \mathbb{R}^{(n+1)\times(n+1)}$ by an upper triangular rectangular matrix $R^n \in \mathbb{R}^{(n+1)\times n}$, all of whose diagonal elements are strictly positive, and whose last line is zero:

$$\bar{H}^n = Q^n R^n, \quad (Q^n)^{-1} = (Q^n)^T, \quad R^n_{ij} = 0, \forall i > j. \tag{6.40}$$

Furthermore, this decomposition is unique if the matrix $\bar{H}^n$ is invertible. For details pertaining to the justification of these results, we refer to Ciarlet (1989).

We shall now carry out this factorisation progressively, i.e. at each index j iteration of step 2 of the algorithm, when a new column of $\bar{H}^n$ appears.

At the outset, $\bar{H}^1$ is a column vector $(\alpha, \beta)^T$, with $\beta = h_{21}$. Let us assume $h_{21} \neq 0$, and set

$$c_1 = \frac{\alpha}{\sqrt{\alpha^2 + \beta^2}}, \; s_1 = -\frac{\beta}{\sqrt{\alpha^2 + \beta^2}}, \quad F_2^1 = \begin{pmatrix} c_1 & -s_1 \\ s_1 & c_1 \end{pmatrix}.$$

Then F_2^1 is a rotation matrix in the plane (e_1^2, e_2^2) and $F_2^1 \bar{H}^1 = R^1$, when $R^1 = (\sqrt{\alpha^2 + \beta^2}, 0)^T$ is an upper triangular matrix of order 2×1, whose only diagonal element is strictly positive, since the last line is zero.

Let us assume we have carried out j iterations of the algorithm, so that

$$\left(F_{j+1}^j F_{j+1}^{j-1} \dots F_{j+1}^2 F_{j+1}^1\right) \bar{H}^j = R^j,$$

where R^j is an upper triangular $(j+1) \times j$ matrix, all of whose diagonal elements are strictly positive, the last line of R^j being zero, and where each of the matrices F_{j+1}^k is square of order $j+1$, its restriction to the plane $(e_k^{j+1}, e_{k+1}^{j+1})$ being an orthogonal rotation, whereas its restriction to the orthogonal of this plane is the identity. In other words, we have:

$$F^k_{j+1} = \begin{bmatrix} 1 & & & & & & 0 \\ & 1 & & & & & \\ & & \cdots & & & & \\ & & & c_k & -s_k & & \\ & & & s_k & c_k & & \\ & & & & & \cdots & \\ 0 & & & & & & 1 \end{bmatrix}_{(j+1)\times(j+1)} \begin{matrix} \\ \\ \\ \longrightarrow \text{ row } k \\ \longrightarrow \text{ row } k+1 \\ \\ \\ \end{matrix} .$$

At step $j+1$, a new column and a new row appear in the matrix $\bar{H}^{j+1}$. We start by carrying out the preceding transformations (i.e. multiplications on the left by $F^j_{j+2} F^{j-1}_{j+2} \ldots F^2_{j+2} F^1_{j+2}$) on the newly created column of $\bar{H}^{j+1}$. We note that the coefficient with index $(j+2, j+1)$, which is nothing else but $\beta = h_{(j+2)(j+1)}$, is not affected by these transformations. At the end of this computation, we then have:

$$\left(F^j_{j+2} F^{j-1}_{j+2} \ldots F^2_{j+2} F^1_{j+2}\right) \bar{H}^{j+1} = \tilde{R}^{j+1},$$

where $\tilde{R}^{j+1}$ is a $(j+2) \times (j+1)$ matrix, whose last two lines are zero, except for their last coefficients, i.e.

$$\tilde{R}^{j+1} = \begin{bmatrix} & & & & x \\ & & & & x \\ & R^j & & & x \\ & & & & x \\ & & & & \alpha \\ 0 & 0 & 0 & 0 & \beta \end{bmatrix}, \quad \text{with } R^j = \begin{bmatrix} x & x & x & x \\ & x & x & x \\ & & x & x \\ & 0 & & x \\ 0 & 0 & 0 & 0 \end{bmatrix}_{(j+1)\times j},$$

where 'x' denotes elements of the matrix that are a priori non-zero. Let us assume $\beta = h_{(j+2)(j+1)} \neq 0$; we shall multiply this matrix on the left by a rotation matrix so as to zero out the element under the diagonal. To do this, let us set

$$c_{j+1} = \frac{\alpha}{\sqrt{\alpha^2 + \beta^2}}, \quad s_{j+1} = -\frac{\beta}{\sqrt{\alpha^2 + \beta^2}},$$

$$F^{j+1}_{j+2} = \begin{bmatrix} 1 & & & & & & \\ & 1 & & & 0 & & \\ & & \cdots & & & & \\ & & & 1 & & & \\ & & & & 1 & & \\ & 0 & & & & c_{j+1} & -s_{j+1} \\ & & & & & s_{j+1} & c_{j+1} \end{bmatrix}_{(j+2)\times(j+2)} .$$

Then we have

$$F_{j+2}^{j+1}\left(F_{j+2}^{j}...F_{j+2}^{2}F_{j+2}^{1}\right)\bar{H}^{j+1} = R^{j+1},$$

where

$$R^{j+1} = F_{j+2}^{j+1}\tilde{R}^{j} = \begin{bmatrix} & & & & & x \\ & & & & & x \\ & & R^{j} & & & x \\ & & & & & x \\ & & & & & x \\ & & & & & \sqrt{\alpha^2+\beta^2} \\ 0 & 0 & 0 & 0 & 0 & 0 \end{bmatrix},$$

is a $(j+2)\times(j+1)$ upper triangular matrix, whose last line is zero and all of whose diagonal elements are strictly positive.

We have thus proven by induction the existence of this construction, and after step n we have (6.40) with:

$$Q^{n} = (F_{n+1}^{1})^{-1}(F_{n+1}^{2})^{-1}...(F_{n+1}^{n-1})^{-1}(F_{n+1}^{n})^{-1}. \tag{6.41}$$

We can now solve the minimisation problem (6.38)–(6.39). Indeed, since Q^n is orthogonal, we have:

$$J^{n}(z) = ||\,||r^{0}||e_{1}^{n+1} - Q^{n}R^{n}z\,|| = ||\,g^{n} - R^{n}z\,||, \tag{6.42}$$

$$\text{with } g^{n} = (Q^{n})^{-1}\,||r^{0}||\,e_{1}^{n+1}. \tag{6.43}$$

But since the last line of the matrix R^n is identically zero, we can now find a vector y^n in $\mathbb{R}^n$ such that $J^n(y^n) = 0$: to do this, it suffices to solve the following upper triangular system

$$R^{n}y^{n} = P^{n}(g^{n}), \tag{6.44}$$

where P^n is the projection from $\mathbb{R}^{n+1}$ onto $\mathbb{R}^n$. From a programming point of view, this system is easily solved by a 'forward' solve as soon as the right hand side g^n is known. But according to (6.43), (6.41), we have

$$g^{n} = (Q^{n})^{-1}||r^{0}||e_{1}^{n+1} = F_{n+1}^{n}F_{n+1}^{n-1}...F_{n+1}^{2}F_{n+1}^{1}\,(||r^{0}||e_{1}^{n+1}), \tag{6.45}$$

which shows that g^n is easily computed, *as the algorithm progresses*, by multiplying the right hand side by all the above elementary rotations.

Last, let us end with a remark pertaining to the *computation of the residual* at step n, for which *it is not necessary to know x^n explicitly*. Indeed, we have, according to (6.36) and (6.42)

$$||r^{n}|| = J^{n}(y^{n}) = |(e_{n+1}^{n+1}, g^{n})|, \tag{6.46}$$

the last equality being linked directly to the choice (6.44) of y^n. Last, let us note that the Euclidean norms of these residuals are linked together by the following recurrence relation:

$$||r^n|| = |s^n|.\quad ||r^{n-1}||, \text{with}\quad s^n = -k^n h_{(n+1)n}, k^n > 0. \tag{6.47}$$

Let us now state some of the convergence properties of this algorithm.

6.2.3.3 Properties of the GMRES algorithm

The algorithm can only 'degenerate' if it has already converged; more precisely we have:

Proposition 6.7 *The approximate solution at step n of the algorithm is exact (i.e. $x^n = x$, x solving $Ax = b$), if and only if one of the following equivalent conditions is satisfied:*
(i) *the algorithm stops at step n,*
(ii) $\hat{v}^{n+1} = 0$,
(iii) $h_{(n+1)n} = 0$.

PROOF That (ii) and (iii) are equivalent is trivial, and it is clear that (iii) implies (i). Conversely, let us assume that $h_{(n+1)n} \neq 0$; then from the construction of the previous section, we can define the rotation F^n_{n+1}. Since all diagonal elements of the upper triangular matrix $R^n = F^n_{n+1}\tilde{R}^n$ are strictly positive, this matrix is invertible, and we can solve the system (6.44). Thus the algorithm is well defined at step n.

Let us now prove that if (iii) holds, then x^n is the exact solution. Let us assume that n is the first index j for which $h_{(j+1)j} = 0$, which proves on the one hand that the vectors $(v^1, ..., v^n)$ are independent, and on the other hand that the algorithm does degenerate at step n, since we cannot construct v^{n+1}. According to (6.39) and the computations leading to (6.36) we have, denoting the residual $b - A(x^0 + V^n y)$ by $r^n(y)$:

$$J^n(y) = ||r^n(y)|| = ||r^0 - AV^n y||.$$

However, since $h_{(n+1)n}$ is zero, we now have $AV^n = V^n H^n$, which implies:

$$J^n(y) = ||r^0 - V^n H^n y|| = ||\, ||r^0||v^1 - V^n H^n y\, || = ||V^n(||r^0||e^n_1 - H^n y)||.$$

Because V^n is an orthogonal matrix, we deduce that:

$$J^n(y) = ||\, ||r^0||e^n_1 - H^n y\, ||.$$

If we assume that H^n is invertible, then the above equality shows that J^n is minimum for $y = y^n = (H^n)^{-1}(||r^0||e^n_1)$, and in that case, $J^n(y^n) = ||r^n(y^n)|| = 0$, which proves that $x^n = x^0 + V^n y^n$ is a solution.

It remains to prove that H^n is invertible, and to do this it suffices to prove, since A is invertible, that any eigenvalue λ of H^n is an eigenvalue of A. Let u be an eigenvector of H^n associated with the eigenvalue λ; by definition $H^n u = \lambda u$, which implies $V^n H^n u = \lambda V^n u$. Since $AV^n = V^n H^n$, we conclude that the vector $v = V^n u$ satisfies $Av = \lambda v$; but this vector is non-zero because V^n has full rank, which proves that v is an eigenvector associated with the eigenvalue λ for the matrix A, and the property is proven.

To end the proof, it remains to show that if n is the first index j for which x^j is the exact solution, then $h_{(n+1)n} \neq 0$. By definition of the residual, n is the first index j for which $r^j = 0$, and according to (6.47), this means that s_n is zero, then that $h_{(n+1)n}$ is also zero. □

6.2.3.4 GMRES(m,C) algorithm

We conclude from the previous result that the GMRES algorithm is 'well posed', and that it converges, as algorithm 3, in at most $m = N$ iterations. However, for large values of m ($m \geq 10$), the memory required (of the order of mN), and the computer time for the orthogonalisation process (in $O(m^2N)$) can become prohibitive. Whence the idea to arbitrarily stop the iterations at an index m fixed in advance, and to restart, until the residual becomes sufficiently small. We shall denote by GMRES(m) the algorithm thus defined, whose Fortran implementation is given in the Appendix.

We also note that because the orthogonalisation process generates round-off errors, it could make the algorithm unstable if the matrix A is ill conditioned. So the search for a preconditioning matrix is important, both for the stability of the method (for large values of m), and for its convergence speed (for small values of m). Let us denote by C the preconditioning matrix and by GMRES(m, C) the preconditioned GMRES algorithm, with preconditioning matrix C, and with the dimension of the Krylov subspace limited by m; this general algorithm is written in the following way:

Algorithme 4 (GMRES(m,C))

0 initialisation: choose x^0 and the preconditioning matrix C (it must be positive definite); compute the initial residual $r^0 = C^{-1}(b - Ax^0)$; if $r^0 = 0$ the sought solution is x^0 so stop; otherwise, choose the dimension m of the Krylov subspace, the accuracy ϵ of the stopping test, then

1 set $v^1 = r^0/||r^0||$;

2 iterate over $j \in [1, m]$:

$$n = j$$

$$h_{ij} = (C^{-1}Av^j, v^i), \quad 1 \leq i \leq j, \tag{6.48}$$

$$\widehat{v}^{j+1} = C^{-1}Av^j - \sum_{i=1}^{j} h_{ij}v^i, \tag{6.49}$$

$$h_{(j+1)j} = ||\widehat{v}^{j+1}||, \tag{6.26}$$

$$\text{if } h_{(j+1)j} = 0, \text{go to } \mathbf{3}; \text{ otherwise}$$
$$v^{j+1} = \frac{\widehat{v}^{j+1}}{h_{(j+1)j}}. \tag{6.27}$$

3 Compute the approximate solution:

$$x^n = x^0 + V^n y^n, \tag{6.37}$$
$$\text{where } y^n \in \mathbb{R}^n \text{ satisfies}: \ \forall y \in \mathbb{R}^n, J^n(y^n) \leq J^n(y), \tag{6.38}$$
$$\text{with } J^n(y) = \|\, \|r^0\| e_1^{n+1} - \bar{H}^n y \,\|, \tag{6.39}$$

where we write $e_1^{n+1} = (1, 0, ..., 0) \in \mathbb{R}^{n+1}$, and where the rectangular matrix $\bar{H}^n$ is formed with the coefficients h_{ij} according to (6.33). This matrix satisfies:

$$C^{-1}AV^n = V^n \bar{H}^n. \tag{6.50}$$

4 Compute the residual $r^n = C^{-1}(b - Ax^n)$; if $\| r^n \| \leq \epsilon$, stop, and x^n is the sought approximate solution; otherwise, set $x^0 = x^n$, $r^0 = r^n$ and go back to step **1**.

Remark 6.8 For safety's sake, it is recommended to choose a maximum number *nit* of restarts. The solution of the minimisation problem (6.38)–(6.39) and the residual computation are carried out as explained in the previous section; in particular it is not necessary, contrary to the above expression for r^n, to compute x^n to decide whether or not to stop the algorithm. □

Remark 6.9 A possible choice for the preconditioning matrix is for instance the diagonal of A. The storage is on the order of mN. In practice, m is chosen around 10, and so is *nit*, and the algorithm is still very efficient. □

Remark 6.10 We note that this algorithm is easily vectorisable since most of the computations are in the matrix-vector product. Also, the matrix A itself only occurs through this product with a vector, so it not necessary to store it. Of course, the version without storage is more costly in computer time than that with storage. □

For more details on the GMRES method, we refer the reader to the paper by Saad–Schultz (1986).

6.3 ONE EXAMPLE OF A NON-LINEAR PROBLEM

Non-linear partial differential equations occur very frequently in concrete problems from physics. We have already alluded to them in Chapter 1 with an example from heat radiation.

There is no systematic method for this type of problems. We shall present here a few tools from the theory of non-linear systems of equations. We shall apply them to a concrete example, namely that of transsonic flows.

6.3.1 *Transsonic Flows*

The veolcity u of a fluid in a domain Ω is written as $u = \nabla\varphi$ where the potential φ is a solution of ($\Gamma = \partial\Omega$)

$$\nabla \cdot (\rho\nabla\varphi) = 0 \text{ in } \Omega, \tag{6.51}$$

$$\left(\rho\frac{\partial\varphi}{\partial n}\right)|_\Gamma = g, \tag{6.52}$$

the density ρ being given by

$$\rho = \left(1 - (\nabla\varphi)^2\right)^{5/2}. \tag{6.53}$$

After discretisation with P^1 finite elements, the problem is reduced to the solution of the system with unknown $\Phi = \{\varphi_i\}_1^{nv}$ defined by

$$\int_\Omega \left[\left(1 - |\nabla\varphi_h|^2\right)^{5/2} \nabla\varphi_h \cdot \nabla w^j\right](\vec{x})d\vec{x} = \int_\Gamma (gw^j)(\vec{x})d\gamma(\vec{x}), \quad \forall j = 1, \ldots, nv, \tag{6.54}$$

with

$$\varphi_h = \sum_{i=1}^{nv} \varphi_i w^i, \tag{6.55}$$

the functions w^i being the usual basis functions in P^1 approximation, i.e. the continuous functions on $\bar{\Omega}$, linear on each triangle of the triangulation, and whose value at vertex q^i is given by:

$$w^i(q^j) = \delta_{ij}. \tag{6.56}$$

6.3.2 *A Fixed Point Algorithm*

Equation (6.54) has the form:

$$F(\Phi) = 0. \tag{6.57}$$

Let us set $G(\Phi) = \Phi + wF(\Phi)$, where w is an arbitrary real number; then problem (6.57) is equivalent to

$$G(\Phi) = \Phi, \tag{6.58}$$

which amounts to searching for the fixed points of the non-linear functional G. Let us consider the following algorithm ($\|.\|$ denotes the usual Euclidean norm in $\mathbb{R}^{nv}$):

Algorithm 5 (fixed point)
0 choose an initialisation Φ^0 for Φ and an accuracy $\varepsilon > 0$;
1 loop over integers $m \geq 0$
compute $\Phi^{m+1} = G(\Phi^m)$;
2 stopping test of the above loop: if $\|\Phi^{m+1} - G(\Phi^{m+1})\| < \varepsilon$, stop, because Φ^{m+1} is the sought approximate solution to (6.57); otherwise continue.

This algorithm converges under certain conditions summarised in the following proposition.

Proposition 6.11 *Let us assume that the functional G is continuously differentiable. Then, if the differential of G, which we denote by G′, is bounded by a constant C, with* $0 \leq C < 1$, *i.e. if*

$$\forall \Phi \in \mathbb{R}^{ns}, \quad \|G'(\Phi)\| \leq C < 1, \tag{6.59}$$

the above algorithm converges towards a solution of (6.57) and (6.58).

PROOF By (6.59), we have

$$\|\Phi^{n+1} - \Phi^n\| = \|G(\Phi^n) - G(\Phi^{n-1})\| \leq C\|\Phi^n - \Phi^{n-1}\|,$$

which gives by an immediate induction:

$$\|\Phi^{n+1} - \Phi^n\| \leq C^n\|\Phi^1 - \Phi^0\|.$$

Then by summation we obtain

$$\begin{aligned}\|\Phi^{m+p} - \Phi^m\| &\leq \sum_{n=m}^{m+p-1} \|\Phi^{n+1} - \Phi^n\| \leq \left(\sum_{n=m}^{m+p-1} C^n\right) \|\Phi^1 - \Phi^0\| \\ &\leq \frac{C^m}{1-C} \|\Phi^1 - \Phi^0\|,\end{aligned}$$

and this inequality shows that $\{\Phi^m\}$ is a Cauchy sequence in $\mathbb{R}^{nv}$; so it converges to a limit denoted by Φ^*. By definition and because of the continuity of G, we deduce that

$$0 = \Phi^{m+1} - G(\Phi^m) \to 0 = \Phi - G(\Phi) = wF(\Phi),$$

which shows that Φ^* is solution of (6.58), and also, since $w \neq 0$, of (6.57). □

Let us try to apply this method to solve our example. We shall iteratively generate vectors φ^m and scalars ρ^m by 'semi-linearizing' problem (6.51)–(6.53) in the following way

$$(\varphi^m, \rho^m) \to (\varphi^{m+1}, \rho^{m+1}), \quad \text{with}$$

$$\nabla \cdot (\rho^m \nabla \varphi^{m+1}) = 0 \text{ in } \Omega, \quad \left(\rho^m \frac{\partial \varphi^{m+1}}{\partial n}\right)|_\Gamma = g, \tag{6.60}$$

$$\rho^{m+1} = \left(1 - (\nabla\varphi^{m+1})^2\right)^{5/2}, \tag{6.61}$$

which gives, with obvious notation, the following discrete problem:

$$\int_\Omega \left[\left(1 - |\nabla\varphi_h^m|^2\right)^{5/2} \nabla\varphi_h^{m+1} \cdot \nabla w^j\right](\vec{x})\mathrm{d}\vec{x} = \int_\Gamma (gw^j)(\vec{x})\mathrm{d}\gamma(\vec{x}),\ \forall j = 1, \ldots, nv. \tag{6.62}$$

The link between this formulation and algorithm 5 is far from being obvious. In fact, there are multiple ways for applying the fixed point algorithm to a non-linear problem and (6.62) is one of them (constructing G from F is another one). It is not easy to show that the method works. Formally, we can write (6.62) in the form

$$A(\Phi^m)\Phi^{m+1} = b, \tag{6.63}$$

where the matrix $A(\Phi^m) = ((A(\Phi^m)_{ij}))$ and the right hand side $b = (b_i)$ are defined by:

$$A(\Phi^m)_{ij} = \int_\Omega \left[\left(1 - |\nabla\varphi_h^m|^2\right)^{5/2} \nabla w^i \cdot \nabla w^j\right](\vec{x})\mathrm{d}\vec{x}, \tag{6.64}$$

$$b_i = \int_\Omega (gw^i)(\vec{x})\mathrm{d}\gamma(\vec{x}).$$

Now equation (6.63) can be written

$$\Phi^{m+1} = G(\Phi^m), \text{ with } \quad G(\Phi^m) = A(\Phi^m)^{-1}b. \tag{6.65}$$

It remains to compute the differential mapping ($\Phi \to A(\Phi)$), so as to check the conditions of the proposition; we leave this point as an exercise to the reader.

6.3.3 *The non-linear GMRES Algorithm*

The non-linear GMRES algorithm is a 'quasi-Newton' method proposed by Brown and Saad (1990) to solve non-linear systems of the type

$$F(x) = 0, \quad x \in \mathbb{R}^N, \tag{6.66}$$

where F is a vector valued function. The iterative process in a quasi-Newton method is of the type

$$x_{n+1} = x_n - (A_n)^{-1}F(x_n), \tag{6.67}$$

where the matrix A_n is an approximation to the differential of F at the point x_n, denoted by $F'(x_n)$. To avoid having to compute this differential, we approximate it by a difference quotient along one direction, that is, assuming δ is a small positive parameter, we set:

$$\forall y,\ A_n y = D_\delta F(x_n\ ;y) = \frac{F(x_n+\delta y)-F(x_n)}{\delta} \cong F'(x_n)y. \tag{6.68}$$

Once we have this approximation, we must then solve the system $A_n y = -F(x_n)$. To do this, we consider, as in the GMRES algorithm, the Krylov subspace

$$K_m(A_n) = \text{Span}\ \{r_0, A_n r_0 \ldots, A_n^{m-1} r_0\}, \tag{6.69}$$

with r_0=$-F(x_n) - A_n x_0$, x_0 being an initialisation of the sought solution x.

Now, instead of defining the approximate solution by relation (6.67), the non-linear GMRES algorithm consists of writing

$$x_{n+1} = x_n + z_n, \text{where } z_n \in K_m(A_n) \text{satisfies}: \tag{6.70}$$

$$\forall z \in K_m(A_n), J_n(x_n+z_n) \leq J_n(x_n+z), \tag{6.71}$$

$$\text{with } J_n(v) = ||F(v)||, \tag{6.72}$$

the index m being the dimension of the Krylov subspace, chosen in advance. We note that we have, to first order:

$$\forall z \in K_m(A_n), \quad F(x_n+z) \cong F(x_n) + A_n z,$$

which shows that the minimisation problem (6.70)–(6.72) approximately amounts to minimising the residual in the linear GMRES method that we could use to solve $A_n y = -F(x_n)$, with procedure (6.67).

Algorithm 6 (non-linear GMRES(m,C)):

0 initialisation: choose x_0, the preconditioning matrix C (it must be symmetric and positive definite), the dimension m for the Krylov subspace, the accuracy ϵ, the stopping test and the small parameter δ; set $n = 0$;

1 compute $\widehat{v}_n^1 = -C^{-1}(F_n + A_n x_n)$, where $F_n = F(x_n)$ and $A_n y = D_\delta F(x_n\ ;y)$, then set $v_n^1 = \widehat{v}_n^1/||\widehat{v}_n^1||$;

2 iterations on $j \in [1, m]$:

$$n_m = j$$

$$h_{ij}^n = (C^{-1} A_n v_n^j, v_n^i), \quad 1 \leq i \leq j,$$

$$\widehat{v}_n^{j+1} = C^{-1} A_n v_n^j - \sum_{i=1}^{j} h_{ij}^n v_n^i,$$

$$h_{(j+1)j}^n = ||\widehat{v}_n^{j+1}||,$$

if $h_{(j+1)j}^n = 0$, go to **3**; otherwise

$$v_n^{j+1} = \frac{\widehat{v}_n^{j+1}}{h_{(j+1)j}^n}.$$

3 compute the approximate solution:

$$\begin{aligned}&x_{n+1} = x_n + z_n, \quad \text{where } z_n \in K_{n_m}(A_n) \text{ satisfies :}\\&\forall z \in K_{n_m}(A_n), J_n(x_n + z_n) \leq J_n(x_n + z),\\&\text{with } J_n(v) = ||F(v)||,\end{aligned}$$

4 If $||F(x_{n+1})|| < \epsilon$, then stop; otherwise change n into $n + 1$ and go back to step **1**.

APPENDIX 6.A

Fortran Subroutine for Solving a Linear System with an Invertible, but not Necessarily Symmetric, Matrix *A*, by the Linear GMRES(m) Algorithm.

```
C  File gmres.for
C
C  -----------------------------------------------------
         subroutine gmres(nv,x,eps,nkr,nit)
C  -----------------------------------------------------
C  This subroutine solves the linear system Ax = b
C  by the GMRES(nkr) method, i.e. with a dimension
C  nk of the Krylov space limited to nkr, which requires
C  several successive restarts, whose number is
C  arbitrarily limited to nit. The stopping test of the
C  method bears on the residual Euclidean norm.
C
C  This method is based on the construction, according to
C  the Arnoldi method, of an orthonormal basis of the
C  Krylov space basis which is stored in array v;
C  more precisely, v(i,j) holds the ith componentof the
C  jth basis vector.
C
C  Simultaneously, we build array h, which at the outset is
C  defined by h(i,j) = (Av_i,v_j); the jth column of array
C  h then represents the jth column of the Hessenberg matrix
C  H̄^nk of order (nk + 1) x nk, which is
C  then factorised in the form Q^nk R^nk, with Q^nk
C  an orthogonal matrix of order (nk + 1) x (nk + 1) and
C  R^nk an upper triangular matrix of order (nk + 1) x nk,
C  with a zero last row, and all diagonal elements
C  are strictly positive.
C
C  In a second stage, h(.,j) holds the index j column
C  of the upper triangular matrix R^nk, after having
C  carried out the jth rotation (the cosine of its angle
C  is denoted by cos, its sine by sin) whose aim is to
C  zero out the coefficient with index (j + 1)j in the
C  Hessenberg matrix H̄^kr.
C
```

```
C  Array r contains the right hand side
C  g^nk = (Q^nk)^T||r^0||e_1^nk+1 of the
C  minimisation problem with unknown y^nk.
C  Then, this array r holds the solution y^nk that
C  is computed by solving an upper triangular system
C  with matrix R^nk. This operation is done in
C  subroutine soln.
C
C  The approximate solution at step nk of the algorithm is
C  stored in array x, and this array holds at the outset
C  the initialisation x^0 of the solution. This operation
C  is also carried out in subroutine soln.
C
C  We then compute the 'residual vector' that we need to
C  initialiSe the next it step; its norm is simultaneously
C  used as a stopping test.
C
C  Remark 1:
C  ---------
C  We have chosen here to construct the basis of the
C  Krylov space first, then to factor the Hessenberg
C  matrix.
C
C  Remark 2:
C  ---------
C  Since both the matrix A of size nv*nv, and the right
C  hand side b of size nv of the linear system to be
C  solved only occur in computing the initial residual,
C  then in the construction of the basis vectors, we have not
C  stored them in this program.
C  They occur in only a subroutine that we called
C  'pmat(nv,x,ax,ind)': for a given vector x(nv)
C  (stored in array x(nv)), and for a fixed flag ind,
C  the subroutine pmat computes and stores in array ax:
C  - the product A*x if ind is 0,
C  - the linear combination b-A*x if ind is 1.
C  This subroutine depends on the storage format of the
C  matrix A under consideration, and we do not give it
C  here. It is to be written by the user depending on his
C  particular needs.
C
C
C          Program written by Pascal Joly
C          (Laboratoire d'Analyse Numerique,
C          Universite Pierre et Marie Curie),
C          modified by Brigitte Lucquin (Juin 95).
C
C       PARAMETER (nvMax=400,nkrMax=50)
        DIMENSION x(nvMax),v(nvMax,nkrMax),
     >            h(nkrMax,nkrMax),r(nkrMax)
        DIMENSION aux(nvMax),aux1(nvMax)
```

```
      CHARACTER*20 filnam
      DATA zero/1.e-8/
      rec = zero
C
C  ...initialisation
C
C       ind=1
      CALL pmat(nv,x,aux,ind)
      res= prods(nv,aux,aux)
      res= sqrt(res)
      IF(res .LT. rec)GOTO 4
      rec = max(rec,res*eps)
C
C  ...iterations start
C
      DO 1 it=1,nit
        nk=nkr
C
C  ...Arnoldi orthogonalisation
C
      DO 10 i=1,nv
        aux(i) = aux(i)/res
        r(i)=0.
10    CONTINUE
      r(1) = res
C
      DO 11 j=1,nkr
        DO 21 i=1,nv
          v(i,j) = aux(i)
21        CONTINUE
        ind=0
        CALL pmat(nv,aux,aux1,ind)
        DO 22 i=1,nv
          aux(i)=aux1(i)
22        CONTINUE
        DO 23 i=1,j
          tem= prods(nv,aux1,v(1,i))
          h(i,j) = tem
          DO 24 il=1,nv
            aux(il)=aux(il)-tem*v(il,i)
24          CONTINUE
23        CONTINUE
        dem= prods(nv,aux,aux)
        dem=sqrt(dem)
        h(j+1,j) = dem
        IF (dem .LT. rec) THEN
          nk=j
          GOTO 5
        ENDIF
        DO 25 i=1,nv
          aux(i) = aux(i)/dem
```

```
25              CONTINUE
11          CONTINUE
C
C   ...triangularisation and modification of right hand side
C
5           CONTINUE
            DO 31 i=1,nk
              ip=i+1
              hii=h(i,i)
              hipi=h(ip,i)
              gamma=hii*hii+hipi*hipi
              gamma=sqrt(gamma)
              gamma=1./gamma
              cos=hii*gamma
              sin=-hipi*gamma
              DO 41 j=i,nk
                hij=h(i,j)
                hipj=h(ip,j)
                h(i,j) = cos*hij - sin*hipj
                h(ip,j) = sin*hij + cos*hipj
41          CONTINUE
            raux=r(i)
            r(i)=cos*raux
            r(ip)=sin*raux
31          CONTINUE

C
C   ...solution of upper triangular system
C
            CALL soln(nk,nv,r,h,v,x)
C
C   ...residual computation
C
            ind=1
            CALL pmat(nv,x,aux,ind)
            res= prods(nv,aux,aux)
            res= sqrt(res)
            IF(res .LT. rec) THEN
              GOTO 4
            ELSEIF(it.GE.nit) THEN
              write(*,*) 'no convergence'
              GOTO 6
            ENDIF
1         CONTINUE
C
C   ...output results
C
4         CONTINUE
            write(*,*) 'name of file for storing'
            write(*,*) 'the approximate solution'
            read(*,'(a)') filnam
```

```
        CALL writeF(filnam,nv,x)
6       CONTINUE
C
C
      END
C
C
C*********************************
C
C       LIBRARY
C
C*********************************
C
C       INCLUDE 'utilP1.for'
C
C-----------------------------------------
        SUBROUTINE soln(nk,nv,r,h,v,x)
C-----------------------------------------
C  This subroutine solves the upper triangular
C  system that lets us compute the solution of the
C  minimisation problem in the GMRES algorithm.
C  The approximate solution at step nk is computed
C  and stored in vector x.
C
        PARAMETER (nvMax=400,nkrMax=50)
        DIMENSION x(nvmax),v(nvmax,nkrmax),
     >            h(nkrmax,nkrmax),r(nkrmax)
C
        DO 1 i=nk,1,-1
          tem=r(i)/h(i,i)
          r(i)=tem
          DO 2 j=i-1,1,-1
            r(j)=r(j)-h(j,i)*tem
2         CONTINUE
1       CONTINUE
C
        DO 3 i=1,nk
          tem=r(i)
          DO 4 il=1,nv
            x(il)=x(il)+tem*v(il,i)
4         CONTINUE
3       CONTINUE
        END
C
C ----------------------------------------------
        FUNCTION prods(n,x,y)
C ----------------------------------------------
C  This function computes the inner product
C  of two vectors x and y of size n.
C
      DIMENSION x(n),y(n)
```

```
      som = 0.
      DO 1 i=1,n
1       som = som + x(i)*y(i)
      prods = som
      END
```

APPENDIX 6.B

The C Program for Solving a Linear Non-symmetric PDE by Gauss Factorisation of the Linear System.

```
/* file femgauss.c

This program solves a second order elliptic PDE by P1
finite elements and Gauss factorisation.
The triangulation must be in file mesh.dta and the
parameters and data of the PDE in file param.dta. These 2
files can be created by freefem with option 'saveall'.
Solves
   alpha u + (u1,u2)grad u- div(mat[rho] grad u) = f,
   u|_{ng!=0 && u0!=0} = u0,(if u0=0 change it to
   u0=eps<<1)
   (beta u + n^T mat[rho]grad u)|_{ng!=0} = g
with all data P^1 */

#include <stdio.h>
#include <stdlib.h>
#include <math.h>
                /* derivatives of the basis functions */
#define
dwya(i,k)(-(q[me[k][next[i]]].x-q[me[k][next[i+1]]].x)/2)
#define
dwxa(i,k)((q[me[k][next[i]]].y-q[me[k][next[i+1]]].y)/2)
#define dwy(i,k) (dwya(i,k)/area[k]) /* dw^i/dx on T^k */
#define dwx(i,k) (dwxa(i,k)/area[k]) /* dw^i/dy on T^k */
#define sqr(x) ((x)*(x))
#define norm(x,y)(sqrt(sqr(x)+sqr(y)))
#define mmax(a,b)(a>b?a:b)
#define mmin(a,b)(a<b?a:b)
#define abss(a)(a >= 0 ? a : -(a))

#define penal 1.0e10

typedef struct{ float x,y;} rpoint;
typedef int triangle[3];
typedef struct{
   int np, nt;
   rpoint *rp;
   triangle *tr;
```

```
   int* ngt;
   int* ng;
 }triangulation;

 /* prototypes */
 int      initFEM(triangulation* t);
 float    pdeian(float *, float *, float *, float *, float *,
   float *, float *, float *, float *, float *, float *,
   float *, float *, int, int);
 float    gaussband(float *, float *, int, long, int, float);
 void rhsPDE(float* fw, float* f, float* g);
 int      buildarea();
 void pdemat (float *a, float* alpha, float *rho11,
     float *rho12, float *rho21, float *rho22,
     float* u1, float* u2, float* beta);
 int readtriangulation( triangulation *, char * );
 int savefct(float *, int , char *);
 int readparam(triangulation* , char *);

 /* global variables */
 triangulation t;

 int next[5]={1,2,0,1,2};        /* i+1 mod 3 */
 char*    pathT = "mesh.dta"; /* mesh file */
 char*    pathP = "param.dta";  /* param coef PDE file */
 char* pathU = "sol.dta";    /* result file */
   float* area; /* triangles area */
   float* a;  /* system matrix */
   float* u;  /* solution */
   float* f;  /* sources */
   float* g;  /* Neumann */
   float* u0; /* Dirichlet */
   float* beta; /* Order 0 coef in Fourier */
   float* alpha;  /* Order 0 coeff in PDE */
   float* rho11;  /* Viscosity matrix */
   float* rho12;
   float* rho21;
   float* rho22;
   float* u1; /* convection velocity */
   float* u2;
   int factorize = 1; /* to factor the matrix */
/* Copy of the triangulation */
  int nv, nt ;
   rpoint* q;
   triangle* me;
   int* ng;
   int* ngt;
   int bdth;
void rhsPDE( float* fw, float* f, float* g)
/* ---------------------------------------------
  Computes the linear system right hand side
```

```
  fw(j) = ∫Ω fw^j + ∫ng<>0 gw^j + penal|ng<0 u0^j
  OTHER INPUT next,area,q,me,ng,nt,penal
*/
{ int j,k, ir, ir1, ir2, meirk, meirkn ;
  float x1,x2, aux;

for(j=0;j<nv;j++) fw[j] = 0 ;
  for(k=0;k<nt;k++)
     for(ir=0;ir<=2;ir++)
     {
     ir1 = next[ir] ; ir2 = next[ir1] ;
     meirk = me[k][ir] ; meirkn = me[k][ir1] ;
     fw[meirk]
       += (2*f[meirk] + f[meirkn] + f[me[k][ir2]])*area[k]/12;
}
     for(k=0;k<nt;k++)
    for(ir=0;ir<=2;ir++)
{
     ir1 = next[ir] ; ir2 = next[ir1] ;
     meirk = me[k][ir] ; meirkn = me[k][ir1] ;
       if ((ng[meirk] != 0)&&(ng[meirkn] != 0))
       {
            aux = norm(q[meirk].x-q[meirkn].x,q[meirk].y
              -q[meirkn].y)/6;
            x1 = g[meirk] * aux;
     x2 = g[meirkn] * aux;
     fw[meirk] += 2*x1 + x2;
       fw[meirkn] += x1 + 2 * x2;
     }
  }
}

int buildarea(void)
/*-------------------------------------------------
 Computes areas: INPUT traingulation nv,nt,me,q
 RETURNS 1 if an area is <0, 0 otherwise
*/

{ int i,k,ir, err = 0;
float qq[2][3];
for( k=0;k<nt; k++)
{
for(ir=0;ir<=2;ir++)
{
     i=me[k][ir] ; qq[0][ir]=q[i].x; qq[1][ir]=q[i].y ;
}
     area[k]=( (qq[0][1]-qq[0][0])*(qq[1][2]-qq[1][0])
                  -(qq[1][1]-qq[1][0])*(qq[0][2]-qq[0][0]))/2;
     err = (area[k] < 0);
 }
 return err;
```

```
}

void pdemat (float *a, float* alpha, float *rho11,
      float *rho12, float *rho21, float *rho22,
      float* u1, float* u2, float* beta)
------------------- Computes matrix -------------------
{
int k,i,j,mejk, meik, k0, k1, k2;
float rhomean[2][2], alphamean, isii, x1, x2, x3, x4, aux;
long ai, nvl=nv, nvbdt = (2*bdth+1)*nvl;

for(ai=0;ai< nvbdt;ai++) a[ai] = 0.0F ;
for(k=0;k<nt;k++)
for(i=0;i<=2;i++)
{
  meik = me[k][i] ;
  k0 = meik; k1=me[k][next[i]]; k2=me[k][next[i+1]] ;
    x1=rho11[k0]; x2=rho11[k1]; x3= rho11[k2] ;
          rhomean[0][0] = (x1+x2+x3) / 3 ;
    x1=rho12[k0]; x2=rho12[k1]; x3= rho12[k2] ;
          rhomean[0][1] = (x1+x2+x3) / 3 ;
    x1=rho21[k0]; x2=rho21[k1]; x3= rho21[k2] ;
          rhomean[1][0] = (x1+x2+x3) / 3 ;
    x1=rho22[k0]; x2=rho22[k1]; x3= rho22[k2] ;
          rhomean[1][1] = (x1+x2+x3) / 3 ;
    x1 = alpha[k0]; x2 = alpha[k1]; x3 = alpha[k2];
          alphamean = ( x1 +x2 +x3 ) / 3 ;
  for(j=0;j<=2;j++)
 {
      mejk = me[k][j];
      isii = i==j ? 1.0F/6.0F : 1.0F / 12.0F ;
      ai = nvl*(meik-mejk+bdth)+mejk;
          aux = dwxa(i,k) * dwx(j,k);
          x1 = rhomean[0][0] * aux;
          aux = dwya(i,k) * dwx(j,k);
          x2 = rhomean[1][0] * aux;
          aux = dwxa(i,k) * dwy(j,k);
          x3 = rhomean[0][1] * aux;
          aux = dwya(i,k) * dwy(j,k);
          x4 = rhomean[1][1] *aux;
          a[ai] += x1 + x2 + x3 + x4;
          x1 = u1[k0]; x2 = u1[k1]; x3 = u1[k2] ;
          a[ai] += (2*x1 + x2 + x3) * dwxa(j,k) / 12;
          x1 = u2[k0]; x2 = u2[k1]; x3 = u2[k2] ;
          a[ai] += (2*x1 + x2 + x3) * dwya(j,k) / 12
            + alphamean * area[k] * isii;
     if((ng[meik] != 0)&&(ng[mejk] != 0)&&(meik < mejk))
     {
       k0 = meik; k1 = mejk;
     x1 = beta[k0]; x2 = beta[k1];
       x1 = (x1 + x2) *
```

```
            norm(q[meik].x-q[mejk].x,q[meik].y-q[mejk].y)/2;
        a[ai] += x1/6;
        ai = nvl*bdth+mejk; a[ai] += x1/3;
        ai = nvl*bdth+meik; a[ai] += x1/3;
      }
    }
  }
}

/*-----------------------------------------------------*/
float gaussband (float *a, float *x, int n, long bdthl,
                        int first, float eps)
/*-----------------------------------------------------*/
/* Factors (first=1), solves Ay=x with result in x */
/* LU is stored in A; returns smallest pivot * */
/* Any pivot < eps is set to eps */
/*   a[i][j] is stored in a[i-j+bdthl][j] */
/*      =a[n*(i-j+bdthl)+j) where -bdwth <= i-j <= bdthl */
{
int i,j,k;
float s, s1, smin = 1e9;

if (first) /* factorisation */
  for (i=0;i<n;i++)
{
for(j=mmax(i-bdthl,0);j<=i;j++)
{
  s=0; for (k=mmax(i-bdthl,0); k<j;k++)
s += a[n*(i-k+bdthl)+k]*a[n*(k-j+bdthl)+j] ;
  a[n*(i-j+bdthl)+j] -= s ;
}
for(j=i+1;j<=mmin(n-1,i+bdthl);j++)
{
s=0;for (k=mmax(j-bdthl,0);k<i;k++)
     s += a[n*(i-k+bdthl)+k]*a[n*(k-j+bdthl)+j] ;
s1 = a[n*bdthl+i];
if(abss(s1) < smin) smin=abss(s1);
if(abss(s1) < eps) s1 = eps;
a[n*(i-j+bdthl)+j] = ( a[n*(i-j+bdthl)+j] - s)/s1;
}
}
  for (i=0;i<n;i++) /* solution */
{
s=0; for (k=mmax(i-bdthl,0);k<i;k++)
     s += a[n*(i-k+bdthl)+k] * x[k];
x[i] = (x[i] - s ) / a[n*bdthl+i] ;
}
  for (i=n-1;i>=0;i--)
{
s=0; for (k=i+1; k<=mmin(n-1,i+bdthl);k++)
     s += a[n*(i-k+bdthl)+k] * x[k];
```

```
x[i] -= s ;
}
return smin;
}

int initFEM(triangulation* t)
{ int i,k, baux;
long matsize;
  nv = t->np; nt = t->nt;
  q =t->rp; me = t->tr;
  ng = t->ng; ngt =t->ngt;
bdth = 0;
for( k=0;k<nt;k++)
for(i=0;i<=2;i++)
{
baux = abss((me[k][i] - me[k][next[i]]));
bdth = (bdth > baux)? bdth : baux;
}
  area = (float*)calloc(nt, sizeof(float));
matsize = (long)nv*(2*(long)bdth+1);
a = (float*)calloc(matsize, sizeof(float));
return buildarea();
}


===================I/O functions=====================
int readtriangulation( triangulation *t, char *path )
{
  int i, j, nv, nt;
  FILE *data;

  if(data=fopen(path, "r"), !data) return 2;
  fscanf( data, "%d %d \n", &t->np, &t->nt );
  nv = t->np; nt = t->nt;
  t->rp = (rpoint*)calloc(nv, sizeof(rpoint));
  t->tr = (triangle*)calloc(nt, sizeof(triangle));
  t->ng = (int*)calloc(nv, sizeof(int));
  t->ngt = (int*)calloc(nt, sizeof(int));
  for( i=0; i<t->np; i++ )
     fscanf( data, "%f %f %d\n", &t->rp[i].x,
                          &t->rp[i].y, &t->ng[i] );
  for( i=0; i<t->np; i++ )
  {
     for( j=0; j<3; j++ )
     {
       fscanf( data, "%d", &t->tr[i][j] );
       t->tr[i][j]--; /* Fortran standard
     }
     fscanf( data, "%d\n", &t->ngt[i] );
  }
```

```
  fclose(data);
  return 0;
}

int savefct(float *f, int nv, char *path)
{
  int i;
  FILE *data;

  if(data=fopen(path, "w"), !data) return 2;
  fprintf(data, "%d\n", nv);
  for(i=0; i<nv; i++) fprintf(data, "%f\n", f[i]);
  fclose(data);
  return 0;
}

int readparam(triangulation* t, char *path)
{
  int i, nv = t->np;
  FILE *data;

  if(data=fopen(path, "r"), !data) return 2;
  fscanf(data, "%d %d", &nv,&i);/* i=1 if scalar eq */
  if(nv != t->np) return 3;
  beta = (float*)calloc(nv, sizeof(float));
  g = (float*)calloc(nv, sizeof(float));
  u0 = (float*)calloc(nv, sizeof(float));
  f = (float*)calloc(nv, sizeof(float));
  alpha = (float*)calloc(nv, sizeof(float));
  u1 = (float*)calloc(nv, sizeof(float));
  u2 = (float*)calloc(nv, sizeof(float));
  rho11 = (float*)calloc(nv, sizeof(float));
  rho22 = (float*)calloc(nv, sizeof(float));
  rho12 = (float*)calloc(nv, sizeof(float));
  rho21 = (float*)calloc(nv, sizeof(float));
  for(i=0; i<nv; i++)
  fscanf(data, "%f %f %f %f %f %f %f %f %f %f %f",
     &f[i],
     &g[i],
     &u0[i],
     &alpha[i],
     &beta[i],
     &u1[i],
     &u2[i],
     &rho11[i],
     &rho12[i],
     &rho21[i],
     &rho22[i]);
  fclose(data);
  u = (float*)calloc(nv, sizeof(float));
  return (u==0);
```

```
}

void main( )
/*-----------
Solves alpha u + (u1,u2)grad u- div(mat[rho] grad u) = f,
            u|_{ng!=0 && u0!=0} = u0,
            (beta u + n^T mat[rho]grad u)|_{ng!=0} = g
with f P1
*/
{
     int i; long ai, bdthl;
     float err;

  if(readtriangulation( &t, pathT))
  { print f('error reading triangulation\n'); exit(0);}
  initFEM(&t); bdthl =bdth;
  if(readparam(&t,pathP) != 0)
     { print f('error reading parameters\n'); exit(0);}
if(factorize)
{
pdemat(a,alpha,rho11,rho12,rho21,rho22, u1, u2, beta);
for(i=0;i<nv;i++)
if(u0[i] !=0 ){ ai = nv*bdthl+i; a[ai] += penal;}
}
rhsPDE(u, f, g);
for(i=0;i<nv;i++)
if(u0[i] !=0) u[i] += u0[i]*penal;
err = gaussband(a,u, nv, bdthl, factorize,1.0/penal);
printf ('smallest pivot = %f \n',err);
savefct(u,nv, pathU);
}

/* TEST OF femgauss.c */
/*-------------------
1.   Choose a solution, for example ue=sin(x+y)
Choose PDE params to find this solution.

2.   Solve with freefem. Here the freefem program will be:
     changewait;
     nv:=40;
border(1,0,2*pi,2*nv) begin x:=3*cos(t);
                                  y:=2*sin(t);end;
border(2,0,2*pi,nv) begin x:= cos(-t); y:= sin(-t); end;
buildmesh(nv*nv);
savemesh('mesh.dta');

ue = sin(x+y);
p = ue; nx = -x; ny =- y; dxue = cos(x+y);
c = 0.2; a1 = y; a2 = x; b=1;
nu = 1; nu11 = 1; nu22 = 2; nu21 =0.3; nu12 =0.4;
dnuue=dxue*(nu*(nx+ny) +
```

```
(nu11 + nu12)*nx + (nu21+ nu22)*ny);
g = ue*c+dnuue;
f = b*ue+dxue*(a1+a2) +ue*(2*nu+nu11+nu12+nu21+nu22);

solve(u) begin
onbdy(1) u = p;
onbdy(2) id(u)*c + dnu(u) = g;
pde(u) id(u)*b + dx(u)*a1 + dy(u)*a2
-laplace(u)*nu - dxx(u)*nu11 - dxy(u)*nu12
     - dyx(u)*nu21 - dyy(u)*nu22 =f;
end;

plot(u-ue);

3.   Check that u-ue is small.
4.   To generate data for femgauss.c use the same
freefem program with 'solve' replaced by

     saveall ('param.dta',u)


5.   The solution computed by femgauss.c is in 'sol.dta'
     check that it is the right one by executing the
     following freefem program
           loadmesh('mesh.dta');
           load('sol.dta',u);
           plot(u);
           plot(sin(x+y));
           plot(u-sin(x+y));
*/
```

7 Evolution Problems: Finite Differences in Time

Summary

We introduce here the finite difference method for approximating the three types of equations set forth in Chapter 1. We present the principle of the method on the example of a simple elliptic equation, then we shall focus our study on parabolic and hyperbolic equations. The generalisation to several space dimensions is described in the case of the heat equation. If the spatial mesh is not uniform, it might be more advisable to use the finite element method, or the finite volume method, which we shall briefly introduce in this chapter.

7.1 THE FINITE DIFFERENCE METHOD

7.1.1 Introduction

Let us go back to the case of a rod heated at both ends, which we studied in the Introduction to this book. The temperature φ of this rod, taken to be a line segment of unit length, solves the variational problem:

$$-\frac{d^2\varphi}{dx^2}(x) + c(x)\varphi(x) = f(x), \quad 0 < x < 1, \tag{7.1}$$

$$\varphi(0) = g(0), \quad \varphi(1) = g(1). \tag{7.2}$$

If the function c takes positive values, this problem admits a unique solution. If c is zero, the exact solution to (7.1)–(7.2) is given by

$$\varphi(x) = \int_0^1 G(x,y)f(y)dy + g(0) + x(g(1) - g(0)), \tag{7.3}$$

where G is the Green function, defined by

$$G(x,y) = \begin{cases} (1-x)y, & \text{if } 0 \le y \le x, \\ (1-y)x, & \text{if } x \le y \le 1. \end{cases} \tag{7.4}$$

Apart from this particular case, and in any case when the problem is set in dimension $d \geq 1$, the solution cannot be computed explicitly, and a discretisation is required, so that we can give as accurate an approximation as possible.

Contrary to the finite element method, the finite difference method consists in *approximating the derivation operator* by a discrete operator. This approach is easily understood if we notice that, for small h, we have (for instance):

$$\frac{\partial \varphi}{\partial x_i}(x_1, \cdots, x_d) \simeq \frac{1}{h}[\varphi(x_1, \cdots, x_i + h, \cdots, x_d) - \varphi(x_1, \cdots, x_d)]. \tag{7.5}$$

This remark suggests that we 'replace' all continuous derivative operators by difference quotients (whence the name 'difference', 'finite'coming from the fact that the parameter h, though chosen arbitrarily small, has a fixed non-zero value). The *a posteriori* justification of this approximation results from using simple Taylor formulae.

We shall now study some examples of discrete operators obtained this in way, then deduce a approximation to the one-dimensional problem (7.1)–(7.2).

7.1.2 Examples of Difference Operators

We shall work in dimension d and, to simplify notations, we write φ for $\varphi(x_1, \cdots, x_d)$ and $\varphi(x_i + h)$ for $\varphi(x_1, \cdots, x_i + h, \cdots, x_d)$. Let us introduce the following linear operators

$$D_i^+ \varphi = \frac{1}{h}(\varphi(x_i + h) - \varphi), \tag{7.6}$$

$$D_i^- \varphi = \frac{1}{h}(\varphi - \varphi(x_i - h)), \tag{7.7}$$

$$D_i^o \varphi = \frac{1}{h}\left(\varphi\left(x_i + \frac{h}{2}\right) - \varphi\left(x_i - \frac{h}{2}\right)\right). \tag{7.8}$$

The operator D_i^o is called a 'centred operator' in direction i, whereas the other two are 'non-centred': forward for the first one, backward for the second one.

The very definition of the derivative shows that $D_i^+ \varphi$ and $D_i^- \varphi$ tend towards $\frac{\partial \varphi}{\partial x_i}$ when h goes to 0. For the centred operator, this stems from taking the difference between the following two Taylor expansions:

$$\varphi\left(x_i + \frac{h}{2}\right) = \varphi(x_i) + \frac{h}{2}\frac{\partial \varphi}{\partial x_i}(x_i + \theta_1(h)), \quad \lim_{h \to 0} \theta_1(h) = 0,$$

$$\varphi\left(x_i - \frac{h}{2}\right) = \varphi(x_i) - \frac{h}{2}\frac{\partial \varphi}{\partial x_i}(x_i + \theta_2(h)), \quad \lim_{h \to 0} \theta_2(h) = 0.$$

We say these operators are consistent approximations to $\frac{\partial\varphi}{\partial x_i}$. If furthermore the error $|D_i\varphi - \frac{\partial\varphi}{\partial x_i}|$ thus committed is bounded, up to a multiplicative constant, by h^p, the approximation is said to be *consistent of order p*.

EXERCISE What is the consistency order of the above approximations ? □

EXERCISE Show that $[\varphi(x_i + h) - \varphi(x_i - h)]/h$ is not a consistent approximation of $\frac{\partial\varphi}{\partial x_i}$ because this difference quotient tends to $2\frac{\partial\varphi}{\partial x_i}$ when $h \to 0$. □

With a view towards solving our initial problem, we shall now define an approximation for the second derivatives.

Proposition 7.1 *If φ is four times continuously differentiable in the interval $[x_i - h, x_i + h]$,*

$$D_i^o D_i^o \varphi = D_i^+ D_i^- \varphi = D_i^- D_i^+ \varphi = \frac{1}{h^2}[\varphi(x_i + h) - 2\varphi + \varphi(x_i - h)] \tag{7.9}$$

is a consistent, second order, approximation to $\frac{\partial^2\varphi}{\partial x_i^2}$.

PROOF First of all we have

$$D_i^o(D_i^o\varphi) = \frac{1}{h}\left(D_i^o\varphi|_{x_i+\frac{h}{2}} - D_i^o\varphi|_{x_i-\frac{h}{2}}\right) =$$
$$\frac{1}{h}\left(\frac{\varphi(x_i + h) - \varphi}{h} - \frac{\varphi - \varphi(x_i - h)}{h}\right) = \frac{1}{h^2}[\varphi(x_i + h) - 2\varphi + \varphi(x_i - h)].$$

Next, an analogous computation shows that $D_i^o\varphi = D_i^+ D_i^- \varphi = D_i^- D_i^+ \varphi$, which proves (7.9).

Now, if φ is of class C^4 on $[x_i - h, x_i + h]$, we can write the following Taylor expansions

$$\varphi(x_i + h) = \varphi + h\frac{\partial\varphi}{\partial x_i} + \frac{h^2}{2}\frac{\partial^2\varphi}{\partial x_i^2} + \frac{h^3}{6}\frac{\partial^3\varphi}{\partial x_i^3} + \frac{h^4}{24}\frac{\partial^4\varphi}{\partial x_i^4}(\xi^+), \quad \xi^+ \in]x_i, x_i + h[$$
$$\varphi(x_i - h) = \varphi - h\frac{\partial\varphi}{\partial x_i} + \frac{h^2}{2}\frac{\partial^2\varphi}{\partial x_i^2} - \frac{h^3}{6}\frac{\partial^3\varphi}{\partial x_i^3} + \frac{h^4}{24}\frac{\partial^3\varphi}{\partial x_i^4}(\xi^-), \quad \xi^- \in]x_i - h, x_i[,$$

and adding them gives

$$D_i^o D_i^o \varphi = \frac{\partial^2\varphi}{\partial x_i^2} + \frac{h^2}{24}\left[\frac{\partial^4\varphi}{\partial x_i^4}(\xi^+) + \frac{\partial^4\varphi}{\partial x_i^4}(\xi^-)\right].$$

By the mean value theorem, we deduce the existence of a number $\xi \in]x_i - h, x_i + h[$ such that

$$D_i^o D_i^o \varphi - \frac{\partial^2 \varphi}{\partial x_i^2} = \frac{h^2}{12} \frac{\partial^4 \varphi}{\partial x_i^4}(\xi), \tag{7.10}$$

which shows that the consistency error $|D_i^o D_i^o \varphi - \frac{\partial^2 \varphi}{\partial x_i^2}|$ is bounded by Ch^2, with $12C = \sup_{\xi \in]x_i - h, x_i + h[} \left| \frac{\partial^4 \varphi}{\partial x_i^4}(\xi) \right|$. So the approximation is consistent of order 2. □

By combining in different ways the operators D^+, D^- and D^o, it is possible to construct approximations to a partial derivative of any arbitrary order; some are better than others, meaning that their consistency order is higher.

EXERCISE As an example, we suggest that the reader studies the consistency of the approximation $D_i^+ D_i^+ \varphi$ of $\frac{\partial^2 \varphi}{\partial x_i^2}$. □

7.1.3 *Finite Difference Approximation of Problems (7.1)–(7.2)*

We partition the segment $[0,1]$ into $N+1$ intervals of length $h = \delta x = 1/(N+1)$, and we define the $N+2$ *subdivision points*, or *nodes*, of this regular mesh by $x_i = ih$, $i \in \{0, ..., N+1\}$.

The discrete problem will be to find an approximation ψ_i to $\varphi(x_i)$ at each internal (i.e. $x_i, 1 \leq i \leq N$) node of the mesh, since, by virtue of 7.2, the solution is known at the ends $x_0 = 0$ and $x_{N+1} = 1$ of the interval. These different values ψ_i are solutions of the discrete problem

$$-\frac{1}{h^2}[\psi_{i+1} - 2\psi_i + \psi_{i-1}] + c(x_i)\psi_i = f(x_i), \quad 1 \leq i \leq N, \tag{7.11}$$

$$\psi_0 = g(0), \quad \psi_{N+1} = g(1), \tag{7.12}$$

which we call a *finite difference scheme* for problem (7.1)–(7.2). If we denote by Ψ_h the unknown vector with entries $(\psi_1, ..., \psi^n)^T$, this problem (7.11)–(7.12) can be written in matrix form

$$A_h \Psi_h = b_h, \tag{7.13}$$

where the symmetric matrix A_h and the right hand side b_h are given by

$$A_h = \frac{1}{h^2} \begin{bmatrix} 2 + c(x_1)h^2 & -1 & & & 0 \\ -1 & 2 + c(x_2)h^2 & -1 & & \\ & & \cdots & & \\ & & -1 & 2 + c(x_{N-1})h^2 & -1 \\ 0 & & & -1 & 2 + c(x_N)h^2 \end{bmatrix},$$

$$b_h = \begin{pmatrix} f(x_1) + \frac{1}{h^2} g(0) \\ f(x_2) \\ \dots \\ f(x_{N-1}) \\ f(x_N) + \frac{1}{h^2} g(1) \end{pmatrix}. \tag{7.14}$$

More generally, we shall adopt the following definition:

Definition 7.2 Let $L\varphi = 0$ be a partial differential equation and let $L_h \Psi_h = 0$ be a finite difference scheme of the above kind for approximating this problem; we shall call the *consistency error* of the scheme, the vector ε_h defined by

$$\varepsilon_h = L_h \varphi_h, \tag{7.15}$$

where φ_h is the projection on the mesh of the *exact solution* φ of the continuous problem, i.e. if the mesh is formed by the N points x_i, then φ_h is the vector with entries $(\varphi(x_1), ..., \varphi(x_N))^T$. The scheme is said to be *consistent* if this vector ε_h tends to 0 with h. This requires that we have defined a vector norm on $\mathbb{R}^N$. Let us take, for example, the norm $||.||_\infty$ defined by $||X||_\infty = \sup_{i=1}^{N} |x_i|$, $X = (x_1, ..., x_N)^T$. We shall say that the *scheme is of order p* in the l^∞ norm if there is a real positive constant C such that

$$||\varepsilon_h||_\infty \leq Ch^p. \tag{7.16}$$

According to Proposition 7.1, we deduce the following result:

Corollary 7.3 If the exact solution of problem (7.1)–(7.2) is of class C^4 on $[0, 1]$, the scheme (7.11)–(7.12) is consistent of order 2.

EXERCISE Still using only the three points x_{i-1}, x_i and x_{i+1}, it is possible to construct fourth order approximations to the first and second derivatives. Indeed, show that, if φ is of class C^6, then

$$\varphi'(x_{i+1}) + 4\varphi'(x_i) + \varphi'(x_{i-1}) = \frac{3}{h} [\varphi(x_{i+1}) - \varphi(x_{i-1})] + 0(h^4),$$

$$\varphi''(x_{i+1}) + 10\varphi''(x_i) + \varphi''(x_{i-1}) = \frac{12}{h^2} [\varphi(x_{i+1}) - 2\varphi(x_i) + \varphi(x_{i-1})] + 0(h^4),$$

By linearly combining the equations $-\varphi''(x_k) = f_k$, $k \in \{i+1, i, i-1\}$, this enables us, for instance, to obtain the following scheme for the Laplacian in one dimension,

$$-\frac{12}{h^2} [\psi_{i+1} - 2\psi_i + \psi_{i-1}] = f(x_{i+1}) + 10f(x_i) + f(x_{i-1}).$$

Show that this scheme is fourth order in the l^∞ norm □

Still regarding problem (7.1)–(7.2), two questions then arise:

1: does the discrete problem admit *a unique solution*?
2: is the method *convergent*, that is does it hold that

$$||\Psi_h - \varphi_h||_\infty \to 0, \quad \text{when } h \to 0? \tag{7.17}$$

It is easy to give an affirmative answer to the first question, since a simple computation shows that the matrix A_h is positive definite; indeed, if $X = (x_1, ..., x_N) \in \mathbb{R}^N$, we have

$$X^T A_h X = x_1^2 + (x_2 - x_1)^2 + ... + (x_N - x_{N-1})^2 + x_N^2 + h^2 \sum_{i=1}^{N} c(x_i) x_i^2,$$

and this quantity is positive, because of the positiveness assumption on c, and can only be zero if all the x_i are zero.

The answer to the second question is also positive. This stems from the *consistency* of the scheme and from *stability* results due to the monotonicity of the matrix A_h. We shall not prove these results here, as we reserve this stability study for the case of *parabolic and hyperbolic problems*, which are the main goal of this chapter. For the case of elliptic operators, such results are proved in Ciarlet (1989), to which we refer the reader .

Remark 7.4 Since the matrix A_h is symmetric and positive definite, several methods are possible for actually solving system (7.13): Choleski's method, the Gauss–Seidel or conjugate gradient algorithms. Because the matrix is diagonally dominant, Jacobi's method also converges. □

Remark 7.5 The above study can be generalised to elliptic operators in *dimension larger than one*. For example, in the case of the Laplacian with Dirichlet boundary conditions on a rectangular domain in the plane,

$$-\Delta\varphi = f \text{ in } \Omega =]0, L_1[\times]0, L_2[, \tag{7.18}$$

$$\varphi = g \text{ on } \Gamma = \partial\Omega, \tag{7.19}$$

the mesh is made up of small elementary rectangles of length h_i along each axis x_i, $i \in \{1,2\}$, and, in two dimensions, the scheme can be written, for the internal node of the mesh with index (i, j) :

$$-\frac{1}{(h_1)^2}[\psi_{i+1,j} - 2\psi_{i,j} + \psi_{i-1,j}] - \frac{1}{(h_2)^2}[\psi_{i,j+1} - 2\psi_{i,j} + \psi_{i,j-1}] = f(ih_1, jh_2), \tag{7.20}$$

where $\psi_{ij} \simeq \varphi(ih_1, jh_2)$. The matrix of the resulting linear system is again symmetric and positive definite; it is pentadiagonal and block-tridiagonal, with each diagonal block itself a tridiagonal matrix. We shall have the opportunity to come back to this point when we treat the heat equation in two dimensions. This scheme is called a 'five-point scheme' for the Laplacian (cf. Figure 7.1). □

Remark 7.6 A last remark to conclude: how can we handle *Neumann boundary conditions?* Mathematically, the answer is much less clear than it was for the finite

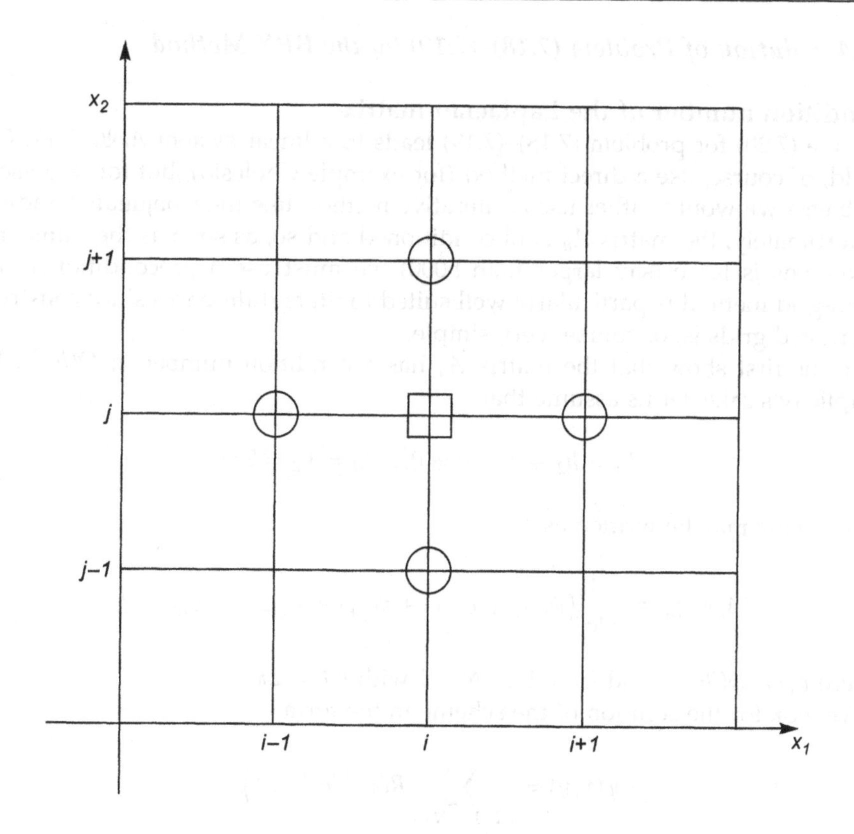

Figure 7.1 The five-point scheme for the Laplacian

element method, for which this type of boundary conditions was taken into account naturally by the variational formulation. It can be useful, for more complicated operators than the mere Laplacian, to combine this variational formulation to 'ad-hoc' quadrature formulae, so as to find the 'right scheme' near the boundary [Lucquin and Pironneau (1997)].

For the one-dimensional problem

$$-\varphi''(x) = f(x), \quad 0 < x < 1, \tag{7.21}$$

$$-\varphi'(0) = g(0), \quad \varphi'(1) = g(1), \tag{7.22}$$

we can for instance propose, keeping the same notation as above, and obtain the following approximation of the boundary condition

$$\psi_0 - \psi_1 = hg(0), \quad \psi^n - \psi_{N-1} = hg(1). \tag{7.23}$$

However, the accuracy of the global scheme is only h near the boundary.

By a clever linear combination (determined thanks to Taylor's formula) involving additional interior nodes, it is possible to improve the accuracy of the boundary condition approximation. □

7.1.4 Solution of Problem (7.18)–(7.19) by the BPX Method

Condition number of the Laplacian matrix

Scheme (7.20) for problem (7.18)–(7.19) leads to a linear system $A_h\Psi_h = F_h$. One could, of course, use a direct method (for example Choleski), but for large-scale problems we would rather use an iterative method like the Conjugate Gradient. Unfortunately, the matrix A_h is ill conditioned and so, as soon as the number of unknowns is large (say larger than 5000) we must use a preconditioner. The multi-grid method is particularly well suited to finite differences since constructing nested grids is, of course, very simple.

Let us first show that the matrix A_h has a condition number in $O(h^{-2})$. For simplicity's sake, let us assume that

$$h_1 = h_2 \equiv h, \quad g = 0, \quad L_1 = L_2 = 2\pi.$$

The scheme may be written as

$$(A_h\Psi_h)_{ij} \equiv -\frac{1}{h^2}\left(\psi_{i+1,j} + \psi_{i-1,j} + \psi_{i,j+1} + \psi_{i,j-1} - 4\psi_{i,j}\right) = f_{i,j}$$

where $\psi_{i,j} \approx \varphi(ih, jh)$ and $i, j = 1, \ldots, N-1$ with $Nh = 2\pi$.

We look for the solution of the scheme in the form

$$\psi(x,y) = \sum_{k,l=1,\ldots,N-1} Re\left(u^{k,l}e^{i(kx+ly)}\right).$$

By substituting $u^{k,l}e^{i(kx+ly)}$ in the scheme, we obtain:

$$-\frac{1}{h^2}\left(u^{k,l}e^{i(kih+ljh)}\left[e^{ikh} + e^{-ikh} + e^{ilh} + e^{-ilh} - 4\right]\right) = f_{i,j}$$

So, if $\hat{f}^{k,l}$ is the discrete Fourier transform of f, that is

$$f(x,y) \approx \sum_{k,l=1,\ldots,N-1} Re\left(\hat{f}^{k,l}e^{i(kx+ly)}\right).$$

then $u^{k,l} = \hat{f}^{k,l}h^2/(4\sin^2(kh/2) + 4\sin^2(lh/2))$ is the solution of the scheme, because

$$e^{ikh} + e^{-ikh} + e^{ilh} + e^{-ilh} = 2\cos(kh) + 2\cos(lh) = 4 - 4\sin^2\frac{kh}{2} - 4\sin^2\frac{lh}{2}.$$

For ψ to satisfy the boundary conditions, we must take a linear combination of functions that gives only sine functions:

$$\psi(x,y) = \sum_{k,l=0,\ldots,N} u^{k,l}\sin(kx)\sin(ly)$$

and this is only possible if f itself has the same form, that is if $f(x,y) = -f(2\pi - x, y) = -f(x, 2\pi - y)$.

This computation also shows that $v^{k,l}(x,y) = \sin(kx)\sin(ly)$ is an eigenvector of the scheme for the eigenvalue $h^{-2}(4\sin^2(\frac{kh}{2}) + 4\sin^2(\frac{lh}{2}))$ because

$$(A_h v^{k,l})_{i,j} = \frac{1}{h^2}(4\sin^2(kh/2) + 4\sin^2(lh/2))v^{k,l}(ih, jh).$$

Since we have $(N-1)^2$ solutions of this form and the matrix A_h has size $(N-1)^2 \times (N-1)^2$ we have all the eigenvalues. The smallest one corresponds to $k = l = 1$ and the largest one to $k = l = \pi/h - 1$

The condition number of A_h, i.e. the ratio of its largest to its smallest eigenvalue is now

$$\text{cond } (A_h) = \frac{\sin^2(\frac{\pi}{2} - \frac{h}{2})}{\sin^2 \frac{h}{2}} \approx \frac{4}{h^2}.$$

The BPX preconditioning Let us consider a multigrid mesh with K levels obtained by subdividing each rectangle of the initial finite difference grid into four identical rectangles. At level n the average size of the rectangles is denoted by k_h. Here $k_h = 2\pi/2^n$.

We solve the system $A_h \Psi_h = F_h$ by the preconditioned conjugate gradient method with a preconditioning matrix C. The algorithm (4.13)–(4.17) from Chapter 4 requires us to be able to compute $C^{-1}b$ for a given vector b. Let us denote by b_h^k the P^1 function (i.e. piecewise linear and continuous) equal to $b_{i,j}$ at the vertex i, j of the triangular mesh obtained by subdividing each finite difference rectangle into two triangles by a diagonal, always the same (cf. Chapter 2, Figure 2.9). We set

$$C^{-1}b = \sum_{k=1}^{K} \sum_{i=1}^{N_k} \frac{\int_\Omega b_h^k w_i^k dx}{\int_\Omega \nabla w_i^k \cdot \nabla w_i^k dx} w_i^k$$

where the sum is over all basis functions (hat functions w_i^k) of level k, then over all K levels. Bramble *et al* (1990) have shown that $C^{-1}A$ has a condition number in $O(1)$.

One should handle the coarsest level in a particular way, by solving the problem exactly. The right formula is then (the coarse level is now level 0)

$$C^{-1}b = A_0^{-1}Q_0 b + \sum_{k=1}^{K} \sum_{i=1}^{N_k} \frac{\int_\Omega b_h^k w_i^k dx}{\int_\Omega \nabla w_i^k \cdot \nabla w_i^k dx} w_i^k$$

where A_0 is the restriction of A_h to the coarse basis functions and Q_0 is the projection L^2 on the coarse space.

Remark The method can also be applied with a finite element discretisation as soon as we have a sequence of nested grids. The result on the condition number is independent of the space dimension.

7.2 FINITE DIFFERENCE SCHEMES FOR LINEAR EVOLUTION PROBLEMS

The main goal of this chapter is to define an approximation for *parabolic and hyperbolic* problems, then study its properties. In this type of equation, one of the variables, called the 'time variable', is singled out, contrary to elliptic equations. The other variables will be called space variables; they lie in the whole of $\mathbb{R}^d$ or in a domain in $\mathbb{R}^d$. Such problems are called *evolution problems* in time, as the solution at time $t \geq 0$ is determined from values at time $t = 0$, which we call 'initial conditions'.

We shall define a numerical approximation, using finite differences for the time variable, for these evolution problems, and we shall assume that their space dependence is limited to linear partial differential operators.

7.2.1 *Introduction*

Let us consider the following Cauchy problem

$$\frac{\partial \varphi}{\partial t}(t) = A(\varphi(t)), \quad 0 < t \leq T, \tag{7.24}$$

$$\varphi(t = 0) = \varphi^0, \tag{7.25}$$

where A is a differential operator, assumed linear and independent of the t variable; for example, A can be the Laplacian, (with boundary conditions if it does not act on the whole space). It is clear that a possible solution φ of problem (7.24)–(7.25) also depends on a space variable x that we have deliberately left out of the equations, on the one hand to simplify the notation, but mostly to emphasise the role of the time variable.

Let us assume that this problem has a classical solution $(t \to \varphi(t))$ in some function space, and let us then denote by $S(t)$ the operator defined by

$$S(t)\varphi^0 = \varphi(t), \tag{7.26}$$

where $\varphi(t)$ is the solution at time t of problem (7.24)–(7.25).

The approximation using *finite differences in time* of this problem consists in partitioning the interval under consideration $[0, T]$ into M subintervals of length $k = \Delta t = T/M$, then, given an approximation ψ^n of $\varphi(t^n)$, $t^n = n\delta t$, in defining an approximation ψ^{n+1} of the exact solution

$$\varphi(t^{n+1}) = S(k)\varphi(t^n) \tag{7.27}$$

of problem (7.24)–(7.25) at the next time step as

$$\psi^{n+1} = G(k)\psi^n, \tag{7.28}$$

where $G(k)$ is the 'discretised in space counterpart' of the operator $S(k)$, in particular meaning that it depends on a space discretisation parameter denoted by $h = \delta x$, assumed to be as small as we wish. In the same way, what we denote by ψ^n is actually a vector, each of whose entries corresponds to its value at a node of the mesh, but, as for the continuous problem, we have not displayed this spatial dependence explicitly.

Naturally, this iterative construction procedure requires the knowledge of the approximate solution at the initial time, and we shall simply define it as

$$\psi^0 = \varphi(t = 0) = \varphi^0. \tag{7.29}$$

What are the properties of such a scheme? How can we measure the error? Building upon the definition of consistency given in the previous section, we shall first answer the second question.

7.2.2 *Consistency of Scheme (7.28)–(7.29)*

By analogy with the notation used in Definition 7.2, we denote by L the continuous operator

$$L = \frac{\partial}{\partial t} - A, \tag{7.30}$$

and we define the discrete operator L^k ($k = \delta t$) on sequences $\Psi^k = (\psi^n)_{n\geq 0}$, $\psi^0 = \varphi^0$ in the following way: $L^k\Psi^k$ is the sequence defined by

$$L^k\Psi^k = \left((L^k\Psi^k)^n \right)_{n\geq 1}, \quad (L^k\Psi^k)^n = \frac{\psi^n - G(k)\psi^{n-1}}{k}, \tag{7.31}$$

since according to (7.28), we have:

$$\frac{\psi^n - \psi^{n-1}}{k} - \frac{G(k)\psi^{n-1} - \psi^{n-1}}{k} = 0.$$

Let us recall that this operator also depends *on the space discretisation step* $h = \delta x$ that implicitly occurs in the discrete operator $G(k)$.

More generally, let L be a partial differential operator depending on time and space. We shall take the following definition for a Cauchy problem associated with this operator:

Definition 7.7 Let φ be the solution of a Cauchy problem associated with operator L ($L\varphi = 0$) and let Ψ^k ($k = \delta t$) be that of the associated discrete problem $L^k\Psi^k = 0$ (with the above notation, and assuming that both these problems have a unique solution). We shall call *consistency error* the quantity

$$\mathcal{E}^k = L^k\varphi^k, \tag{7.32}$$

where φ^k is the projection on the mesh of the exact solution φ of problem (7.24)–(7.25). This consistency error is a vector $\mathcal{E}^k = \left(\varepsilon^n\right)_{n\geq 1}$, each of whose components depends on $h = \delta x$ and on $k = \delta t$. The scheme is called *consistent* if for all n, $\varepsilon^n \to 0$, as $k \to 0$ and $h \to 0$, for a given choice of norm $||.||$ in space. If moreover we have,

$$\forall n, \quad ||\varepsilon^n|| = 0(k^p) + O(h^q), \tag{7.33}$$

the scheme is said to be of *order p in time and q in space* (for this norm).

We note that, by virtue of (7.31) and (7.27), we have:

$$\varepsilon^n = \frac{S(k) - G(k)}{k} \varphi(t^{n-1}). \tag{7.34}$$

The notion of consistency enables us to measure the error produced by *approximating the continuous operator* by a discrete operator. It can be computed on the *exact solution* of the continuous problem, thanks to a Taylor expansion. However, this will not be enough to let us prove the convergence of the scheme. Another notion is required, that of stability, which we shall now define.

7.2.3 *Stability of Scheme (7.28)–(7.29)*

By an immediate induction from (7.28)–(7.29), we obtain

$$\psi^n = G(k)^n \psi^0, \tag{7.35}$$

i.e. the approximate solution at time t^n is defined as a function of the initial condition through the operator $G(k)^n$.

Stability will force these operators to remain bounded when $k = \delta t \to 0, n \to \infty$, with the product $n\delta t$ remaining bounded by the final time T. The idea is that *there can be no growth over time*: the approximate solution must remain bounded, despite the accumulation of discretisation and round-off errors.

Definition 7.8 The scheme (7.28)–(7.29) is said to be *stable* if $G(k)^n$ remains *uniformly bounded* for all $k = \delta t$, n satisfying:

$$0 \leq k \leq k^*, \quad 0 \leq nk \leq T;$$

or, in other words, if there exist a positive constant $C(T)$ (independant of h) such that

$$\forall n, \forall k \in]0, k^*], \, 0 \leq nk \leq T, \quad \text{we have: } ||[G(k)]^n|| \leq C(T), \tag{7.36}$$

for a given choice of norm in space.

We shall come back later, using concrete examples, to the practical way of checking stability. We shall now see how this notion is an essential feature in the convergence of the scheme.

7.2.4 *Convergence of Scheme (7.28)–(7.29)*

The question we now ask is the following: in what sense will the approximate solution $\Psi^k = (\psi^n)_{n\geq 0}$ ($k = \delta t$) defined by (7.28)–(7.29) 'tend' towards the exact solution of the initial problem (7.24)–(7.25) when the space step $h = \delta x$ and the time step $k = \delta t$ both go to 0? The answer lies in the following theorem, usually called the 'Lax Equivalence Theorem'.

Theorem 7.9 If scheme (7.28)–(7.29) is stable and consistent, then it is convergent, that is the the error $e^n = \varphi(t^n) - \psi^n$ at time t^n goes to zero when the time and space steps both go to 0 (for the norm used in Definitions 7.1 and 7.2), with the constraint $0 \leq n\delta t \leq T$.

PROOF Let us denote by E^k the error 'vector' whose index n entry is the error e^n. By the definition of ψ^0, we have $e^0 = 0$. Furthermore, since $L^k\Psi^k = 0$, we obtain, with notation as in (7.32),

$$L^kE^k = L^k\varphi^k = \mathcal{E}^k,$$

and this means, according to (7.31), that

$$\left(L^kE^k\right)^n = \frac{e^n - G(k)e^{n-1}}{k} = \varepsilon^n,$$

or that:

$$e^n = G(k)e^{n-1} + k\varepsilon^n.$$

By induction it follows that:

$$e^n = [G(k)]^n e^0 + k\sum_{i=1}^{n} [G(k)]^{n-i}\varepsilon^i;$$

then, as the error at the initial time e^0 is zero, we obtain the following estimate, for all integers n and all time steps $k = \delta t$ satisfying the constraint $0 \leq nk \leq T$,

$$\|e^n\| \leq nkC \max_{i\in\{1,\dots,n\}} \|\varepsilon^i\|,$$

because of the stability condition (7.36). We deduce that

$$\|e^n\| \leq TC \max_{\in\{1,\dots,n\}} \|\varepsilon^i\|,$$

and because the scheme is consistent, this goes to 0 with the space and time steps. □

Remark 7.10 All the above estimates are for a given choice of norm in space, and for a discretisation scheme (finite differences, finite elements,...) yet to be determined. □

Remark 7.11 We could also consider the case of an equation of the type $L\varphi = f$, with a source term f depending only on time; this would not change the above stability analysis, since this additional term would only affect the consistency error. The convergence theorem remains valid in that case. □

Now that we have defined all these notions, we shall apply them to the study of some classical examples for the approximation of certain 'model' evolution problems, starting with the example of the heat equation.

7.3 *THE HEAT EQUATION*

We present and analyse here several schemes to approximate the heat equation, using finite differences in both time and space. We start our study with the case of only one space variable in the whole space $\mathbb{R}$:

$$\frac{\partial\varphi}{\partial t} - \frac{\partial^2\varphi}{\partial x^2} = 0, \quad x \in \mathbb{R}, \quad 0 < t \leq T, \tag{7.37}$$

$$\varphi(x,0) = \varphi^0(x). \tag{7.38}$$

This problem is mathematically well-posed, and has a number of properties (Brezis, 1987), among them the *maximum principle* that we state without proof:

$$\text{if } \varphi^0 \geq 0, \quad \text{then} \quad 0 \leq \varphi(.,t) \leq \sup_{x\in\mathbb{R}} \varphi^0. \tag{7.39}$$

We shall denote by $\psi_i^n \simeq \varphi(x_i, t^n)$ the approximate solution taken at time $t^n = n\delta t$ ($n \in \{0,\dots,M\}, M\delta t = T$) and at point $x_i = i\delta x$ ($i \in \mathbb{Z}$). To simplify the notation we shall frequently write (k,h) for $(\delta t, \delta x)$.

7.3.1 *An Explicit Scheme*

We approximate the continuous operator

$$L = \frac{\partial}{\partial t} - \frac{\partial^2}{\partial x^2} \tag{7.40}$$

by the operator discretised *in space and time*, denoted by $\bar{L}$ for simplicity's sake, defined on sequences $\bar{\psi} = (\psi_i^n)_{i\in\mathbb{Z}, n\in\{0,\dots,M\}}$ by:

$$\bar{L}\bar{\psi} = \left((\bar{L}\bar{\psi})_i^n \right)_{i\in\mathbb{Z}, n\in\{1,\dots,M\}}, \tag{7.41}$$

$$\forall i \in \mathbb{Z}, \forall n \in \{0,\dots,M-1\}, \quad (\bar{L}\bar{\psi})_i^{n+1} = \frac{\psi_i^{n+1} - \psi_i^n}{k} - \frac{\psi_{i+1}^n - 2\psi_i^n + \psi_{i-1}^n}{h^2}.$$

Then the discrete problem is written as:

$$\bar{L}\bar{\psi} = 0, \tag{7.42}$$

$$\forall i \in \mathbb{Z}, \quad \bar{\psi}_i^0 = \varphi^0(x_i). \tag{7.43}$$

This scheme is *fully explicit*, i.e. given the approximate solution at step n (ψ_i^n known, $\forall i \in \mathbb{Z}$), we obtain the approximate solution at step $n+1$ from the very simple relation

$$\forall i \in \mathbb{Z}, \quad \psi_i^{n+1} = \psi_i^n + \frac{k}{h^2}\left(\psi_{i+1}^n - 2\psi_i^n + \psi_{i-1}^n\right), \tag{7.44}$$

which shows, in particular, that problem (7.42)–(7.43) admits a unique solution. This scheme is called the *forward Euler scheme.*

7.3.1.1 Convergence in the l^∞ norm in space

As far as the consistency error is concerned, we have the following result:

Proposition 7.12 *If the solution of the continuous problem (7.37)–(7.38) is C^2 in time and C^4 in space, then the scheme (7.41)–(7.42) is consistent and of order 1 in time and 2 in space (for the norm $\|.\|_\infty$ in space).*

PROOF The consistency error is defined as $\bar{\varepsilon} = \bar{L}\bar{\varphi} = \bar{L}\bar{\varphi} - L\varphi$, where $\bar{\varphi}$ is the projection on the mesh of the exact solution φ at each node of the mesh in both time and space. It is a doubly indexed sequence $\bar{\varepsilon} = (\varepsilon_i^n)_{i\in\mathbb{Z}, n\in\{1,\ldots,M\}}$ defined by:

$$\begin{aligned}\varepsilon_i^{n+1} = &\left[\frac{\varphi(x_i, t^{n+1}) - \varphi(x_i, t^n)}{k} - \frac{\partial\varphi}{\partial t}(x_i, t^n)\right] \\ &- \left[\frac{\varphi(x_{i+1}, t^n) - 2\varphi(x_i, t^n) + \varphi(x_{i-1}, t^n)}{h^2} - \frac{\partial^2\varphi}{\partial x^2}(x_i, t^n)\right].\end{aligned}$$

The consistency error is thus the sum of two errors, the first one linked to the discretisation of the time derivative operator, and the second one, relative to the spatial derivative, that has already been estimated in Proposition 7.1. A simple Taylor expansion up to second order now gives, because of (7.10):

$$\varepsilon_i^{n+1} = \frac{k}{2}\frac{\partial^2\varphi}{\partial t^2}(x_i, \tau^n) - \frac{h^2}{12}\frac{\partial^4\varphi}{\partial x^4}(\xi_i, t^n),$$

with $\tau^n \in]t^n, t^{n+1}[$ and $\xi_i \in]x_{i-1}, x_{i+1}[$. Thus, modulo the smoothness hypotheses in the statement of the theorem, we deduce that there exist two positive constants C_1 and C_2 such that, for all indices $i \in \mathbb{Z}$ et $n \in \{0, \ldots, M-1\}$, we have

$$|\varepsilon_i^{n+1}| \leq C_1 k + C_2 h^2,$$

which shows that the scheme is consistent, of first order in time and second order in space; the above constants are defined by:

$$C_1 = \sup_{(x,t)\in\mathbb{R}\times[0,T]} \left|\frac{\partial^2\varphi}{\partial t^2}(x,t)\right|, \quad C_2 = \sup_{(x,t)\in\mathbb{R}\times[0,T]} \left|\frac{\partial^4\varphi}{\partial x^4}(x,t)\right|. \qquad \square$$

Let us now study the stability of the scheme in the l^∞ norm in space.
Let us set:

$$\lambda = \frac{k}{h^2} \geq 0; \tag{7.45}$$

equality (7.44) then becomes

$$\forall i \in \mathbb{Z}, \quad \psi_i^{n+1} = \lambda\psi_{i+1}^n + (1-2\lambda)\psi_i^n + \lambda\psi_{i-1}^n, \tag{7.46}$$

i.e. ψ_i^{n+1} is a linear combination of ψ_{i+1}^n, ψ_i^n and ψ_{i-1}^n. We then note that, if the following condition is satisfied

$$0 \leq \lambda = \frac{k}{h^2} \leq \frac{1}{2}, \tag{7.47}$$

all the coefficients of this linear combination are positive, and their sum is 1, so that

$$\forall i \in \mathbb{Z}, \quad |\psi_i^{n+1}| \leq ||\Psi^n||_\infty,$$

if we denote by ψ^n the vector in $\mathbb{R}^{\mathbb{Z}}$ whose entries are ψ_i^n. By induction, we obtain

$$\forall n \geq 0, \quad ||\Psi^n||_\infty \leq ||\Psi^0||_\infty,$$

and this proves the l^∞ stability of the scheme.
We have just proved the following result:

Proposition 7.13 *Under condition (7.47), scheme (7.41)–(7.43) is stable for the norm $||.||_\infty$ in space.*

Propositions 7.12 and 7.13 completed by Theorem 7.9 allow us to conclude that scheme (7.41)–(7.43) is convergent under condition (7.47).

Remark 7.14 By induction, we note that, if Ψ^0 is positive (meaning that all its components are positive), then this is also true for vector Ψ^n, still assuming hypothesis (7.47). We find here a discrete version of the maximum principle (7.39) seen at the beginning of this section. □

Remark 7.15 The stability condition (7.47) we found is seen here as a *sufficient* condition for stability. Is it also necessary? To answer this question, we shall study stability in a different context. This is what we do in the next section. □

7.3.1.2 Von Neumann stability

To make the practical study of stability simpler, we shall change (7.41)–(7.42) into a 'continuous' in space version written as:

$$\forall x \in \mathbb{R}, \quad \frac{\psi^{n+1}(x) - \psi^n(x)}{k} - \frac{\psi^n(x+h) - 2\psi^n(x) + \psi^n(x-h)}{h^2} = 0. \tag{7.48}$$

Let us denote by $\hat{\psi}$ the *Fourier transform* of ψ defined by:

$$\hat{\psi}(\xi) = \int_{-\infty}^{+\infty} e^{-i\xi x}\psi(x)\mathrm{d}x. \tag{7.49}$$

Let us recall that:

$$\hat{\hat{\psi}} = 2\pi\psi, \quad ||\hat{\psi}||_2 = \sqrt{2\pi}\,||\psi||_2, \tag{7.50}$$

if we denote by $||.||_2$ the norm in the space $L^2(\mathbb{R})$; the second property is called 'Plancherel's theorem'. Let us Fourier transform equation (7.48); after using a change of variables to observe that

$$\int_{-\infty}^{+\infty} e^{-i\xi x}\varphi(x+h)\mathrm{d}x = e^{i\xi h}\int_{-\infty}^{+\infty} e^{-i\xi y}\varphi(y)\mathrm{d}y, \tag{7.51}$$

we obtain

$$\hat{\psi}^{n+1}(\xi) = \hat{\psi}^n(\xi) + \frac{k}{h^2}(e^{ih\xi}\hat{\psi}^n(\xi) - 2\hat{\psi}^n(\xi) + e^{-ih\xi}\hat{\psi}^n(\xi)). \tag{7.52}$$

This relation can be written

$$\hat{\psi}^{n+1}(\xi) = a(\xi)\hat{\psi}^n(\xi), \tag{7.53}$$

where the factor $a(\xi)$, called the *amplification factor*, is the real number defined by:

$$a(\xi) = 1 + \frac{k}{h^2}(e^{ih\xi} - 2 + e^{-ih\xi}) = 1 - 4\frac{k}{h^2}\sin^2\frac{h\xi}{2}. \tag{7.54}$$

By induction, we obtain

$$\hat{\psi}^n(\xi) = [a(\xi)]^n\hat{\psi}^0(\xi) = [a(\xi)]^n\hat{\varphi}^0(\xi). \tag{7.55}$$

If the following condition is satisfied

$$||a||_\infty = \sup_{\xi\in\mathbb{R}} |a(\xi)| \leq 1, \tag{7.56}$$

we have, by using Plancherel's relation

$$||\psi^n||_2 = \frac{1}{\sqrt{2\pi}}||\hat{\psi}^n||_2 \leq \frac{1}{\sqrt{2\pi}}||\hat{\varphi}^0||_2 = ||\varphi^0||_2,$$

which proves the stability of the scheme for the norm $||.||_2$. Condition (7.56) thus appears as a sufficient condition for the stability of scheme (7.48) in the norm $||.||_2$. Is the condition necessary?

Before we answer this question, let us try to link this condition with the one we found previously. We remark that $a(\xi)$ is always less than 1, so that (7.56) is equivalent to $\forall \xi \in \mathbb{R}$, $a(\xi) \geq -1$, and this relation is equivalent to (7.47).

In Figure 7.2, we have plotted the graph of the function $A(t) = a(\xi)$, $t = h\xi$, for three different values of the ratio k/h^2: the curve with diamond-shaped symbols corresponds to $k/h^2 = 1/4$, that with crosses corresponds to the limit case $k/h^2 = 1/2$, and last the solid line curve corresponds to $k/h^2 = 1$. As the theoretical study predicts, condition (7.56) is not satisfied in the last case, whereas it is satisfied in the other cases.

We shall now prove the following result, in a slightly more general framework than is needed for the scheme (7.48):

Proposition 7.16 *A scheme of the type $\psi^n = G(k)^n\varphi^0$ $(k = \delta t)$, where the Fourier transform of the operator $G(k)$ is a multiplication operator by a scalar function a, and is stable in the $L^2(\mathbb{R})$ norm (we also say 'von Neumann stable') if and only if the following stability condition, called the 'von Neumann condition', is satisfied:*

$$\exists C \geq 0, \exists (\delta t)^*, \quad \text{such that: } \forall \delta t \in]0, (\delta t)^*], \forall \xi \in \mathbb{R}, |a(\xi)| \leq 1 + C\delta t. \tag{7.57}$$

To prove this result, we shall use a lemma that we present here in the general vector case (i.e. in $\mathbb{R}^d$ with $d \geq 1$), because we will find it useful later. Before we do that, we shall recall some definitions and results from matrix algebra [Ciarlet (1989), Lascaux and Théodor (1987), Schatzman (1991)].

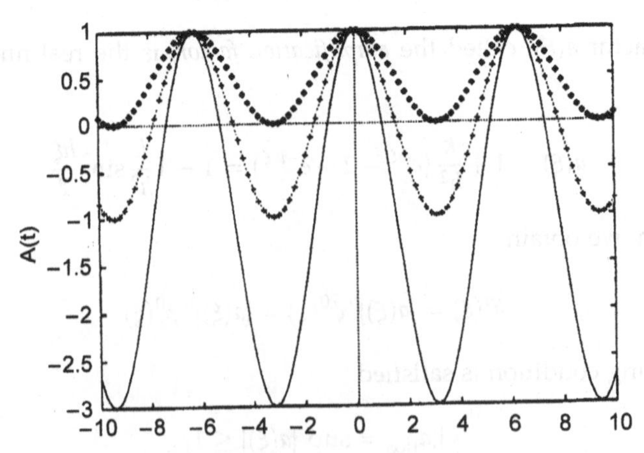

Figure 7.2 Graph of function $A(t) = a(\xi)$, $t = h\xi$

Let B be a matrix with real or complex entries, of size $d \times d$, and let $B^* = \bar{B}^T$ be its conjugate transpose matrix; we say B is a *normal* matrix if it commutes with B^*, i.e. if $BB^* = B^*B$. Real symmetric matrices are normal. The *spectral radius* of matrix B, denoted by $\rho(B)$, is the largest modulus of all the eigenvalues of B. For any vector norm, it is possible to define a matrix norm, called the *subordinate matrix norm* to this vector norm, through the relation:

$$||B|| = \sup_{x \in \mathbb{R}^d,\ x \neq 0} \frac{||Bx||}{||x||};$$

for such a matrix norm, we have:

$$\forall B, \quad \rho(B) \leq ||B||.$$

In particular, the matrix norm subordinate to the Euclidean norm $||.||_2$ is defined by:

$$||B||_2 = \sqrt{\rho(BB^*)}.$$

If the matrix B is normal, its Euclidean norm is equal to its spectral radius.

After these reminders, we now state the following classical result, a proof of which can be found, for example, in Lascaux (1976), Richtmyer and Morton (1967).

Lemma 7.17 Let F be the operator defined over $L^2(\mathbb{R}^d)$ by: $\forall U \in \mathbb{R}^d$, $F(U) = AU$, where, for all $\xi \in \mathbb{R}^d$, $A(\xi)$ is a matrix; this linear operator has norm (we write L^2_d for $L^2(\mathbb{R}^d)$)

$$||F||_{\mathcal{L}(L^2_d, L^2_d)} = \sup_{\xi \in \mathbb{R}} ||A(\xi)||_2. \tag{7.58}$$

PROOF OF PROPOSITION 7.16 Let us consider a scheme of the form $\psi^n = G(k)^n \varphi^0$, written, after Fourier transform, as $\hat{\psi}^n = F(\hat{\varphi}^0)$, where the operator F is the multiplication operator by the scalar function a^n. If we use first Plancherel's Theorem and then Lemma 7.17 in the scalar case (i.e. $d = 1$), we deduce

$$||[G(k)]^n||_{\mathcal{L}(L^2(\mathbb{R}_x), L^2(\mathbb{R}_x))} = ||F||_{\mathcal{L}(L^2(\mathbb{R}_\xi), L^2(\mathbb{R}_\xi))} = ||a^n||_\infty = ||a||^n_\infty,$$

with the notation: $||a||_\infty = \sup_{\xi \in \mathbb{R}} |a(\xi)|$.

Let us assume that the scheme is stable in $L^2(\mathbb{R})$. By definition, the norms of the operators $G(k)^n$, as operators from $L^2(\mathbb{R})$ into $L^2(\mathbb{R})$, remain bounded independently of n, for all integers n and for time steps $k = \delta t \in]0, k^*]$ satisfying the constraint $nk \leq T$. There exist a positive constant C_1 (we may assume $C_1 > 1$) such that $||a||^n_\infty \leq C_1$. This holds in particular for the integer $n = n_0$, where n_0 is half the integer part of the ration $T/k \geq 1$. Thus, we have

$$||a||_\infty \leq C_1^{\frac{2k}{T}} \leq 1 + \frac{C_1^{\frac{2k^*}{T}} - 1}{k^*} k,$$

or (7.57) with $C = (C_1^{2k^*/T} - 1)/k^*$.

Conversely, if a satisfies condition (7.57), then

$$\|a\|_\infty^n \leq (1 + Ck)^n \leq (e^{Ck})^n \leq e^{CT},$$

for all integers n and all time steps $k = \delta t$ such that $nk \leq T$, which proves the L^2 stability of the scheme and ends the proof. □

Remark 7.18 The term $C\delta t$ in (7.57) allows an exponential growth in time of the numerical solution (while noting that this growth is actually limited by the fact that the definition of stability is only concerned with solutions in finite time T).

In general, this condition is often replaced by the more restrictive condition (7.56), called the 'strict von Neumann condition'; this is true, in particular, if we know that the exact solution has no exponential behaviour in time, as is precisely the case here. Under this condition, the scheme in Proposition 7.16 is 'strictly stable': the numerical solution does not grow faster than the exact solution, and we have stability in the limit case $T = +\infty$. From a practical point of view, it is also preferable to use condition (7.56); indeed, even though the theoretical condition (7.57) is true in the limit $\delta t \to 0$, because the discretisation steps have a non-zero finite value, the numerical solution may grow appreciably during the time iterations, harming the effective stability of the scheme.

However, there are situations [Richtmyer and Morton (1967)] where the condition (7.57) is theoretically indispensable; this would for instance be the case if equation (7.37) featured a dissipation term like $b\varphi$ discretised explicitly (i.e. by $b\psi_i^n$ at point x_i). □

One shows easily that the scheme (7.48) is always consistent, of order 1 in time and 2 in space, and this eventually allows us to state the final convergence result, whose proof is analogous to that of Theorem 7.9.

Theorem 7.19 *Under the stability condition (7.47), the scheme (7.48) is convergent for the $L^2(\mathbb{R})$ norm; it is of order 1 in time and 2 in space.*

Remark 7.20 This Fourier transform method is an undeniably practical tool for studying the stability of schemes, that also gives necessary and sufficient stability conditions: we shall use it as often as possible. However, its major drawback is that it is difficult to generalise to the case of a boundary value problem, and that it is limited to partial differential equations with constant coefficients discretised on a uniform mesh. □

Remark 7.21 The stability condition (7.47) imposes a particularly severe constraint on the time step, since it must be of the order of the square of the space step (and even less!), and this means that in order to reach a given final time T, we shall have to iterate the algorithm a large number of times, which leads to prohibitive computer time. This fully explicit scheme, even though it is very simple to program, is thus not a very good scheme. We shall now propose others, and study their properties somewhat more rapidly. □

7.3.2 *Towards Implicit Schemes*

When using scheme (7.41)–(7.43) to determine the approximate solution at step $n+1$ from that at step n, the Laplacian was computed at time step n. This had the advantage of making the programming at each time step quite simple, but made it very costly because of condition (7.47) on the time step. Conversely, what happens if the Laplacian is taken at time step $n+1$?

7.3.2.1 Study of the fully implicit scheme

The operator L defined by (7.40) is now approximated by the operator $\bar{L}$, discrete in time and space, that is defined on sequences $\bar{\psi} = (\psi_i^n)_{i\in\mathbb{Z},n\in\{0,\dots,M-1\}}$ by

$$\bar{L}\bar{\psi} = \left((\bar{L}\bar{\psi})_i^n \right)_{i\in\mathbb{Z},n\in\{1,\dots,M\}}, \tag{7.59}$$

$$\forall i \in \mathbb{Z},\ \forall n \in \{0,\dots,M-1\}, \quad (\bar{L}\bar{\psi})_i^{n+1} = \frac{\psi_i^{n+1} - \psi_i^n}{k} - \frac{\psi_{i+1}^{n+1} - 2\psi_i^{n+1} + \psi_{i-1}^{n+1}}{h^2},$$

where h still denotes the space step and k is the time step. The resulting scheme $\bar{L}\bar{\psi} = 0$ is said to be *implicit because,* when we know the approximate solution at time t^n, the solution at the next time step (given by relation (7.59)) is not determined in a simple way, since it requires *solving a non-diagonal linear system*. We shall come back to this point in more details when we treat the case of the heat equation in a bounded domain in $\mathbb{R}$. Before we do that, we shall study the stability properties of such a scheme.

Analogously to (7.48), the 'continuous in space' version of this scheme is written as:

$$\forall x \in \mathbb{R}, \quad \frac{\psi^{n+1}(x) - \psi^n(x)}{k} - \frac{\psi^{n+1}(x+h) - 2\psi^{n+1}(x) + \psi^{n+1}(x-h)}{h^2} = 0, \tag{7.60}$$

which enables us to study its von Neumann stability. The Fourier transform $\hat{\psi}$ of ψ is now a solution of

$$\hat{\psi}^{n+1}(\xi) = \hat{\psi}^n(\xi) + \frac{k}{h^2}(\hat{\psi}^{n+1}(\xi)e^{ih\xi} - 2\hat{\psi}^{n+1}(\xi) + e^{-ih\xi}\hat{\psi}^{n+1}(\xi));$$

in other words

$$\hat{\psi}^{n+1}(\xi) = b(\xi)\hat{\psi}^n(\xi), \tag{7.61}$$

where the amplification factor $b(\xi)$ is now defined by:

$$b(\xi) = \frac{1}{1 + 4\frac{k}{h^2}\sin^2\frac{h\xi}{2}}. \tag{7.62}$$

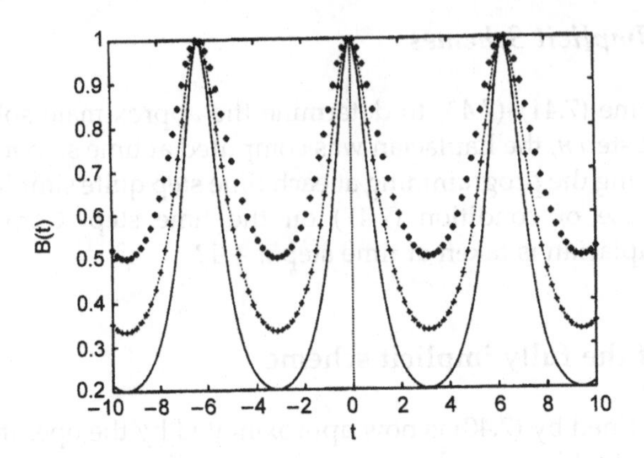

Figure 7.3 Graph of function $B(t) = b(\xi)$, $t = h\xi$

We notice that

$$\forall h \geq 0, \forall k \geq 0, \quad 0 \leq b(\xi) \leq 1, \tag{7.63}$$

which allows us to obtain by induction and Plancherel's theorem the following inequalities

$$||\psi^n||_{L^2} \leq ||\psi^{n-1}||_{L^2} \leq \ldots \leq ||\psi^1||_{L^2} \leq ||\varphi^0||_{L^2}, \tag{7.64}$$

showing that the scheme is stable, without any restrictive condition on the time step; we say the scheme is *unconditionally stable*. We have just proven

Proposition 7.22 *The fully implicit scheme is unconditionally von Neumann stable.*

In Figure 7.3, we have plotted the graph of the function $B(t) = b(\xi)$, $t = h\xi$, for the three values of the ratio k/h^2 already considered in Figure 7.2: the curve with the diamond shaped symbols corresponds to $k/h^2 = 1/4$, that with crosses is the limit case $k/h^2 = 1/2$, and the solid line curve corresponds to $k/h^2 = 1$. Relation (7.63) is satisfied in all three cases.

EXERCISE By doing a Taylor expansion around point (x, t^{n+1}), show that the consistency error is the same as that of the explicit scheme. □

This scheme, called the *backward Euler scheme* is very robust, but it is still only of first order in time. We shall now try to construct other stable schemes, hopefully more accurate, by using a linear combination with the explicit scheme.

7.3.2.2 A little implicit, a little explicit

Let θ be a fixed parameter in $[0, 1]$; we define the θ-scheme by

$$\frac{\psi_i^{n+1} - \psi_i^n}{k} - \mathcal{D}_i^+\mathcal{D}_i^- \left(\theta\Psi^{n+1} + (1-\theta)\psi^n\right)) = 0, \tag{7.65}$$

where ψ^n denotes the vector in $\mathbb{R}^{\mathbb{Z}}$ with entries ψ_i^n, and where $\mathcal{D}^+$ and $\mathcal{D}^-$ are non-centred operators in space defined on sequences $U = (u_i)_{i\in\mathbb{Z}}$ of $\mathbb{R}^{\mathbb{Z}}$ by a relation analogous to definitions (7.6) and (7.7) for functions, that is:

$$\mathcal{D}_i^+ U = \frac{1}{h}(u_{i+1} - u_i), \quad \mathcal{D}_i^- U = \frac{1}{h}(u_i - u_{i-1}); \tag{7.66}$$

by analogy with (7.9), we have obviously:

$$\mathcal{D}_i^+ \mathcal{D}_i^- U = \mathcal{D}_i^- \mathcal{D}_i^+ U = \frac{u_{i+1} - 2u_i + u_{i-1}}{h^2}. \tag{7.67}$$

For $\theta = 0$, this scheme is the explicit scheme, whereas for $\theta = 1$, it is the implicit scheme. We shall see that $\theta = 1/2$ plays a particular role: in that case the scheme is called *the Crank–Nicolson scheme.*

By Fourier transform in space, the scheme

$$\forall x \in \mathbb{R}, \quad \frac{\psi^{n+1}(x) - \psi^n(x)}{k} - \theta\,\frac{\psi^{n+1}(x+h) - 2\psi^{n+1}(x) + \psi^{n+1}(x-h)}{h^2} - (1-\theta)\,\frac{\psi^n(x+h) - 2\psi^n(x) + \psi^n(x-h)}{h^2} = 0, \tag{7.68}$$

becomes

$$\hat{\psi}^{n+1}(\xi) = c(\xi)\hat{\psi}^n(\xi),$$

and the amplification factor $c(\xi)$ is now equal to:

$$c(\xi) = \frac{1 - 4(1-\theta)\frac{k}{h^2}\sin^2\frac{h\xi}{2}}{1 + 4\theta\frac{k}{h^2}\sin^2\frac{h\xi}{2}}. \tag{7.69}$$

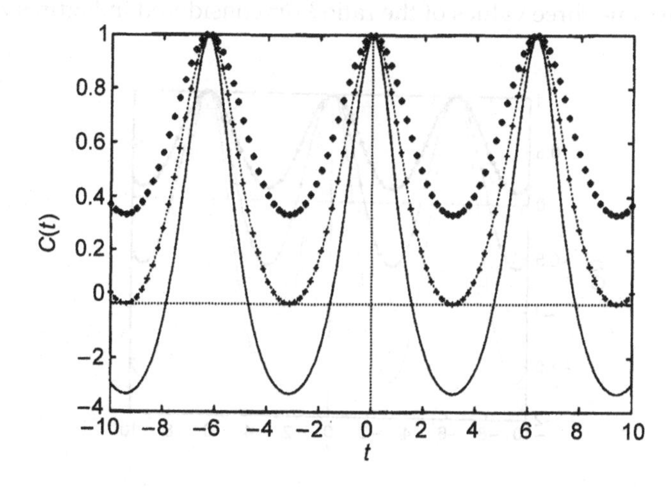

Figure 7.4 Graph of $C(t) = c(\xi)$, $t = h\xi$, for $\theta = 1/2$

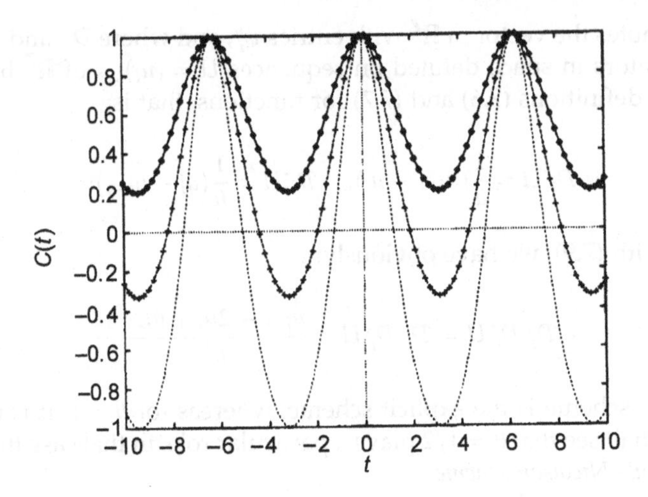

Figure 7.5 Graph of $C(t) = c(\xi)$, $t = h\xi$, for $\theta = 1/4$

We notice that for all real values of ξ we have $c(\xi) \leq 1$, so that the scheme is von Neumann stable if and only if: $\forall \xi \in \mathbb{R}$, $c(\xi) \geq -1$. This last property may be written

$$\forall \xi \in \mathbb{R}, \quad 2 - 4(1-2\theta)\frac{k}{h^2}\sin^2\frac{h\xi}{2} \geq 0,$$

and this relation is always satisfied for $\theta \geq 1/2$ (but this is not a necessary and sufficient condition). We thus have

Proposition 7.23 *For $\theta \geq 1/2$, the θ-scheme is unconditionally von Neumann stable.*

In Figures 7.4, 7.5 and 7.6, we have plotted the graph of the function $C(t) = c(\xi)$, $t = h\xi$, for the same three values of the ratio k/h^2 considered in Figures 7.2 and 7.3

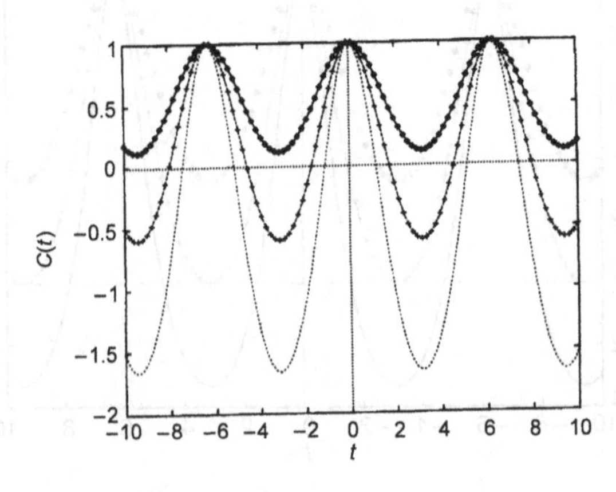

Figure 7.6 Graph of $C(t) = c(\xi)$, $t = h\xi$, for $\theta = 1/8$

above (the symbol captions are the same as in those figures), for three different values of the parameter θ: first for $\theta = 1/2$, then for $\theta = 1/4$, and last for $\theta = 1/8$. Figures 7.4 and 7.6 confirm the theoretical results; for $\theta = 1/4$, the scheme is stable if $k/h^2 \leq 1$, and this is precisely the case on Figure 7.5.

Let us now show a consistency property peculiar to the *Crank–Nicolson* scheme. The other schemes have the same accuracy characteristics as the former schemes.

Proposition 7.24 *If the solution of the continuous problem (7.37)–(7.38) is sufficiently smooth (C^3 in time and C^4 in space), the Crank-Nicolson scheme ($\theta = 1/2$) is consistent of order 2 in time and space.*

PROOF The proof follows from a Taylor expansion of the exact solution φ of (7.37)–(7.38) at point $x_i = ih$ in space and $t^n + k/2$ ($t^n = nk$) in time.

For simplicity's sake, let us denote by $\tilde{\varphi}(x_i, .)$ the function of the sole time variable defined by the following difference quotient:

$$\tilde{\varphi}(x_i, t) = \frac{\varphi(x_{i+1}, t) - 2\varphi(x_i, t) + \varphi(x_{i-1}, t)}{h^2}.$$

According to Proposition 7.1, we already know that if φ is of class C^4 in space, we have for any time t:

$$\tilde{\varphi}(x_i, t) = \frac{\partial^2 \varphi}{\partial x^2}(x_i, t) + 0(h^2). \tag{7.70}$$

The consistency error $\bar{\varepsilon} = \bar{L}\bar{\varphi}$, where $\bar{\varphi}$ is the projection of the exact solution φ at each of the nodes of the mesh in time and space, is a doubly indexed sequence $\bar{\varepsilon} = (\varepsilon_i^n)_{i\in\mathbb{Z}, n\in\{1,\ldots,M\}}$ that can be defined by

$$\begin{aligned} \varepsilon_i^{n+1} = & \left[\frac{\varphi(x_i, t^{n+1}) - \varphi(x_i, t^n)}{k} - \frac{\partial \varphi}{\partial t}(x_i, t^{n+1/2})\right] \\ & - \left[\frac{1}{2}\left(\tilde{\varphi}(x_i, t^{n+1}) + \tilde{\varphi}(x_i, t^n)\right) - \frac{\partial^2 \varphi}{\partial x^2}(x_i, t^{n+1/2})\right], \end{aligned}$$

since $(L\varphi)(., t^{n+1/2}) = 0$. The consistency error is thus the sum of two errors $\varepsilon_i^{n+1} = (\varepsilon_t)_i^{n+1} + (\varepsilon_x)_i^{n+1}$, one

$$(\varepsilon_t)_i^{n+1} = \frac{\varphi(x_i, t^{n+1}) - \varphi(x_i, t^n)}{k} - \frac{\partial \varphi}{\partial t}(x_i, t^{n+1/2}),$$

linked to the discretisation of the time derivative operator, whereas the other

$$(\varepsilon_x)_i^{n+1} = -\frac{1}{2}\left[\tilde{\varphi}(x_i, t^{n+1}) + \tilde{\varphi}(x_i, t^n)\right] + \frac{\partial^2 \varphi}{\partial x^2}(x_i, t^{n+1/2}),$$

is linked to the discretisation of the spatial operator.

Let u now be any function of the t variable, assumed of class C^3; we have the Taylor expansions:

$$u(t+k) = u\left(t+\frac{k}{2}\right) + \frac{k}{2}\frac{\partial u}{\partial t}\left(t+\frac{k}{2}\right) + \frac{k^2}{8}\frac{\partial^2 u}{\partial t^2}\left(t+\frac{k}{2}\right) + O(k^3), \tag{7.71}$$

$$u(t) = u\left(t+\frac{k}{2}\right) - \frac{k}{2}\frac{\partial u}{\partial t}\left(t+\frac{k}{2}\right) + \frac{k^2}{8}\frac{\partial^2 u}{\partial t^2}\left(t+\frac{k}{2}\right) + O(k^3).$$

If we apply these expansions to $u = \varphi(x_i, .)$, we obtain by subtracting that:

$$(\varepsilon_t)_i^{n+1} = O(k^2).$$

According to (7.70), the error ε_x can be written as

$$(\varepsilon_x)_i^{n+1} = -\frac{1}{2}\left[\frac{\partial^2\varphi}{\partial x^2}(x_i, t^{n+1}) + \frac{\partial^2\varphi}{\partial x^2}(x_i, t^n)\right] + \frac{\partial^2\varphi}{\partial x^2}\left(x_i, t^{n+1/2}\right) + O(h^2).$$

Then it suffices to apply (7.71) to the function $u = \frac{\partial^2\varphi}{\partial x^2}(x_i, .)$ to deduce that

$$(\varepsilon_t)_i^{n+1} = O(k^2) + O(h^2),$$

and this ends the proof. □

Remark 7.25 If equation (7.37) has a source term f that only depends on time, this term only occurs in the consistency error, and not in stability (cf. Remark 7.11). In order to keep second order accuracy in time, we must consider the scheme

$$\frac{\psi_i^{n+1} - \psi_i^n}{k} - D_i^+ D_i^- (\theta\Psi^{n+1} + (1-\theta)\psi^n) = \frac{f(t^{n+1}) + f(t^n)}{2}, \tag{7.72}$$

because

$$\frac{f(t^{n+1}) + f(t^n)}{2} = f(t^n + \frac{k}{2}) + O(k^2),$$

and this is again true by (7.71). □

7.3.3 *A Three-level Scheme*

Let us go back to the simple example of the explicit scheme. The question we ask is the following: could we not improve the discretisation in time by taking a second order approximation of the operator $\frac{\partial}{\partial t}$?

A rather natural choice consists, for example, in considering the *Richardson scheme*:

$$\forall n \geq 1, \quad \frac{\psi_i^{n+1} - \psi_i^{n-1}}{2k} - \frac{\psi_{i+1}^n - 2\psi_i^n + \psi_{i-1}^n}{h^2} = 0. \tag{7.73}$$

As above, we require at the initial time: $\psi_i^0 = \varphi(x_i), \forall i$. It is clear that we cannot use (7.73) to define the approximate solution at the next step; we shall then use a first order scheme, of the explicit type, to compute Ψ^1.

Proposition 7.26 *Richardson's scheme is, modulo enough smoothness for the solution of (7.37)–(7.38), second order in time and space; unfortunately, it is always unstable.*

PROOF We leave it as an exercise to the reader to compute the consistency error of this scheme. Let us study the von Neumann stability. Again using the Fourier transform, we have:

$$\hat{\psi}^{n+1}(\xi) = \mathrm{d}(\xi)\hat{\psi}^n(\xi) + \hat{\psi}^{n-1}(\xi), \tag{7.74}$$

with

$$\mathrm{d}(\xi) = -8\frac{k}{h^2}\sin^2\frac{h\xi}{2}.$$

Equation (7.74) can be written in matrix form

$$\hat{X}^{n+1}(\xi) = A(\xi)\hat{X}^n(\xi), \tag{7.75}$$

where we denote by X^n the vector with entries $(\psi^{n+1}, \psi^n)^T$, and A the *amplification matrix:*

$$A(\xi) = \begin{bmatrix} \mathrm{d}(\xi) & 1 \\ 1 & 0 \end{bmatrix}. \tag{7.76}$$

Since A is symmetric, its eigenvalues are real. They are solutions of the characteristic equation $\lambda^2 - \lambda \mathrm{d}(\xi) - 1 = 0$. But the product of the roots is -1, so there must exist values of ξ for which one of the eigenvalues has absolute value strictly larger than 1 (they cannot both have absolute value 1, because for $\lambda = 1$ or -1, the above characteristic polynomial is not identically zero). We deduce that there are ξ values for which the spectral radius of $A(\xi)$ is strictly larger than 1, and since this matrix is normal, this contradicts the stability condition of the following lemma (we cannot find a positive bounded constant C such that (7.77) holds): Richardson's scheme is thus always unstable. □

For systems, we have the stability result summed up in the following way:

Lemma 7.27 *A necessary condition for a scheme whose Fourier transform is written in the form (7.75) to be stable is that there exist two positive constants C and k^* such that:*

$$\forall \xi \in \mathbb{R}, \forall k \in]0, k^*[, \quad \rho(A(\xi)) \leq 1 + Ck. \tag{7.77}$$

If the matrix $A(\xi)$ is normal for all ξ, this von Neumann condition is a sufficient stability condition.

PROOF We sketch the proof which is analogous to that of Proposition 7.16. Using Lemma 7.17 in the case $d = 2$ and Plancherel's Theorem, we show, thanks to the following relations

$$[\rho(A(\xi))]^n = \rho[(A(\xi))^n] \leq ||(A(\xi))^n||_2,$$

that (7.77) is necessary for a scheme whose Fourier transform is written in the form (7.75) to be stable. Conversely, if the matrix $A(\xi)$ is normal, we have

$$||(A(\xi))^n||_2 \leq ||A(\xi)||_2^n = [\rho(A(\xi))]^n;$$

if, moreover, the condition (7.77) is satisfied, we have, for all integers n such that $nk \leq T$,

$$||(A(\xi))^n||_2 \leq (1 + Ck)^n \leq e^{CT},$$

which proves that the scheme is stable. □

Remark 7.28 As we already said about the scalar case in Remark 7.18, the condition (7.77) on the spectral radius of the matrix A is usually replaced by the more restrictive condition

$$\forall \xi \in \mathbb{R}, \quad \rho(A(\xi)) \leq 1, \tag{7.78}$$

called the *strict von Neumann condition.* □

For schemes whose Fourier transform is of the form (7.75), with a not necessarily normal matrix A, it is possible to give sufficient stability conditions; this is the topic of the following exercise.

EXERCISE Extend the above analysis to non normal matrices, starting with the case of diagonalisable matrices i.e.

$$A(\xi) = P(\xi)D(\xi)P^{-1}(\xi),$$

where D is a diagonal matrix, and P is an invertible matrix. Show that if there exist a positive constant C such that

$$\sup_{\xi \in \mathbb{R}} \left(||P(\xi)||_2 ||P^{-1}(\xi)||_2 \right) \leq C, \tag{7.79}$$

then condition (7.78) is a sufficient stability condition.

In the general case, the matrix $A(\xi)$ can only be put in triangular form, i.e. it can be written in the form $A(\xi) = P(\xi)T(\xi)P^{-1}(\xi)$, with T a triangular matrix. Show then, under condition (7.79), that the more restrictive relation

$$\forall \xi \in \mathbb{R}, \quad \rho(A(\xi)) < 1, \tag{7.80}$$

is a *sufficient* von Neumann stability condition. It is clear that, condition (7.79) is rarely checked in practice, and that we are content with studying the matrix spectral radius.

Apply this result to study the stability of the implicit scheme defined by:

$$\forall n \geq 2, \quad \frac{\frac{3}{2}\psi_i^{n+1} - 2\psi_i^n + \frac{1}{2}\psi_i^{n-1}}{k} - \frac{\psi_{i+1}^{n+1} - 2\psi_i^{n+1} + \psi_{i-1}^{n+1}}{h^2} = 0.$$

Show that this scheme, called *Gears's scheme,* is second order in time and space and that it is unconditionally stable.

Study the stability of the following *Du Fort–Frankel scheme:*

$$\forall n \geq 2, \quad \frac{\psi_i^{n+1} - \psi_i^{n-1}}{2k} - \frac{\psi_{i-1}^n - \psi_i^{n-1} - \psi_i^{n+1} + \psi_{i+1}^n}{h^2} = 0.$$

Show that this explicit scheme is only consistent if $k/h \to 0$ when h and k go to zero. Conclude. □

One can find in Richtmyer and Morton (1967) other sufficient stability conditions; we state, without proof, the two main results, starting with the case of diagonalisable matrices. Let us recall that, given a matrix P whose columns are the components of a set of vectors, the Gram determinant of this set, denoted by Δ^2, is the determinant of the matrix P^*P; a necessary and sufficient condition for the vectors in this set to be linearly independent is that this Gram determinant be strictly positive.

Let us again consider schemes whose Fourier transform is written in the form (7.75) with a not necessarily normal matrix A; we have:

Theorem 7.29 If the matrix A is diagonalisable and if there exist a constant $\delta > 0$ such that

$$\forall \delta t \in]0, (\delta t)^*[, \ \forall \xi, \quad \Delta(\xi) \geq \delta > 0,$$

where $\Delta^2(\xi)$ is the Gram determinant of the normalized eigenvectors of the matrix $A(\xi)$, then the von Neumann condition is a necessary and sufficient stability condition.

Theorem 7.30 If the elements of the matrix A are bounded for all $\delta t \in]0, (\delta t)^*[$ and for all ξ, and if all the eigenvalues $\lambda_i(\xi), i \in \{1, ..., N\}$ of $A(\xi)$, except possibly one of them, are inside the unit disk, i.e.

$$\forall \delta t \in]0, (\delta t)^*[, \ \forall \xi, \quad |\lambda_1(\xi)| \leq 1,$$
$$\forall i \in \{2, ..., N\}, \forall \delta t \in]0, (\delta t)^*[, \ \forall \xi, \quad |\lambda_i(\xi)| \leq \gamma < 1,$$

then the scheme is stable.

7.3.4 *Taking Boundary Conditions into Account*

7.3.4.1 Statement of the problem

In this section we shall be concerned with discretising the heat equation in a single space variable belonging to a bounded interval $\Omega =]0, L[$ in $\mathbb{R}$, with Dirichlet boundary conditions:

$$\frac{\partial \varphi}{\partial t} - \frac{\partial^2 \varphi}{\partial x^2} = 0, \quad x \in \Omega =]0, L[, \quad 0 < t \leq T, \tag{7.81}$$

$$\varphi(x, 0) = \varphi^0(x), \tag{7.82}$$

$$\varphi(x, t) = g(x, t), \quad \text{for } x = 0 \text{ and } x = L. \tag{7.83}$$

We shall assume that the initial condition φ^0 satisfies the boundary conditions (7.83) at $t = 0$, i.e. $\varphi^0(x) = g(x, 0)$ at the points $x = 0$ and $x = L$.

The difference with what we did previously lies in the way we treat the boundary conditions (7.83). The space step is now equal to $h = \delta x = L/(N+1)$, and that means that at any time t^n, there are N unknowns that are the values ψ_i^n of the approximate solution at the internal vertices $x_i = ih$, $i \in \{1, \ldots, N\}$ of the mesh; these values solve a 'discretised' version, as described above, of equation (7.81). At both ends of the intervals, the solution is required to satisfy the boundary conditions (7.83), i.e.

$$\psi_i^n = g(x_i, t^n), \text{ for } i = 0 \text{ and } i = N + 1. \tag{7.84}$$

Two problems now arise:
1:how to analyze the consistency and the stability of the scheme, so as to ensure its convergence, as was the case in the whole space;
2: how to solve the discrete problem from a practical point of view.

We shall answer both questions successively for concrete schemes. As far as the first question is concerned, since the discrete solution satisfies the continuous boundary conditions at each node of the space–time mesh, the consistency error of the global scheme is that of the scheme used to discretise (7.81). It only remains to analyse its stability: this is what we shall now undertake.
The second problem will be studied in Section 7.3.5.

7.3.4.2 Stability by energy inequalities

Because we are in a bounded spatial domain, we can no longer use the Fourier transform to study stability. We shall first study the stability of the continuous problem, whence we shall deduce a method for the discrete problem. We assume, for simplicity's sake, that the Dirichlet condition is homogeneous, i.e. $g = 0$.

At the continuous level, let us multiply equation (7.81) by φ, integrate the resulting equation over Ω, and integrate by parts; we obtain

$$\frac{1}{2}\frac{\mathrm{d}}{\mathrm{d}t}\left(\int_\Omega \varphi^2(x,t)\mathrm{d}x\right) + \int_\Omega \left(\frac{\partial \varphi}{\partial x}\right)^2 (x,t)\mathrm{d}x = 0,$$

which shows, since the second term is non-negative, that:

$$\frac{1}{2}\frac{\mathrm{d}}{\mathrm{d}t}\left(\int_\Omega \varphi^2(x,t)\mathrm{d}x\right) \leq 0.$$

In other words, the norm in $L^2(\mathbb{R})$ of $\varphi(.,t)$ decreases when time increases, because we have:

$$\forall t \geq 0, \quad ||\varphi(.,t)||_{L^2(\mathbb{R})} \leq ||\varphi^0||_{L^2(\mathbb{R})}. \tag{7.85}$$

This inequality, called the *energy inequality*, shows the 'stability in time' of the solution of the continuous problem (it also shows its uniqueness!). Let us note that inequality (7.85) is also valid for an homogeneous Neumann problem.

Remark 7.31 Let us point out that in the case of an non-homogeneous Dirichlet boundary condition, we obtain the same conclusion, after extending (as we did for finite elements) the boundary condition, so as to reduce to a homogeneous problem. □

We shall show on the example of the Crank–Nicolson scheme, how to copy closely this argument at the discrete level, so as to prove an energy inequality of the type

$$\forall n \geq 0,\ n\delta t \leq T, \quad ||\Psi^n||_{l^2} \leq ||\Psi^0||_{l^2}, \tag{7.86}$$

where we have set:

$$\Psi^n = (\psi_0^n, ..., \psi_{N+1}^n), \quad ||\Psi^n||_{l^2} = \sum_{i=0}^{N+1} (\psi_i^n)^2. \tag{7.87}$$

It is indeed clear, according to (7.35) and (7.36), that (7.86) proves the stability of the scheme for the l^2 norm in space.

The scheme we shall study is defined by (7.65) with $\theta = 1/2$ and (7.84), i.e.

$$\frac{\psi_i^{n+1} - \psi_i^n}{k} - \frac{1}{2}\mathcal{D}_i^+\mathcal{D}_i^-\left(\Psi^{n+1} + \Psi^n\right) = 0, \quad i \in \{1, ..., N\}, \tag{7.88}$$

$$\psi_i^n = 0, \text{ for } i = 0 \text{ and } i = N+1. \tag{7.89}$$

Proposition 7.32 *The Crank–Nicolson scheme (7.88)–(7.89) is unconditionally stable for the l^2 norm in space.*

As in the continuous case, the proof rests on an 'discrete integration by parts formula' that we shall prove first.

Lemma 7.33 Let $U = (u_i)_{i\in\{0,...,N\}}$ and $V = (v_i)_{i\in\{0,...,N\}}$ be two sequences indexed by $\{0, ..., N\}$; then:

$$\sum_{i=1}^{N} \mathcal{D}_i^+ U\, v_i = -\sum_{i=1}^{N+1} u_i\, \mathcal{D}_i^- V + \frac{1}{h}\,(u_{N+1}v_{N+1} - u_1 v_0). \tag{7.90}$$

In particular, if $v_0 = v_{N+1} = 0$, we have:

$$\sum_{i=1}^{N} \mathcal{D}_i^+ U\, v_i = -\sum_{i=1}^{N+1} u_i\, \mathcal{D}_i^- V. \tag{7.91}$$

PROOF Let us expand the left hand side of (7.90); we obtain successively

$$\begin{aligned}
\sum_{i=1}^{N} \mathcal{D}_i^+ U v_i &= \frac{1}{h}\sum_{i=1}^{N}(u_{i+1} - u_i)v_i = \frac{1}{h}\left(\sum_{i=1}^{N} u_{i+1}v_i - \sum_{i=1}^{N} u_i v_i\right) \\
&= \frac{1}{h}\left(\sum_{i=2}^{N+1} u_i v_{i-1} - \sum_{i=1}^{N} u_i v_i\right) = \frac{1}{h}\left[\sum_{i=1}^{N+1} u_i(v_{i-1} - v_i) - u_1 v_0 + u_{N+1}v_{N+1}\right] \\
&= -\sum_{i=1}^{N+1} u_i \mathcal{D}_i^- V + \frac{1}{h}(-u_1 v_0 + u_{N+1}v_{N+1}),
\end{aligned}$$

which proves (7.90) and ends the proof of the Lemma, (7.91) being a particular case of the above equality. □

PROOF OF PROPOSITION 7.32 Let us multiply equation (7.88) by $\psi_i^{n+1} + \psi_i^n$ and sum over all indices $i \in \{1, ..., N\}$; we obtain:

$$\sum_{i=1}^{N} \frac{(\psi_i^{n+1})^2 - (\psi_i^n)^2}{k} - \frac{1}{2}\sum_{i=1}^{N} \mathcal{D}_i^+\big(\mathcal{D}_i^-(\Psi^{n+1} + \Psi^n)\big)\,(\psi_i^{n+1} + \psi_i^n) = 0. \tag{7.92}$$

By virtue of the boundary conditions (7.89) satisfied by ψ^n and Ψ^{n+1}, we can apply the relation (7.91) to the sequences $V = \Psi^{n+1} + \Psi^n$ and $U = (u_i)_i$, $u_i = \mathcal{D}_i^-(\Psi^{n+1} + \Psi^n)$, which gives us:

$$\sum_{i=1}^{N} \mathcal{D}_i^+\big(\mathcal{D}_i^-(\Psi^{n+1} + \Psi^n)\big)\,(\psi_i^{n+1} + \psi_i^n) = -\sum_{i=1}^{N+1}\big(\mathcal{D}_i^-(\Psi^{n+1} + \Psi^n)\big)^2 \le 0.$$

We thus deduce from (7.92) that

$$\sum_{i=1}^{N} \frac{(\psi_i^{n+1})^2 - (\psi_i^n)^2}{k} \le 0,$$

or

$$\sum_{i=1}^{N}(\psi_i^{n+1})^2 \leq \sum_{i=1}^{N}(\psi_i^n)^2.$$

According to (7.89), we have, using notation as in (7.87):

$$\forall n \geq 0, \quad ||\Psi^{n+1}||_{l^2} \leq ||\Psi^n||_{l^2},$$

which gives, by an easy induction, a slightly stronger result than (7.86), since it allows the case $T = +\infty$; the l^2 stability of the scheme is thus proven. □

We shall now show in detail, using an example, the practical solution of the scheme.

7.3.5 *Practical Implementation of the Implicit Scheme*

The discrete problem associated with problem (7.81)–(7.83) is particularly simple to solve if the scheme is explicit. For the implicit case, and to make the right hand side simpler, we shall consider the example of the fully implicit scheme defined by the discrete operator (7.59). The scheme we study can be written more generally ($f = 0$ in our example):

$$\frac{\psi_i^{n+1} - \psi_i^n}{k} - \frac{\psi_{i+1}^{n+1} - 2\psi_i^{n+1} + \psi_{i-1}^{n+1}}{h^2} = f_i^{n+1}, \quad i \in \{1, ..., N\}, \tag{7.93}$$

$$\psi_i^{n+1} = g(x_i, t^{n+1}), \text{ for } i = 0 \text{ and } i = N+1, \tag{7.94}$$

which means that $X = (\psi_1^{n+1}, ..., \psi_N^{n+1})^T$ solves the linear system

$$AX = b, \tag{7.95}$$

where the matrix A is defined by

$$A = \begin{bmatrix} 2c+1 & -c & \dots & 0 \\ -c & 2c+1 & -c & 0 \\ 0 & -c & 2c+1 & \dots \\ & \dots & \dots & \\ \dots & 0 & -c & 2c+1 \end{bmatrix}, \quad c = \frac{k}{h^2} > 0, \tag{7.96}$$

and the right hand side b is given by:

$$b = \begin{bmatrix} kf_1^{n+1} + \psi_1^n + cg(x_0, t^{n+1}) \\ kf_2^{n+1} + \psi_2^n \\ \dots \\ kf_{N-1}^{n+1} + \psi_{N-1}^n \\ kf_N^{n+1} + \psi_N^n + cg(x_{N+1}, t^{n+1}) \end{bmatrix}. \tag{7.97}$$

We can easily show, as we did in Section 7.1.3, that the matrix A is symmetric and positive definite, which proves that the linear system (7.95), and thus the scheme (7.93)–(7.94) has one and only one solution.

We have at our disposal numerous methods enabling us to actually solve system (7.95); we shall give details for that based on the LU factorisation of the matrix, which is particularly simple here, as A is tridiagonal.

We shall factorise A as a product of two matrices, a lower triangular one, L, and an upper triangular one, U, with unit diagonal; because of the sparse structure of A, these two matrices L and U have only two non-zero diagonals. We thus have:

$$A = LU, \quad L = \begin{bmatrix} d_1 & & & 0 \\ l_1 & d_2 & & \\ & l_2 & d_3 & \\ & & \cdots & \\ 0 & & l_{N-1} & d_N \end{bmatrix}, \quad U = \begin{bmatrix} 1 & u_1 & & 0 \\ & 1 & u_2 & \\ & & \cdots & \\ & & 1 & u_{N-1} \\ 0 & & & 1 \end{bmatrix}. \tag{7.98}$$

The non-zero elements of these matrices are given by:

$$L_{i,i-1} = l_{i-1}, \quad L_{ii} = d_i, \quad U_{i,i+1} = u_i, \quad U_{ii} = 1.$$

With this notation, we have the following relations:

$$A_{ii} = 2c + 1 = (LU)_{ii} = L_{i,i-1}U_{i-1,i} + L_{ii}U_{ii} = u_{i-1}l_{i-1} + d_i, \tag{7.99}$$

$$A_{i,i-1} = -c \quad = (LU)_{i,i-1} = L_{i,i-1}U_{i-1,i-1} = l_{i-1}, \tag{7.100}$$

$$A_{i,i+1} = -c \quad = (LU)_{i,i+1} = L_{i,i}U_{i,i+1} = d_i u_i. \tag{7.101}$$

Once the matrices L and U have been computed, it is easy to solve system (7.95) in two consecutive steps, the first step, called 'forward solve', where we solve the lower triangular system $Lz = b$, then the second step, called 'backward solve', where we solve the upper triangular system $UX = z$. The algorithm is now the following, if we denote by X_i the entries of X and by b_i those of the right hand side b:

LU factorisation algorithm (for solving (7.95))

0 Set $d_1 = 2c + 1, \quad z_1 = b_1/d_1, \quad u_1 = -c/d_1$

1 Loop over $i = 2, \ldots, N$ (factorisation and forward solve)

$l_{i-1} = -c$

$d_i = 2c + 1 - u_{i-1}l_{i-1}$

$u_i = -c/d_i$

$z_i = (b_i - l_{i-1}z_{i-1})/d_i$

2 Initialisation of the backward solve by setting $X_N = z_N$

3 Loop over $j = N-1, N-2, \ldots, 1$ (backward solve)

$X_i = z_i - u_i X_{i+1}$.

We give in an appendix to this chapter the Fortran program to solve the heat equation (7.81)–(7.83) on $]0,1[$, in the homogeneous case (i.e. for $g = 0$), using two

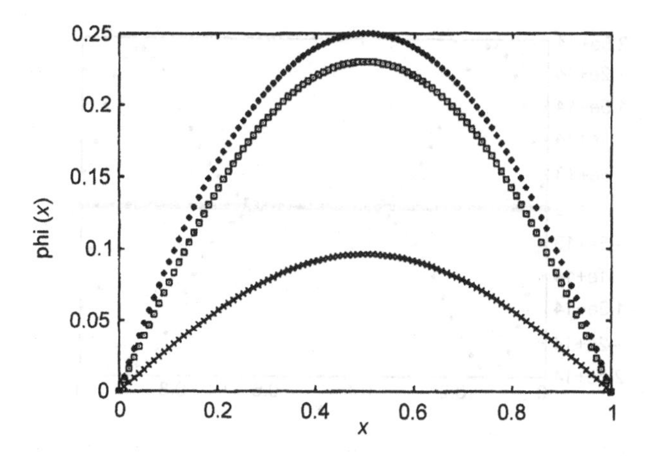

Figure 7.7 Computed solution at different time steps

different schemes: the explicit scheme, and the implicit scheme using the *LU* method as described above. The initial condition is $\varphi^0(x) = x(1 - x)$.

Figure 7.7 plots the solution computed by the explicit scheme at different time steps: the initial time, on the curve with diamond symbols, then the time steps $T = 0.01$ and $T = 0.1$, shown on the curves with square and cross symbols respectively. The space mesh has 100 nodes, the critical time step k_c for the explicit scheme is $k = 5.1 \times 10^{-5}$, and we have taken $k = 10^{-6}$. For clarity's sake, we have not plotted on this figure the results obtained using the implicit scheme; let us just note that, for this latter scheme, we have obtained exactly identical results with only $k = 10^{-3}$!

We have compared, in Figures 7.8 and 7.9, the solutions at time $T = 0.01$ obtained with, either the explicit scheme (curve with diamond symbols), or the implicit scheme (curve with square symbols); the time step is the same for both schemes. Figure 7.8 corresponds to a time step $k = 5 \times 10^{-5} < k_c$, whereas

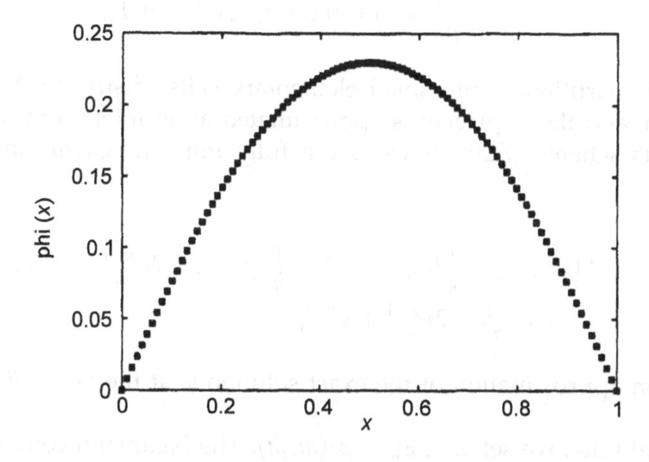

Figure 7.8 Comparison of two schemes: $k = 5 \times 10^{-5}$

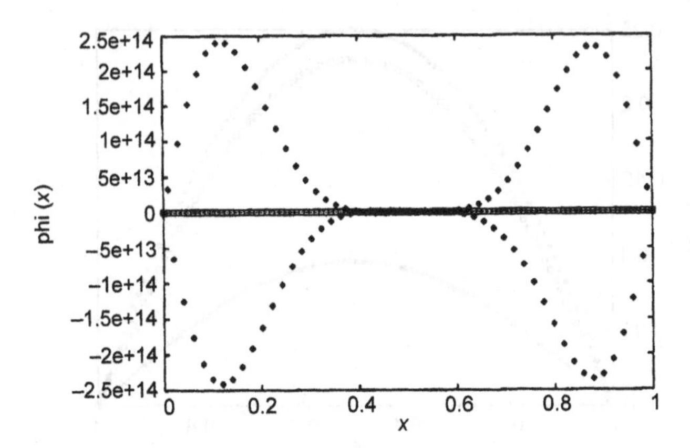

Figure 7.9 Comparison of two schemes: $k = 6 \times 10^{-5}$

$k = 6 \times 10^{-5} > k_c$ in Figure 7.9: the stability results are quite striking, as the explicit scheme 'explodes' in the second case (take note that the scales on both figure are different).

7.3.6 *Two-dimensional Case*

Let us consider, for example, the heat equation in a square $]0, L[^2$ of $\mathbb{R}^2$, with a positive viscosity coefficient μ and homogeneous Neumann boundary conditions:

$$\frac{\partial \varphi}{\partial t} - \mu(\frac{\partial^2 \varphi}{\partial x_1^2} + \frac{\partial^2 \varphi}{\partial x_2^2}) = f \text{ in }]0, L[^2 \times]0, T[, \tag{7.102}$$

$$\varphi(x, 0) = \varphi^0(x) \text{ in }]0, L[^2, \tag{7.103}$$

$$\frac{\partial \varphi}{\partial n}(x, t) = 0 \text{ on } \partial(]0, L[^2) \times]0, T[. \tag{7.104}$$

The domain is partitioned into small elementary cells of size $h = 1/(N+1)$ in each direction, and the Laplacian is approximated at an internal mesh point by the five points scheme (7.20). If we use a fully implicit scheme in time, this gives us:

$$\begin{aligned}\forall i, j \in \{1, ..., N\}, \quad & \frac{1}{k}[\psi_{ij}^{n+1} - \psi_{ij}^n] - \frac{\mu}{h^2}[\psi_{i+1,j}^{n+1} - 2\psi_{i,j}^{n+1} + \psi_{i-1,j}^{n+1} \\ & + \psi_{i,j+1}^{n+1} - 2\psi_{i,j}^{n+1} + \psi_{i,j-1}^{n+1}] = f_{ij}^{n+1},\end{aligned} \tag{7.105}$$

where ψ_{ij}^n is an approximation of the exact solution φ at the node (ih, jh) and at time $t = nk$.

At the initial time, we set $\forall i, j$, $\psi_{ij}^0 = \varphi^0(ih, jh)$. The Neumann condition on the boundary of the domain are discretised in the following way:

$$\forall j \in \{1,...,N\}, \quad \psi_{0j}^n = \psi_{1j}^n, \quad \psi_{N-1,j}^n = \psi_{N,j}^n \qquad (7.106)$$
$$\forall i \in \{1,...,N\}, \quad \psi_{i0}^n = \psi_{i1}^n, \quad \psi_{i,N-1}^n = \psi_{iN}^n.$$

The N^2 equations included in (7.105) can be written in matrix form

$$AX = b, \qquad (7.107)$$

where the matrix A is pentadiagonal, block tridiagonal, with each of the blocks an $N \times N$ square matrix;

$$A = \begin{bmatrix} A_1 & B & 0 & \cdots & \cdots & 0 \\ B & A_2 & B & 0 & \cdots & 0 \\ 0 & B & A_2 & B & 0 & 0 \\ \cdots & \cdots & \cdots & \cdots & \cdots & \cdots \\ 0 & \cdots & 0 & B & A_2 & B \\ 0 & \cdots & \cdots & 0 & B & A_1 \end{bmatrix}, \quad B = \begin{bmatrix} b & 0 & \cdots & 0 \\ 0 & b & 0 & \cdots \\ \cdots & \cdots & \cdots & \cdots \\ 0 & \cdots & 0 & b \end{bmatrix}; \qquad (7.108)$$

the diagonal blocks are written

$$A_1 = \begin{bmatrix} a_1 & b & 0 & \cdots & 0 \\ b & a_2 & b & 0 & 0 \\ \cdots & \cdots & \cdots & \cdots & \cdots \\ 0 & 0 & b & a_2 & b \\ 0 & \cdots & 0 & b & a_1 \end{bmatrix}, \quad A_2 = \begin{bmatrix} a_2 & b & 0 & \cdots & 0 \\ b & a_3 & b & 0 & 0 \\ \cdots & \cdots & \cdots & \cdots & \cdots \\ 0 & 0 & b & a_3 & b \\ 0 & \cdots & 0 & b & a_2 \end{bmatrix}, \qquad (7.109)$$

with

$$a_i = \frac{1}{k} + (1+i)\frac{\mu}{h^2}, \quad b = -\frac{\mu}{h^2}. \qquad (7.110)$$

To solve the system (7.107), we could use the Choleski algorithm, or the preconditioned conjugate gradient method. The successive over-relaxation method, although slower, is simpler to program and does not require storing the matrix; we shall briefly recall its principle.

The Gauss–Seidel algorithm for solving a linear system $Ax = f$ of size $N \times N$ is an iterative method, where the approximate solution at step $m+1$, denoted by x^{m+1}, is computed from that at the previous step by the relation:

$$x_i^{m+1} = \left[f_i - \sum_{j<i} A_{ij} x_j^{m+1} - \sum_{j>i} A_{ij} x_j^m \right] / A_{ii}. \qquad (7.111)$$

We note that in this computation, only the entries of the vector x^m with $j > i$ are used, and none of the previous entries, so that it is not necessary to store two vectors x^m and x^{m+1}: only one vector x is sufficient. As the computation proceeds, the entries of this vector, that are those of x^m at the outset, are replaced by those of the vector x^{m+1}. We can thus suppress the index m. Last, the scheme can be improved by introducing a positive relaxation parameter ω, so that the successive over-relaxation **algorithm** can schematically be written as:

loop over m

loop over i

$$x_i^* \leftarrow (f_i - \sum_{j \neq i} A_{ij} x_j)/(A_{ii}), \tag{7.112}$$

$$x_i \leftarrow x_i + \omega(x_i^* - x_i); \tag{7.113}$$

end of the loops

A reasonable choice for the relaxation parameter, in the case of the heat equation, is to take ω of the order of 1.7.

7.3.7 Explicit Runge-Kutta Schemes

Among the possible choices for explicit schemes, we point out the Runge–Kutta schemes, frequently used in the approximation of ordinary differential equations. For example, the second order method for solving $\mathrm{d}\varphi/\mathrm{d}t(t) = F(\varphi(t), t)$ can be written as

$$\frac{1}{k}[\psi^{n+1} - \psi^n] = F(\psi^{n+1/2}, (n+1/2)k), \quad \psi^{n+1/2} = \psi^n + \frac{k}{2}F(\psi^n, nk),$$

which is justified by noting that:

$$\varphi(t^{n+1/2}) = \varphi(t^n) + \frac{k}{2}F(\varphi(t^n), t^n) + O(k^2).$$

This scheme is second order, i.e. the consistency error is in $O(k^2)$.

We can apply this type of scheme to the solution of the heat equation (7.37)–(7.38), because if we discretise this equation only in the space variables, for example by writing the scheme (with obvious notation):

$$\frac{\partial \psi_i}{\partial t}(t) - \mathcal{D}_i^+ \mathcal{D}_i^- \psi(t) = 0, \tag{7.114}$$

we obtain a system of ordinary differential equations only depending on the t variable. The second order Runge-Kutta scheme (7.114) can then be written:

$$\frac{1}{k}[\psi_i^{n+1} - \psi_i^n] = \mathcal{D}_i^+ \mathcal{D}_i^- (\Psi^{n+1/2}), \quad \psi_i^{n+1/2} = \psi_i^n + \frac{k}{2}\mathcal{D}_i^+ \mathcal{D}_i^- (\Psi^n),$$

still denoting by Ψ^n the vector with entries ψ_i^n. Let us analyse the von Neumann stability of this scheme. By a Fourier transform, we obtain, setting

$\lambda = (2k/h^2)\sin^2(h\xi/2) \geq 0$, and with ξ being the dual variable to the space variable x:

$$\hat{\psi}^{n+1}(\xi) = \hat{\psi}^n(\xi) - 2\lambda\hat{\psi}^{n+\frac{1}{2}}(\xi), \quad \hat{\psi}^{n+\frac{1}{2}}(\xi) = (1-\lambda)\hat{\psi}^n(\xi),$$

which gives eventually:

$$\hat{\psi}^{n+1}(\xi) = \hat{\psi}^n(\xi)\,[\,1 - 2\lambda(1-\lambda)\,].$$

The above polynomial in λ being always positive, it will be strictly less than 1, for all ξ, if and only if the stability condition $k < h^2/2$ is satisfied.

Under this stability condition, we thus obtain a convergent scheme of second order in time and space. It is of course possible to extend this analysis to higher order Runge–Kutta schemes.

7.4 THE CONVECTION EQUATION

We leave the framework of parabolic equations to take on that of hyperbolic equations, starting with the study of convection equations, that frequently occur in practice. Let Ω be a subset of $\mathbb{R}^d$ and let $u = (u_1, u_2, ..., u_d)$ be a given vector field defined on Ω; we seek a function φ satisfying the partial differential equation

$$\frac{\partial \varphi}{\partial t} + u \cdot \nabla\varphi = f \text{ in } \Omega\times]0,T[, \tag{7.115}$$

with the initial condition

$$\varphi(x,0) = \varphi^0(x), \quad \forall x \in \Omega; \tag{7.116}$$

we recall the notation:

$$u \cdot \nabla\varphi = \sum_{i=1}^{d} u_i \frac{\partial \varphi}{\partial x_i}.$$

What are the possible boundary conditions for such a problem? It is clear that since the spatial derivative is only of first order, we cannot prescribe the function φ on the whole boundary Γ of Ω (to convince oneself of that fact, it suffices, in one dimension, to compute explicitly the exact solution).

For simplicity's sake, we shall assume that the velocity field u is independent of the variable t. Let us denote by n the exterior normal to Γ and denote by Γ^- the part of the boundary where the velocity field u is incoming i.e.:

$$\Gamma^- = \{x \in \Gamma : \quad u(x)\cdot n(x) < 0\}. \tag{7.117}$$

Mathematically, the problem is well posed if we take as a boundary condition:

$$\varphi(x,t) = g(x), \quad \forall x \in \Gamma^-, \quad \forall t \in]0,T[. \tag{7.118}$$

EXERCISE It is possible to obtain an analytical solution of problem (7.115)–(7.118) by 'integrating' the equation along the characteristic curves of the vector field u. These curves $(t \to X(t))$ are defined by:

$$\frac{\mathrm{d}X}{\mathrm{d}t}(t) = u(X(t),t), \quad X(t=0) = x. \tag{7.119}$$

Show that, if we set $\Phi(t) = \varphi(X(t),t)$, we have:

$$\Phi'(t) = \left(\frac{\partial \varphi}{\partial t} + (u \cdot \nabla)\varphi\right)(X(t),t). \tag{7.120}$$

The function Φ is thus constant; deduce from this the expression of φ. □

We shall now propose several schemes for approximating with finite differences this convection equation, starting by the one-dimensional case, and assuming that the space variable varies over the whole of $\mathbb{R}$.

7.4.1 Lax Scheme

Let us approximate the continuous operator by $D_t^+ + uD_x^-$; we obtain the explicit backward Euler scheme defined by

$$\frac{1}{k}[\psi_i^{n+1} - \psi_i^n] + \frac{u_i}{h}[\psi_i^n - \psi_{i-1}^n] = f_i^n, \tag{7.121}$$

with the usual initial condition ($h = \delta x; x_i = ih$):

$$\forall i, \; \psi_i^0 = \varphi^0(x_i). \tag{7.122}$$

The consistency error of the scheme is in $O(k+h)$. To simplify studying stability, we shall assume that u is constant in time and space. By Fourier transform, we obtain:

$$\hat{\psi}^{n+1}(\xi) - \hat{\psi}^n(\xi) + \frac{ku}{h}\hat{\psi}^n(\xi)[1 - e^{-i\xi h}] = k\hat{f}^n. \tag{7.123}$$

The amplification factor is thus the complex number defined by

$$a(\xi) = 1 - \frac{ku}{h}[1 - e^{-i\xi h}], \tag{7.124}$$

whose modulus square is equal to:

$$\left\{\begin{aligned}|a(\xi)|^2 &= \left[1-\frac{ku}{h}(1-\cos\xi h)\right]^2+\left[\frac{ku}{h}\sin\xi h\right]^2\\ &=1-2\frac{ku}{h}(1-\cos\xi h)+2\left(\frac{ku}{h}\right)^2(1-\cos\xi h)\\ &=1-2\frac{ku}{h}+2\left(\frac{ku}{h}\right)^2+2\frac{ku}{h}\left(1-\frac{ku}{h}\right)\cos\xi h\\ &\le 1 \text{ if } \frac{ku}{h}\le 1 \text{ and } u\ge 0\,.\end{aligned}\right. \tag{7.125}$$

The proposed scheme is thus von Neumann stable under the two conditions:

$$\frac{ku}{h}\le 1 \text{ and } u\ge 0. \tag{7.126}$$

The first condition is the CFL condition (Courant–Friedrich–Lewy). As far as the second condition is concerned, we can note that if $u<0$, scheme (7.121) is *always unstable*; to have stability in that case, we must use a downwind scheme, that is:

$$\frac{1}{k}[\psi_i^{n+1}-\psi_i^n]+\frac{u_i}{h}[\psi_{i+1}^n-\psi_i^n]=f_i^n. \tag{7.127}$$

One can show in the same way that, if u is constant in time and space, this latter scheme is stable under the condition:

$$\frac{k|u|}{h}\le 1 \text{ and } u\le 0. \tag{7.128}$$

EXERCISE Show that the centred scheme

$$\frac{1}{k}\left[\psi_i^{n+1}-\psi_i^n\right]+\frac{u_i}{2h}\left[\psi_{i+1}^n-\psi_{i-1}^n\right]=f_i^n, \tag{7.129}$$

which is quite natural, is *always unstable*. □

EXERCISE Study the stability of the leap-frog scheme defined by

$$\frac{1}{2k}[\psi_i^{n+1}-\psi_i^{n-1}]+\frac{u_i}{2h}[\psi_{i+1}^n-\psi_{i-1}^n]=f_i^n, \tag{7.130}$$

by using the sufficient stability condition (7.80) established in the exercise at the end of Section 7.3.3. □

Remark 7.34 Let us note that the CFL condition we found to be much less restrictive for the time step than the stability condition we had found for the heat equation. This is of course because the problem is here of first order in space, whereas it was of second order for the heat equation. □

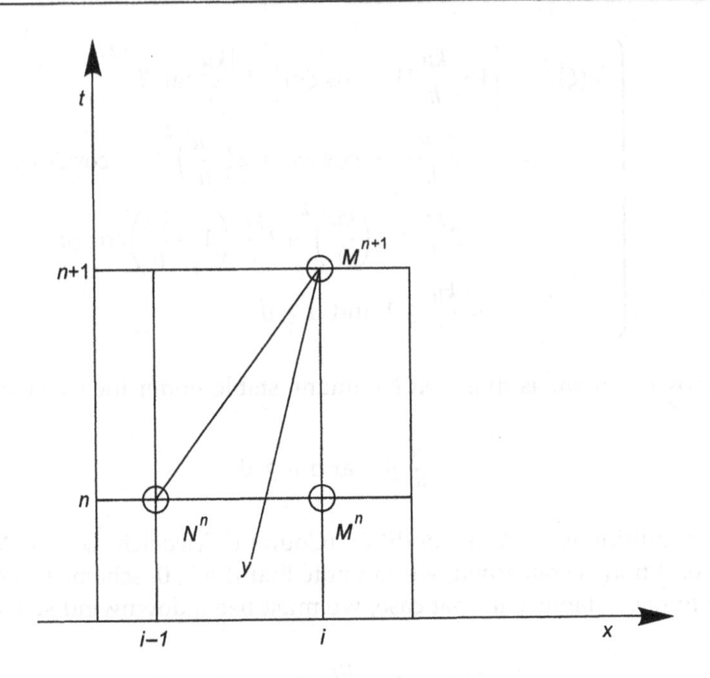

Figure 7.10 Numerical domain of dependence and CFL

Remark 7.35 The CFL condition obtained for the scheme (7.121) can be interpreted graphically in the (x, t) space, as in Figure 7.10. The half-line $M^{n+1}y$ from point M^{n+1} with slope $1/u$ must intersect the line $t = t^n$ at a point of segment N^nM^n, in other words the 'numerical domain of dependence', which is the surface S bounded by the triangle with vertices N^n, M^n and M^{n+1} must contain the 'exact domain of dependence'i.e. the half-line $M^{n+1}y$. We also see that if u is negative, S does not contain this half-line: there is instability. □

7.4.2 *Lax–Wendroff Scheme*

The above scheme is only first order in time and space; we shall construct, using a Taylor expansion, a scheme accurate to second order. We still assume u is constant and we shall set $f = 0$ for simplicity's sake; we have, denoting the exact solution of (7.115)–(7.118) by φ,

$$\varphi(x, t+k) = \varphi(x, t) + k\frac{\partial\varphi}{\partial t}(x, t) + \frac{k^2}{2}\frac{\partial^2\varphi}{\partial t^2}(x, t) + 0(k^3). \tag{7.131}$$

If we express that φ is a solution of (7.115) with $f = 0$, we obtain

$$\begin{aligned}\frac{\partial^2\varphi}{\partial t^2} &= \frac{\partial}{\partial t}\left(-u\frac{\partial\varphi}{\partial x}\right) = -u\frac{\partial}{\partial x}\left(\frac{\partial\varphi}{\partial t}\right)\\ &= -u\frac{\partial}{\partial x}\left(-u\frac{\partial\varphi}{\partial x}\right) = u^2\frac{\partial^2\varphi}{\partial x^2}\end{aligned}, \tag{7.132}$$

so that

$$\varphi(x,t+k) = \varphi(x,t) - ku\frac{\partial\varphi}{\partial x}(x,t) + \frac{k^2}{2}u^2\frac{\partial^2\varphi}{\partial x^2}(x,t) + 0(k^3). \tag{7.133}$$

Using centred finite difference approximations of the space variable derivatives, we then obtain the Lax–Wendroff scheme:

$$\psi_i^{n+1} = \psi_i^n - ku\left(\frac{\psi_{i+1}^n - \psi_{i-1}^n}{2h}\right) + \frac{k^2}{2}\frac{u^2}{h^2}\left[\psi_{j+1}^n - 2\psi_i^n + \psi_{i-1}^n\right]. \tag{7.134}$$

By construction, this scheme is second order in time and space.

EXERCISE Study the von Neumann stability of this scheme; show that it is stable if $\frac{k|u|}{h} \leq 1$. □

EXERCISE Show that, still for u constant, the Crank–Nicolson scheme defined by

$$\frac{1}{k}\left[\psi_i^{n+1} - \psi_i^n\right] + \frac{u}{4h}\left[\psi_{i+1}^{n+1} - \psi_{i-1}^{n+1} + \psi_{i+1}^n - \psi_{i-1}^n\right] = \tfrac{1}{2}(f_i^{n+1} + f_i^n)$$

is marginally unconditionally stable ($|a(\xi)| = 1$), and that it is consistent of second order in time and space. □

EXERCISE What do the stability domains for the Lax and Lax–Wendroff schemes become when they are applied to the following dissipation–convection equation:

$$\frac{\partial\varphi}{\partial t} + u\frac{\partial\varphi}{\partial x} + r\varphi = f \text{ in } \Omega\times]0,T[,$$

by approximating the constant term $r\varphi$ in an implicit way (i.e. at the point with index i by $r\psi_i^{n+1}$)? □

7.4.3 *The Multidimensional Case*

There are no additional difficulties; in two dimensions, the Lax–Wendroff scheme for instance becomes (taking $f = 0$ for simplicity):

$$\begin{cases} \psi_{ij}^{n+1} = \psi_{ij}^n - \dfrac{ku_{ij}}{2h}\left(\psi_{i+1,j}^n - \psi_{i-1,j}^n\right) - \dfrac{kv_{ij}}{2h}\left(\psi_{i,j+1}^n - \psi_{i,j-1}^n\right) \\ \qquad + \dfrac{k^2}{2}\left[u_{ij}^2 D_{x,i}^+ D_{x,i}^+ \Psi_{.j}^n + u_{ij}v_{ij}\left(D_{x,i}^+ D_{y,j}^- + D_{x,i}^- D_{y,j}^+\right)\Psi_{..}^n + v_{ij}^2 D_{y,j}^+ D_{y,j}^+ \Psi_{i.}^n\right], \end{cases}$$

where we have denoted the discrete operators with different notation according to the x and y directions. Index i refers to the first variable x, whereas j is associated with the second variable y: for example $\psi_{.,j}^n$ denotes the singly indexed sequence $(\psi^n_{i,j})_i$ indexed by i whereas $\psi^n_{..}$ is the doubly indexed sequence $(\psi^n_{ij})_{i,j}$ indexed by i and j. Here, the velocity field has two components $\vec{u} = (u, v)$.

In this scheme, the second mixed derivative is approximated by the following seven-point scheme:

$$\frac{\partial^2 \varphi}{\partial x \partial y}(ih, jh) \simeq \frac{1}{h^2}[-\varphi((i-1)h,(j+1)h) + \varphi(ih,(j+1)h) + \varphi((i-1)h, jh)$$
$$-2\varphi(ih, jh) + \varphi((i+1)h, jh) + \varphi(ih,(j-1)h) - \varphi((i+1)h,(j-1)h).$$

Stability analysis is done via a Fourier transform in both variables x and y, assuming the velocity field $\vec{u}$ is constant.

7.5. THE CONVECTION–DIFFUSION EQUATION

Let us consider the following equation

$$\begin{aligned} &\frac{\partial \varphi}{\partial t} - \nu \frac{\partial^2 \varphi}{\partial x^2} + u \frac{\partial \varphi}{\partial x} + r\varphi = 0 \text{ in } \mathbb{R}\times]0, T[, \\ &\varphi(x, 0) = \varphi^0(x), \end{aligned} \tag{7.135}$$

where ν is a real positive parameter, u is a given velocity field and r is a function defined over $\mathbb{R}\times]0, T[$.

7.5.1 Continuous Case

If we multiply equation (7.135) by φ, integrate over the whole space, and integrate by parts, we obtain, (omitting the integration variables to simplify the notation):

$$\frac{1}{2}\frac{\partial}{\partial t}\left(\int_{\mathbb{R}} \varphi^2\right) + \nu \int_{\Omega} \nabla|\varphi|^2 - \int_{\mathbb{R}} \frac{\varphi^2}{2}\frac{\partial u}{\partial x} + \int_{\Omega} r\varphi^2 = 0. \tag{7.136}$$

Of course, this assumes that φ is square integrable and decays to zero at infinity.

The kinetic energy $E = \int_{\mathbb{R}} \varphi^2$ decreases with increasing time as soon as the following condition is satisfied:

$$-\frac{1}{2}\frac{\partial u}{\partial x} + r \geq 0. \tag{7.137}$$

Actually, a simple change of variables shows that this condition is not necessary. Indeed, if we set $\varphi_1 = e^{-\alpha t}\varphi$, equation (7.135) becomes

$$\frac{\partial \varphi_1}{\partial t} + \alpha\varphi_1 - \upsilon \frac{\partial^2 \varphi_1}{\partial x^2} + u \frac{\partial \varphi_1}{\partial x} + r\varphi_1 = 0, \tag{7.138}$$

i.e. it is of the same type as (7.135) except that r is changed to $r+\alpha$. It can be interesting, numerically, to carry out this change of variables, so as to avoid the exponential growth of the solution with time.

7.5.2 *Discretisation*

Let us apply the following Crank–Nicolson type scheme; we obtain:

$$\frac{1}{k}\left[\psi_j^{n+1}-\psi_j^n\right]-\frac{1}{2}\frac{\nu}{h^2}\left[\psi_{j+1}^{n+1}-2\psi_j^{n+1}+\psi_{j-1}^{n+1}\right]-\frac{1}{2}\frac{\nu}{h^2}\left[\psi_{j+1}^n-2\psi_j^n+\psi_{j-1}^n\right]$$
$$+\frac{u_j^{n+1/2}}{2h}\left(\psi_{j+1}^{n+1}-\psi_j^{n+1}\right)+\frac{u_j^{n+1/2}}{2h}\left(\psi_j^n-\psi_{j-1}^n\right)+\frac{r_j^{n+1}}{2}\psi_j^{n+1}+\frac{r_j^n}{2}\psi_j^n=0. \quad (7.139)$$

Let us assume, for simplicity, that all coefficients are constant, and let us perform a stability analysis of this scheme by using a Fourier transform in space; we obtain:

$$\begin{cases}\hat{\psi}^{n+1}(\xi)-\hat{\psi}^n(\xi)+\dfrac{2\nu k}{h^2}\hat{\psi}^{n+1}(\xi)\sin^2\dfrac{\xi h}{2}+\dfrac{2\nu k}{h^2}\hat{\psi}^n(\xi)\sin^2\dfrac{\xi h}{2}\\ +\dfrac{ku}{2h}\hat{\psi}^{n+1}(\xi)(\cos\xi h+i\sin\xi h-1)+\\ \dfrac{-ku}{2h}\hat{\psi}^n(\xi)(\cos\xi h-i\sin\xi h-1)+k\dfrac{r}{2}(\hat{\psi}^{n+1}+\hat{\psi}^n)(\xi)=0\end{cases}$$

The amplification factor can thus be written

$$a(\xi)=\frac{1-\alpha+\gamma-i\beta}{1+\alpha+\gamma+i\beta}, \quad (7.140)$$

with

$$\alpha=\frac{2\nu k}{h^2}\sin^2\left(\frac{\xi h}{2}\right)+k\frac{r}{2},\ \beta=\frac{ku}{2h}\sin(\xi h),\ \gamma=-\frac{ku}{h}\sin^2\left(\frac{\xi h}{2}\right).$$

We have

$$|a(\xi)|^2-1=\frac{-4\alpha(1+\gamma)}{(1+\alpha+\gamma)^2+\beta^2},$$

and α is positive (according to (7.137) we assume $r\geq 0$ since u is constant). Thus, $|a(\xi)|^2-1$ has the same sign as $-(1+\gamma)$. The scheme will be stable if, for any ξ, $1+\gamma(\xi)\geq 0$, or if the following stability condition is satisfied:

$$\frac{ku}{h}\leq 1; \quad (7.141)$$

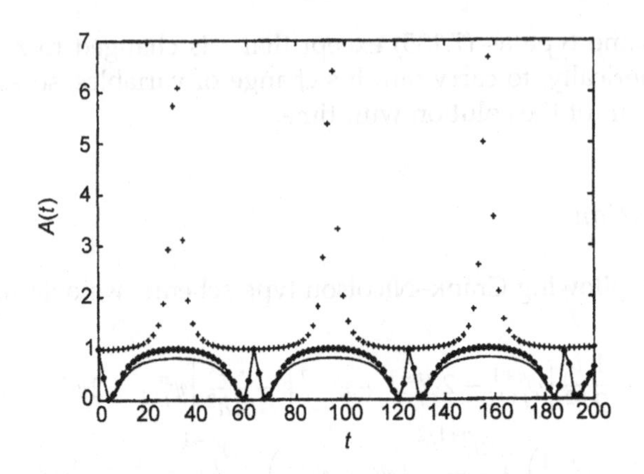

Figure 7.11 Plot of function $A(t) = |a(\xi)|^2$, $t = \xi$

we note that it is always satisfied if u is negative.

In Figure 7.11 we have plotted the function $|a|^2$, for $k = h = 0.1$, and different values of u: the solid curve corresponds to $u = 0$, that with diamond shaped symbols is the limit case $u = 1$, and finally that with + symbols corresponds to $u = 10$. The curve corresponding to $u = -10$ is located under the line with ordinate 1 at $t = 200$. We observe that the stability condition (7.141) is not satisfied for $u = 10$, whereas it is satisfied in the other cases: for $u = 10$, we must decrease the time step so as to obtain a stable scheme.

To compute the order of consistency of the scheme, it suffices to study that of the approximation of the first order term $u(\partial\varphi/\partial x)$, as the other terms, already estimated, give a consistency order of $h^2 + k^2$. We recall that this estimate has been obtained by expanding the exact solution φ in the neighbourhood of the point $(jh, (n+1/2)k)$. By two successive Taylor expansions, one with $n+1$ fixed in the neighbourhood of the point with index $j+1/2$, the other at $j+1/2$ fixed, in the neighbourhood of $t^{n+1/2} = n + 1/2k$, we obtain:

$$\frac{1}{2h}[\varphi((j+1)h,(n+1)k) - \varphi(jh,(n+1)k)] = \tfrac{1}{2}\frac{\partial}{\partial x}\varphi((j+\tfrac{1}{2})h,(n+1)k) + 0(h^2)$$
$$= \frac{1}{2}\frac{\partial}{\partial x}\varphi((j+\tfrac{1}{2})h,(n+\tfrac{1}{2})k) + \frac{k}{4}\frac{\partial^2}{\partial x\partial t}\varphi((j+\tfrac{1}{2})h,(n+\tfrac{1}{2})k) + 0(h^2+k^2). \tag{7.142}$$

In a similar way, we have:

$$\frac{1}{2h}[\varphi(jh,nk) - \varphi((j-1)h,nk)] = \frac{1}{2}\frac{\partial}{\partial x}\varphi((j-\tfrac{1}{2})h,nk) + 0(h^2) \tag{7.143}$$
$$= \frac{1}{2}\frac{\partial}{\partial x}\varphi((j-\tfrac{1}{2})h,(n+\tfrac{1}{2})k) - \frac{k}{4}\frac{\partial^2\varphi}{\partial x\partial t}((j-\tfrac{1}{2})h,(n+\tfrac{1}{2})k) + 0(h^2+k^2).$$

Using the following relations, for $p = n + 1/2$,

$$\frac{1}{2}\frac{\partial\varphi}{\partial x}((j+\tfrac{1}{2})h,pk)+\frac{1}{2}\frac{\partial\varphi}{\partial x}((j-\tfrac{1}{2})h,pk)=\frac{\partial\varphi}{\partial x}(jh,pk)+0(h^2), \qquad (7.144)$$

$$\frac{\partial^2\varphi}{\partial x\partial t}((j+\tfrac{1}{2})h,pk)-\frac{\partial\varphi}{\partial x\partial t}((j-\tfrac{1}{2})h,pk)=\frac{\partial\varphi}{\partial x}(jh,pk)+0(h),$$

we obtain, by adding (7.142) and (7.143)

$$\frac{1}{2h}[\varphi((j+1)h,(n+1)k)-\varphi(jh,(n+1)k)+\varphi(jh,nk)-\varphi((j-1)h,nk)]$$
$$=\frac{\partial\varphi}{\partial x}(jh,(n+\tfrac{1}{2})k)+0(h^2+hk+k^2), \qquad (7.145)$$

which proves that the scheme (7.139) is in $0(h^2+hk+k^2)$: if h and k are of the same order of magnitude, the proposed scheme is of second order in time and space.

EXERCISE Study the stability and the consistency of the following scheme

$$\frac{1}{k}[\psi_j^{n+1}-\psi_j^n]-\frac{1}{2}\frac{\nu}{h^2}[\psi_{j+1}^{n+1}-2\psi_j^{n+1}+\psi_{j-1}^{n+1}]-\frac{1}{2}\frac{\nu}{h^2}[\psi_{j+1}^n-2\psi_j^n+\psi_{j-1}^n]$$
$$+\frac{u_j^{n+1/2}}{2h}(\psi_j^{n+1}-\psi_{j-1}^{n+1})+\frac{u_j^{n+1/2}}{2h}(\psi_{j+1}^n-\psi_j^n)+\frac{r_j^{n+1}}{2}\psi_j^{n+1}+\frac{r_j^n}{2}\psi_j^n=0.$$

Note in particular that the stability condition can be written here

$$-\frac{ku}{h}\leq 1,$$

and this condition is always satisfied if u is positive. From the two schemes above construct a second order in time and space scheme. □

7.6 THE WAVE EQUATION

Let us consider the wave equation:

$$\frac{\partial^2\varphi}{\partial t^2}-c^2\frac{\partial^2\varphi}{\partial x^2}=0 \text{ in }]0,L[\times]0,T[, \qquad (7.146)$$

$$\varphi(x,0)=\varphi^0(x),\frac{\partial\varphi}{\partial t}(x,0)=g(x),\varphi(0,t)=\varphi(L,t)=0. \qquad (7.147)$$

This problem admits a unique solution that has the usual shape of a wave (depending on the data φ^0 and g) propagating with speed c. If c is constant, the operator factors in the form

$$\frac{\partial^2}{\partial t^2}-c^2\frac{\partial^2}{\partial x^2}=\left(\frac{\partial}{\partial t}-c\frac{\partial}{\partial x}\right)\left(\frac{\partial}{\partial t}+c\frac{\partial}{\partial x}\right),$$

so, at least formally, equation (7.146) decomposes into a system of two convection equations:

$$\frac{\partial \psi}{\partial t} - c\frac{\partial \psi}{\partial x} = 0,$$
$$\frac{\partial \varphi}{\partial t} + c\frac{\partial \varphi}{\partial x} = \psi.$$

This remark enables us, in particular, to construct numerical schemes from those studied previously for the convection equation.

Another possibility consists of introducing the auxiliary variables u and v defined by

$$u = \frac{\partial \varphi}{\partial t}, \quad v = c\frac{\partial \varphi}{\partial x}, \tag{7.148}$$

so that is suffices to solve the coupled system

$$\frac{\partial u}{\partial t} = c\frac{\partial v}{\partial x}, \quad \frac{\partial v}{\partial t} = c\frac{\partial u}{\partial x}. \tag{7.149}$$

Several schemes in this spirit are proposed, for instance, in Euvrard (1990) and Richtmyer and Morton (1967).

We can also discretise directly equation (7.146). Let us consider for example the centred scheme defined by

$$\frac{1}{k^2}\left(\psi_j^{n+1} - 2\psi_j^n + \psi_j^{n-1}\right) - \frac{c^2}{h^2}\left(\psi_{j+1}^n - 2\psi_j^n + \psi_{j-1}^n\right) = 0, \tag{7.150}$$

which is obviously consistently of second order in time and space. We note that this scheme is equivalent to the scheme

$$\begin{cases} \dfrac{1}{k}\left(u_j^{n+1} - u_j^n\right) = \dfrac{c}{h}\left(v_{j+1/2}^n - v_{j-1/2}^n\right), \\ \dfrac{1}{k}\left(v_{j-1/2}^{n+1} - v_{j-1/2}^n\right) = \dfrac{c}{h}\left(u_j^{n+1} - u_{j-1}^{n+1}\right), \end{cases} \tag{7.151}$$

if we identify u_j^n with $(\psi_j^n - \psi_j^{n-1})/k$ and $v_{j-1/2}^n$ with $c(\psi_j^n - \psi_{j-1}^n)/h$. By analogy with the notation of Section 7.3.3, we denote by X^n the vector with entries $(u^n, v^n)^T$. Then we can write the system obtained by Fourier transform of the scheme (7.151) in matrix form. The amplification matrix can be written, with i denoting the square root of -1,

$$A(\xi) = \begin{bmatrix} 1 & ia(\xi) \\ ia(\xi) & 1 - a^2 \end{bmatrix}, \quad a(\xi) = 2\frac{ck}{h}\sin\left(\frac{\xi h}{2}\right). \tag{7.152}$$

This matrix is not normal, and has as its characteristic equation

$$\lambda^2 - \lambda(2 - a^2(\xi)) + 1 = 0,$$

whose discriminant is $a^2(\xi)(a^2(\xi) - 4)$. If $ck/h < 1$, this discriminant is non-positive, and the eigenvalues of $A(\xi)$ are complex conjugate, with modulus 1, so that $\rho(A(\xi)) = 1$. We check that $A(\xi)$ is diagonalisable in all cases (i.e. whether a is zero or not) and the scheme is thus stable.

Remark 7.36 We note a slight difficulty if we study directly the stability of scheme (7.150). By Fourier transforming in the space variables, we obtain

$$\hat{\psi}^{n+1}(\xi) - 2\hat{\psi}^n(\xi) + \hat{\psi}^{n-1}(\xi) = -4\left(\frac{ck}{h}\right)^2 \hat{\psi}^n(\xi) \sin^2\left(\frac{\xi h}{2}\right);$$

or, if we denote by X^n the vector with entries $(\psi^{n+1}, \psi^n)^T$

$$\hat{X}^{n+1}(\xi) = B(\xi)\hat{X}^n(\xi), \quad B(\xi) = \begin{bmatrix} b(\xi) & -1 \\ 1 & 0 \end{bmatrix},$$

with

$$b(\xi) = 2\left[1 - 2\left(\frac{ck}{h}\right)^2 \sin^2\left(\frac{\xi h}{2}\right)\right].$$

The eigenvalues λ of B are solutions of:

$$\lambda^2 - b(\xi)\lambda + 1 = 0.$$

The reduced discriminant of this equation is written as:

$$\Delta' = 4\left(\frac{ck}{h}\right)^2 \sin^2\left(\frac{\xi h}{2}\right)\left[\left(\frac{ck}{h}\right)^2 \sin^2\left(\frac{\xi h}{2}\right) - 1\right].$$

For values of ξ such that $\sin\left(\frac{\xi h}{2}\right) \neq 0$, there are no difficulties: if $\left|\frac{ck}{h}\right| < 1$, the above discriminant is always strictly negative and the matrix $B(\xi)$ is diagonalisable because it has two distinct eigenvalues; also these eigenvalues both have modulus 1.

On the other hand, for values of ξ such that $\sin\left(\frac{\xi h}{2}\right)$ is zero, the matrix is not diagonalisable, and has a double eigenvalue, which renders condition (7.80) invalid (let us note that these very values of ξ make the above matrix $A(\xi)$ equal to the identity matrix, which itself is perfectly diagonalisable!). Does that mean that the scheme is unstable? It actually depends on the vector norm chosen for the vector X^n, and therein lies the ambiguity. It is clear that if $X^1 = (\psi^1, \psi^0)^T$, then $X^n = (\psi^0 + (n+1)(\psi^1 - \psi^0), \psi^0 + n(\psi^1 - \psi^0))^T$, so that if we take the norm $||X^n|| = \max(|\psi^{n+1}|, |\psi^n|)$, the scheme if clearly unstable. But if we choose the norm (cf. Godunov and Ryabenkii (1987))

$$||X^n|| = \max\left(|\psi^{n+1}|, |\frac{\psi^{n+1} - \psi^n}{k}|\right),$$

for all integers n satisfying $(n+1)k \leq T$, then we have

$$||X^n|| \leq 2 \max(1, T)||X^0||.$$

Let us note that this norm involves the (discrete) time derivative of the solution, whereas this derivative does not occur in the scheme (7.150), but it does in scheme (7.151) by the very choice of the variable u.

In practice, we shall usually be content to check the stability condition (7.78), without worrying about the delicate choice of the norm, knowing that one of two things can happen: if this condition is not satisfied, we cannot hope to have stability for any 'reasonable' choice of norm, and conversely, if it is, we can hope that there exist a 'reasonable' norm for which we have stability... □

EXERCISE In the same spirit, we could study the scheme defined by

$$\frac{1}{k^2}\left(\psi_j^{n+1} - 2\psi_j^n + \psi_j^{n-1}\right) - \mathcal{D}_j^+\mathcal{D}_j^-\left(\theta\Psi^{n+1} + (1-2\theta)\Psi^n) + \theta\Psi^{n-1}\right) = 0,$$

where we have used the notation of Section 7.3.2.2, θ being a parameter in the interval $[0, 1/2]$. We can show, in particular, that for $\theta \in [1/4, 1/2]$, this scheme is unconditionally stable, with the nuances alluded to in the previous remark. What is the stability condition in the case $\theta \in [0, 1/4[$? □

7.7 FINITE DIFFERENCES IN TIME AND FINITE ELEMENTS IN SPACE

In two or more dimensions, as soon as the computational domain does not have its boundary piecewise parallel to the coordinate axes, the finite difference method (in the space variables) is no longer practical: we must transform the partial differential equation into local coordinates before discretising it, and this is a lenghthy, and sometimes painful, computation. It may then be preferable to change the discretisation method in space and to choose, for instance, a finite element or a finite volume method: this is what we do in this section and in the next.

Let us consider, for instance, the following convection-diffusion equation:

$$\begin{aligned}
&\frac{\partial \varphi}{\partial t} + \nabla \cdot (u\varphi) - \nu\Delta\varphi = f \text{ in } \Omega\times]0, T[, \\
&\varphi(x, 0) = \varphi^0(x) \text{ in } \Omega, \\
&\varphi(x, t) = 0 \text{ on } \Gamma\times]0, T[,
\end{aligned} \tag{7.153}$$

where $u = u(x, t)$ is a given velocity field, and ν is a strictly positive constant.

To fix ideas, let us discretise the time derivative by using a fully implicit finite difference scheme. We then inductively define a sequence of functions

$(x \rightarrow \phi^n(x))_n$ ($\phi^n \simeq \varphi(.,t^n)$) only depending on the space variable x: we start with the initialisation $\phi^0 = \varphi^0$, then, knowing ϕ^n, we compute ϕ^{n+1} by solving the following boundary value problem ($k = \delta t > 0$):

$$\begin{aligned} &\text{find } \phi^{n+1} \text{ solving} \\ &\frac{\phi^{n+1} - \phi^n}{k} + \nabla \cdot (u^{n+1}\phi^{n+1}) - \nu\Delta\phi^{n+1} = f^{n+1} \text{ in } \Omega, \\ &\phi^{n+1} = 0 \text{ on } \Gamma, \end{aligned} \tag{7.154}$$

with the notation: $u^{n+1} = u(.,t^{n+1}), f^{n+1} = f(.,t^{n+1})$.

We thus obtain a sequence of boundary value problems, to which we may apply the finite element technique as explained in the beginning of this book. To do this, we start with the continuous problem (in space) (7.154), and write its variational formulation as:

$$\text{find } \phi^{n+1} \in H_0^1(\Omega) \text{ such that } \forall w \in H_0^1(\Omega), \text{we have: } \mathcal{A}(\phi^{n+1}, w) = l(w), \tag{7.155}$$

where $\mathcal{A}$ and l are respectively the bilinear and linear forms defined over $H_0^1(\Omega)$ by:

$$\begin{aligned} \mathcal{A}(v,w) &= \int_\Omega (vw)(x)\mathrm{d}x - k\int_\Omega (u^{n+1}v \cdot \nabla w)(x)\mathrm{d}x + \nu k \int_\Omega (\nabla v \cdot \nabla w)(x)\mathrm{d}x, \\ l(w) &= k\int_\Omega (f^{n+1}w)(x)\mathrm{d}x + \int_\Omega (\phi^n w)(x)\mathrm{d}x. \end{aligned} \tag{7.156}$$

Using Green's formula in the second integral defining $\mathcal{A}$, we have:

$$\mathcal{A}(w,w) = \int_\Omega w^2(x)\mathrm{d}x + \frac{k}{2}\int_\Omega [(\nabla \cdot u^{n+1})w^2](x)\mathrm{d}x + \nu k \int_\Omega |(\nabla w)(x)|^2 \mathrm{d}x,$$

so that if $1 + (k/2)(\nabla \cdot u^{n+1}) \geq 0$, the bilinear form $\mathcal{A}$ is elliptic on $H_0^1(\Omega)$ and problem (7.155) admits a unique solution (let us note that the above condition is always satisfied if u is a divergence-free field).

We then proceed to the discretisation by covering Ω with triangles with vertices q^i; let us call Ω_h the computational domain defined as the union of all these triangles. Let $w^1, \ldots, w^{nv}$ be the usual basis functions for the P^1 finite element approximation ($w^i(q^j) = \delta_{ij}$). At every time step t^{n+1}, we are led to solve the above variational problem in the subspace V_h of $H_0^1(\Omega)$ generated by all the functions w^i associated with interior nodes of the mesh, i.e. the space of functions w_h continuous on Ω_h, equal to zero on the boundary of Ω_h whose restriction to each triangle is a linear function. The discrete variational problem is then written:

$$\text{find } \phi_h^{n+1} \in V_h \text{ such that } \forall w_h \in V_h, \text{ we have: } \mathcal{A}_h(\phi_h^{n+1}, w_h) = l_h(w_h), \tag{7.157}$$

where $\mathcal{A}_h$ has the same expression as $\mathcal{A}$, except that we integrate over the computational domain Ω_h, and the linear form l_h is now defined by:

$$l(w_h) = k \int_{\Omega_h} (f^{n+1} w_h)(x) \mathrm{d}x + \int_{\Omega_h} (\phi_h^n w_h)(x) \mathrm{d}x.$$

The numerical technique is then the one that was explained in the first chapters: we expand ϕ_h^{n+1} on the basis of V_h and we express (7.157) by taking for w_h each of the basis functions. This leads us to the solution of a linear system whose solution gives us the components of ϕ_h^{n+1}.

If the solution of the continuous problem (7.153) is sufficiently smooth, which actually depends on the smoothness of the data, it is shown, for example, in Pironneau (1989) that this method is convergent, the error between the exact solution and the approximate solution being $O(h+k)$, if we denote by h the size of the mesh. More generally, the error would be of order r in space, should we have chosen an approximation with P^r finite elements.

It is of course possible to choose other discretisation schemes in time; let us single out, for example, the Crank–Nicolson scheme defined by:

$$\begin{aligned}
&\text{find } \phi_h^{n+1} \in V_h \text{ such that } \forall w_h \in V_h, \text{ we have:}\\
&\int_{\Omega_h} (\phi_h^{n+1} w_h)(x)\mathrm{d}x - k \int_{\Omega_h} \left(\frac{\phi^{n+1} + \phi^n}{2} u^{n+1} \cdot \nabla w_h \right)(x)\mathrm{d}x\\
&+ \nu k \int_{\Omega_h} \left(\nabla \left(\frac{\phi^{n+1} + \phi^n}{2} \right) \cdot \nabla w_h \right)(x)\mathrm{d}x,\\
&= k \int_{\Omega_h} \left(f^{n+\frac{1}{2}} w_h \right)(x)\mathrm{d}x + \int_{\Omega_h} (\phi_h^n w_h)(x)\mathrm{d}x.
\end{aligned} \tag{7.158}$$

The stability of this scheme is easily proven by energy qualities, such as were described in Section 7.3.4.1 (it suffices to take $w_h = \phi^{n+1} + \phi^n$). We can also make the linear system giving the components of ϕ^{n+1} in the basis of V_h explicit and study the eigenvalues of the corresponding matrix [Pironneau (1989)]. Let us note that on a uniform triangular mesh of a rectangular domain, the scheme is the same as the one we would obtain with a finite difference method, and this gives another way to check its stability. Last, this scheme is second order in time, and for a sufficiently regular mesh, we can hope for a second order accuracy in space.

7.8 FINITE DIFFERENCES IN TIME AND FINITE VOLUMES IN SPACE

Let us again consider a convection–diffusion equation (the notation is the same as in the previous section), with, for a change, Neumann boundary conditions:

$$\begin{aligned}
&\frac{\partial \varphi}{\partial t} + \nabla \cdot (u\varphi) - \nu \Delta \varphi = f \text{ in } \Omega \times]0, T[,\\
&\varphi(x, 0) = \varphi^0(x) \text{ in } \Omega,\\
&\frac{\partial \varphi}{\partial n}(x, t) = g(x, t) \text{ on } \Gamma \times]0, T[.
\end{aligned} \tag{7.159}$$

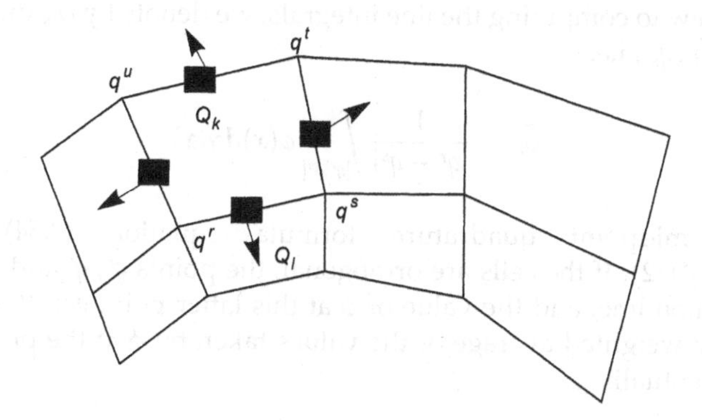

Figure 7.12 A finite volume mesh defined by rectangles

We cover the domain Ω by quadrangles Q_k, so that $\cup_{k\in K} Q_k = \Omega$, and we assume that this mesh is *admissible*, i.e. the intersection between two quadrangles is either empty, or reduced to one point, or to a whole side. We integrate the partial differential equation on any one of those quadrangles and obtain, after using Green's formula,

$$\begin{aligned}&\frac{\partial}{\partial t}\int_{Q_k}\varphi(x,t)\mathrm{d}x+\int_{\partial Q_k}(u\cdot n\varphi)(x,t)\mathrm{d}\gamma(x)+\nu\int_{\partial Q_k}\left(\frac{\partial\varphi}{\partial n}\right)(x,t)\mathrm{d}\gamma(x)\\&=\int_{Q_k}f(x,t)\mathrm{d}x.\end{aligned}\tag{7.160}$$

As above, we can discretise this equation in time, choosing an implicit, explicit, or semi-implicit scheme. If, for instance, we choose an explicit scheme, we are led to compute a sequence of functions ϕ^m defined by the recurrence relation

$$\begin{aligned}&\int_{Q_k}\frac{1}{\delta t}\left[\phi^{m+1}-\phi^m\right](x)\mathrm{d}x+\int_{\partial Q_k}(u^m\cdot n\phi^m)(x)\mathrm{d}\gamma(x)\\&+\nu\int_{\partial Q_k}\left(\frac{\partial\phi^m}{\partial n}\right)(x)\mathrm{d}\gamma(x)=\int_{Q_k}f^m(x)\mathrm{d}x,\end{aligned}\tag{7.161}$$

and initialised by $\phi^0=\varphi^0$.

7.8.1 A Cell Centred Scheme

We shall evaluate each of the above integrals by using only the values of the unknown at the centre of the cells. At step $n+1$, the unknowns of the problem are the values of ϕ^{n+1} at each of those nodes (i.e. the cell centres), so that there are as many unknowns as equations like (7.161).

Let us now describe how to approximate these integrals. To do this, let us use the notation of Figure 7.12, and denote by q^k (resp. q^l) the centre of cell Q_k (resp.

Q_l). With a view to computing the line integrals, we denote by $\tilde{\phi}_{rs}$ the average of ϕ on the edge $[q_r, q_s]$, i.e.

$$\tilde{\phi}_{rs} = \frac{1}{|q^r - q^s|} \int_{]q^r, q^s[} \phi(x) \mathrm{d}\gamma(x). \tag{7.162}$$

Using the midpoint quadrature formula equation (2.54), we have $\tilde{\phi}_{rs} \simeq \phi((q^r + q^s)/2)$. If the cells are orthogonal, the points q^k, q^l and $(q^r + q^s)/2$ all lie on a common line, and the value of ϕ at this latter point can then be approximated by the weighted average of the values taken by ϕ at the points q^k and q^l. This gives eventually:

$$\tilde{\phi}_{rs} \simeq \phi_{rs} = \phi_l \frac{|q^k - \frac{q_r + q_s}{2}|}{|q^k - q^l|} + \phi_k \frac{|q^l - \frac{q_r + q_s}{2}|}{|q^k - q^l|}. \tag{7.163}$$

We shall use this approximation for the average of ϕ over $[q^r, q^s]$ in all cases.

The integrals featured in (7.161) are then computed in the following way:

$$\begin{aligned} &\int_{Q_k} \phi(x) \mathrm{d}x \simeq \phi_k \text{ area } (Q_k) \\ &\int_{]q^r, q^s[} \frac{\partial \phi}{\partial n}(x) \mathrm{d}\gamma(x) \simeq \frac{\phi_l - \phi_k}{|q^k - q^l|} |q^r - q^s| \\ &\int_{\partial Q_k} (un\, \phi)(x) \mathrm{d}x \simeq un_{rs}\, \phi_{rs} + un_{st}\, \phi_{st} + \cdots . \end{aligned} \tag{7.164}$$

In this way, we obtain a numerical scheme for all almost orthogonal meshes. If this is not the case, we must modify the above approximate formulae. For example, to compute the normal derivative of ϕ, we can go back to the definition $\partial\phi/\partial n = \nabla\phi \cdot n$ and compute an approximation $\tilde{\nabla\phi}$ to $\nabla\phi$ by solving the system (with notation as in Figure 7.12):

$$\begin{aligned} (\tilde{\nabla\phi}).(q^r + q^s - q^u - q^t) &= 2(\varphi_{rs} - \varphi_{ut}), \\ (\tilde{\nabla\phi}).(q^s + q^t - q^u - q^r) &= 2(\varphi_{st} - \varphi_{ur}). \end{aligned} \tag{7.165}$$

Stability and convergence results can be obtained for regular meshes by applying the method described above for finite differences. Other direct proofs exist, based on the consistency of the approximate flux (i.e. the line integrals featured in (7.161) and the conservative nature of the scheme (i.e. if ϕ is a smooth function, the approximation of the integral over $Q_k \cap Q_l$ of $\partial\phi/\partial n$ in the equation over Q_k is the opposite to that associated with Q_l).

Although it is in some sense close to both the finite difference and the P^0 finite element method, the finite volume differs from those methods on other points. As the finite element method, it is based on a weak formulation of the partial differential equation, obtained by integrating, but only against the function 1, and the sought solution is not expanded on a basis. The flux approximation uses a 'finite difference' principle, but the resulting scheme is not consistent in the finite difference sense; consistency only plays a role at the integral formulation level, in the flux approximation.

7.8.2 Other Possible Schemes

As we have seen, the principle of the finite volume method is to integrate the equation over a cell, and then to interpolate this integral with function values at the nodes. We have seen the case of quadrangular cells, with nodes at the centre of the cells, the interpolation being linear. Other choices are possible; let us quote, for instance:

- the nodes are the vertices of the quadrangle Q_k and the control cells are the quadrangles obtained by splitting each quadrangle Q_k into four (along the medians) and joining together all the sub-quadrangles thus formed that have a common side;
- the nodes are the midpoints of the quadrangle edges and the cells are obtained by joining together all quadrangles having a common edge;
- in all of the above, we can replace the quadrangles by triangles; for example, the nodes are the vertices of the triangulation and the cell associated with a node is the polygon obtained by joining the medians of the triangles sharing this vertex.

Naturally, for all these choices, it is necessary that the number of unknowns be exactly equal to the number of equations. Despite this, there can be problems. Let us consider, for example, the convection equation

$$\frac{\partial \varphi}{\partial t} + u \cdot \nabla \varphi = 0, \tag{7.166}$$

with: $\nabla \cdot u = 0$. Let us choose quadrangular cells, and define the nodes as the centre of the cells; we have

$$\int_{Q_k} \frac{\partial \varphi}{\partial t} + \int_{\partial Q_k} u \cdot n\varphi = 0. \tag{7.167}$$

We cannot assume that φ is constant on each cell, because the second integral would be zero. This shows the importance of the choice of the interpolation.

These finite volume methods are comparatively recent [Jameson *et al.* (1986)]. They are conceptually simple and easy to implement. The theory is still under development (Eymard and Gallouët (1993), Morton (1996)).

These methods have first been developed for convection equations (Euler equation for fluid flows). Nicolaïdes (1992) extended them to second order equations, after decoupling them into a system of two first order equations, according to the scheme:

$$\Delta \varphi = 0 \quad \leftrightarrow \quad u = \nabla \varphi, \quad \nabla \cdot u = 0.$$

Some of the equations are integrated over triangles, whereas the others are integrated over the Voronoï polygons of the triangulation.

One of the interests of this method is that it makes it possible, contrary to finite difference methods, to treat equations with possibly discontinuous coefficients. Numerical tests show that for an operator like $-\nabla \cdot (\nu \nabla \varphi)$, where ν is a matrix with variable and discontinuous coefficients, the finite volume method gives, for a Neumann boundary condition, better results than the finite element method; on the other hand, the conclusion is reversed for a Dirichlet boundary condition. For a recent survey of this type of methods, the reader is referred to the forthcoming book by Eymard, Gallouët and Herbin (1995), where a rather complete mathematical bibliography can be found.

APPENDIX 7.A

The Fortran Program for Solving the Heat equation with Homogeneous Dirichlet Boundary Condition by two Different Schemes in Time: the Explicit Scheme, and the Implicit Scheme.

```
C file: evolHEAT.for
C
            PROGRAM evolHEAT
C           ----------------
C
C This program solves the evolution problem
C governed by the heat equation in one dimension,
C with homogeneous Dirichlet boundary conditions
C and zero source term, by two methods: explicit
C scheme, implicit scheme and LU factorisation of
C the matrix.
C
C   arrays:
C   ------
C   mesh:
C     x:coordinates of the mesh points
C   solution
C     phi: values of the approximate solution
C   other arrays
C     al,d,u: coefficients of the LU factors of the
C       linear system matrix (implicit case)
C     phin: array for explicit case
C
C
C
            PARAMETER (nMax=1000)
            DIMENSION x(nMax),phi(nMax)
            CHARACTER*20 filnam
C
            write(*,*) 'number of points (min 2 ,max 1000)'
```

```
          read(*,*) n
          dx=1./(n-1.)
          DO 1 i=1,n
            x(i)=(i-1)*dx
1         CONTINUE
          write(*,*) 'chosen scheme: 0=implicit'
          write(*,*)                   '1=explicit'
          read(*,*) ischem
          write(*,*) 'final time'
          read(*,*) Tf
          CALL init(n,x,phi)
          IF(isch.EQ.1) THEN
           dtmax=0.5*dx*dx
           write(*,*) 'enter time step'
           write(*,*) 'attention dtmax=',dtmax
           read(*,*) dt
           m=aint(Tf/dt)
           c=dt/(dx*dx)
           CALL Sexpli(n,m,c,x,phi)
          ELSE
            write(*,*) 'enter time step'
            read(*,*) dt
            m=aint(Tf/dt)
            c=dt/(dx*dx)
            CALL Simpli(n,m,c,x,phi)
          ENDIF
          write(*,*) 'name of file for storing the '
          write(*,*) 'approximate solution'
          read(*,'(a)') filnam
          CALL writeF(filnam,n,phi)
          PAUSE
          END
C
C
C ---------------------------------------------------
          SUBROUTINE init(n,x,phi)
C ---------------------------------------------------
C
          DIMENSION x(*),phi(*)
          DO 1 i=1,n
            xi=x(i)
            phi(i)=xi*(1-xi)
1           CONTINUE
          END
C
C ---------------------------------------------------
          SUBROUTINE Sexpli(n,m,c,x,phi)
C ---------------------------------------------------
C
          PARAMETER (nMax=1000)
          DIMENSION x(*),phi(*),phin(nMax)
```

```
            DO 1 k=1,m
            DO 2 i=2,n-1
              phin(i)=phi(i)+c*(phi(i+1)-2*phi(i)+phi(i-1))
2             CONTINUE
            DO 3 i=2,n-1
              phi(i)=phin(i)
3             CONTINUE
1           CONTINUE
            END
C
C ----------------------------------------------------------------
            SUBROUTINE Simpli(n,m,c,x,phi)
C ----------------------------------------------------------------
            PARAMETER (nMax=1000)
            DIMENSION x(*)
            DIMENSION phi(*)
            DIMENSION al(nMax),d(nMax),u(nMax)
C
C matrix
C
            d2=2*c+1
            d(2)=d2
            u(2)=-c/d2
            DO 1 i=3,n-1
              al(i-1)=-c
              d(i)=2*c+1-u(i-1)*al(i-1)
              u(i)=-c/d(i)
1           CONTINUE
C iterations:  foward-backward solves
C
            DO 2 k=1,m
              phi(2)=phi(2)/d(2)
            DO 3 i=3,n-1
              phi(i)=(phi(i)-al(i-1)*phi(i-1))/d(i)
3             CONTINUE
            DO 4 i=n-2,2,-1
              phi(i)=phi(i)-u(i)*phi(i+1)
4             CONTINUE
2           CONTINUE
            END
```

Part C

COMPLEMENTS ON NUMERICAL METHODS

8 Boundary Integral Methods for the Laplacian

Summary

In this chapter we introduce the boundary integral representation method, or boundary integral method for short: this method can only be applied to operators whose Green's function is known. We shall explain it in the case of the Laplacian in three dimensions and state the results in two dimensions. This method was first introduced by fluid mechanics engineers, but is currently being extended to many other domains beyond aerodynamics such as acoustics and electromagnetism. We shall briefly present the case of the Helmholtz equation. Let us note that the method is sometimes called the singularity method, or panel method: we shall take this opportunity to explain the origin of this name.

8.1 PRELIMINARIES

We set about to solve the model problem of the Laplacian with Dirichlet boundary conditions in a domain Ω in the space $\mathbb{R}^3$ with a sufficiently smooth boundary:

$$-\Delta\varphi = f \quad \text{in } \Omega, \quad \varphi_{|\Gamma} = \varphi_\Gamma. \tag{8.1}$$

The boundary Γ determines a bounded domain Ω_i, and we shall be concerned with the following two cases: $\Omega = \Omega_i$, and the problem is the *interior Dirichlet problem*; and $\Omega = \Omega_e = \mathbb{R}^3 - \bar{\Omega}_i$, and (8.1) is then the *exterior Dirichlet problem*.

The principle of the boundary integral method is based on the search for an *elementary solution, or Green's function,* of the operator. We shall now define this for our case of interest, that is the Laplacian.

8.1.1 Elementary Solution of the Laplacian

In three dimensions the function $(x = (x_1, x_2, x_3) \to 1/|x|)$, with the notation: $|x| = (x_1^2 + x_2^2 + x_3^2)^{1/2}$, is the right candidate up to a multiplicative constant, since it satisfies the following Lemma:

Lemma 8.1 The function defined over the space $\mathbb{R}^3$ minus the origin by $(x \to 1/|x|)$ is harmonic, i.e. it satisfies:

$$\forall x \in \mathbb{R}^3, x \neq 0, \quad \Delta\left(\frac{1}{|x|}\right) = 0. \tag{8.2}$$

Moreover, we have:

$$\forall a > 0, \quad \int_{|x|<a} \Delta\left(\frac{1}{|x|}\right) = -4\pi. \tag{8.3}$$

PROOF Let us start by proving the first point. It follows from a simple algebraic computation

$$\frac{\partial}{\partial x_i}\left(x_1^2 + x_2^2 + x_3^2\right)^{-\frac{1}{2}} = -\frac{1}{2}\left(|x|^2\right)^{-\frac{3}{2}}\frac{\partial}{\partial x_i}(x_i^2) = -\frac{x_i}{\left(|x|^2\right)^{\frac{3}{2}}},$$

so that

$$\begin{aligned}\frac{\partial^2}{\partial x_i^2}\left(\frac{1}{|x|}\right) = \frac{\partial}{\partial x_i}\left[\frac{-x_i}{(|x|^2)^{\frac{3}{2}}}\right] &= -(|x|^2)^{-\frac{3}{2}} - x_i\frac{\partial}{\partial x_i}\left[(|x|^2)^{-\frac{3}{2}}\right] \\ &= -|x|^{-3} + \tfrac{3}{2}x_i|x|^{-5}(2x_i) = -|x|^{-3} + 3x_i^2|x|^{-5},\end{aligned}$$

which gives (8.2) by summing over indices $i \in \{1, 2, 3\}$.

The second point is proved by using a Green's formula. Let us denote by n the outer normal to the sphere with centre at 0 and radius a. Then we have, since $\nabla 1 \equiv 0$:

$$\int_{|x|<a} \Delta\left(\frac{1}{|x|}\right)\mathrm{d}x = \int_{|x|=a} \frac{\partial}{\partial n}\left(\frac{1}{|x|}\right)\mathrm{d}\gamma(x). \tag{8.4}$$

By using spherical coordinates (r, θ, φ), this integral becomes

$$\begin{aligned}\int_{|x|=a} \frac{\partial}{\partial n}\left(\frac{1}{|x|}\right)\mathrm{d}\gamma(x) &= \int_0^{2\pi}\left(\int_{-\frac{\pi}{2}}^{\frac{\pi}{2}} \frac{\partial}{\partial r}\left(\frac{1}{r}\right)_{|r=a} a^2 \cos\varphi \,\mathrm{d}\varphi\right)\mathrm{d}\theta \\ &= -\int_0^{2\pi}\int_{-\frac{\pi}{2}}^{\frac{\pi}{2}} \cos\varphi \,\mathrm{d}\varphi \mathrm{d}\theta = -2\pi[\sin\varphi]_{-\frac{\pi}{2}}^{\frac{\pi}{2}} = -4\pi,\end{aligned}$$

which, together with (8.4), proves (8.3) and ends the proof of the Lemma. □

EXERCISE Show that

$$\forall a > 0, \quad \int_{\{x, |x|<a, x_3 \geq 0\}} \Delta\left(\frac{1}{|x|}\right) = -2\pi; \tag{8.5}$$

this result can obviously be generalised to any half-ball centred at 0. □

Proposition 8.2 *We have*

$$\Delta\left(\frac{1}{|x|}\right) = -4\pi\delta, \tag{8.6}$$

if we denote by δ the Dirac measure at 0, defined over compactly supported continuous functions ψ by: $<\delta,\psi>=\psi(0)$.

PROOF We denote by the same symbol $<.,.>$ the duality bracket between distributions and functions in $\mathcal{D}(\mathbb{R}^3)$, where this space denotes the set of indefinitely differentiable functions with compact support in $\mathbb{R}^3$. Because the function $(x \to 1/|x|)$ is locally integrable, it is a distribution, and we may differentiate it in the space $\mathcal{D}'(\mathbb{R}^3)$ of distributions over $\mathbb{R}^3$.

By the definition of the distributional derivative, we have for any function $\psi \in \mathcal{D}(\mathbb{R}^3)$,

$$\left\langle \Delta\left(\frac{1}{|x|}\right), \psi \right\rangle = \left\langle \frac{1}{|x|}, \Delta\psi \right\rangle = \lim_{\varepsilon\to 0} I_\varepsilon, \quad I_\varepsilon = \int_{|x|>\varepsilon} \frac{1}{|x|}\Delta\psi(x)\mathrm{d}x, \tag{8.7}$$

where the last equality follows from applying Lebesgue's dominated convergence theorem.

Using Green's formula, we obtain, denoting by n the normal to the sphere $|x| = \varepsilon$ oriented towards the exterior of the ball $|x| < \varepsilon$ (pay attention to signs!):

$$\begin{aligned} I_\varepsilon = &\int_{|x|>\varepsilon} \Delta\left(\frac{1}{|x|}\right)\psi(x)\mathrm{d}x - \int_{|x|=\varepsilon} \frac{1}{|x|}\frac{\partial\psi}{\partial n}(x)\mathrm{d}\gamma(x) \\ &+ \int_{|x|=\varepsilon} \frac{\partial}{\partial n}\left(\frac{1}{|x|}\right)(x)\psi(x)\mathrm{d}\gamma(x). \end{aligned} \tag{8.8}$$

According to (8.2), the first integral is zero. A change of variables to spherical coordinates similar to the one above (the Jacobian is equal to $\varepsilon^2\cos(\varphi)$) enables us to show that the absolute value of the second integral is bounded, up to a multiplicative constant, by ε: so, this integral goes to 0 with ε. It remains to show that the third integral tends to $-4\pi\psi(0)$.

The normal n in (8.8) can be written as $n = n(x) = x/|x|$, and $\nabla(1/|x|) = -x/|x|^3$ so that we have:

$$\frac{\partial}{\partial n}\left(\frac{1}{|x|}\right)(x) = -\frac{1}{|x|^2}.$$

It follows successively that

$$\left| \int_{|x|=\varepsilon} \frac{\partial}{\partial n}\left(\frac{1}{|x|}\right)(x)\psi(x)d\gamma(x) + 4\pi\psi(0) \right|$$

$$= \left| \int_{|x|=\varepsilon} \left[\frac{\partial}{\partial n}\left(\frac{1}{|x|}\right)(x)\psi(x) + \frac{1}{\varepsilon^2}\psi(0)\right] d\gamma(x) \right|$$

$$= \frac{1}{\varepsilon^2}\left| \int_{|x|=\varepsilon} [-\psi(x) + \psi(0)]d\gamma(x) \right| \quad \leq \varepsilon \sup_{x\in\mathbb{R}^3} \sum_{1\leq i\leq 3} \left|\frac{\partial\psi}{\partial x_i}(x)\right|,$$

the last inequality follows from a change of variables to spherical coordinates, followed by the mean value inequality. In the limit $\varepsilon \to 0$, we then have, for any function $\psi \in \mathcal{D}(\mathbb{R}^3)$,

$$\lim_{\varepsilon\to 0} I_\varepsilon = -4\pi\psi(0),$$

which, together with (8.7), proves (8.6) in the sense of distributions. □

By an identical argument (that we shall take up again in the proof of Theorem 8.5), we obtain the following Corollary:

Corollary 8.3 Let D be a smooth bounded domain in $\mathbb{R}^3$. For any sufficiently smooth function ψ, with compact support in $\mathbb{R}^3$, let us set:

$$K\psi(x) = \int_D \Delta\left(\frac{1}{|x-y|}\right)\psi(y)dy. \tag{8.9}$$

We have:

$$K\psi(x) = \begin{cases} 0 & \text{if } x \notin D, \\ -4\pi\psi(x) & \text{if } x \in D - \partial D, \\ -2\pi\psi(x) & \text{if } x \in \partial D. \end{cases} \tag{8.10}$$

Remark 8.4 In two dimensions, all the above conclusions remain valid, if we replace the function $-1/(4\pi|x|)$ by $\ln(|x|)/2\pi$. We then obtain the following results:

$$\forall x \in \mathbb{R}^2, x \neq 0, \quad \Delta(\ln(|x|)) = 0, \tag{8.11}$$

$$\forall a > 0, \quad \int_{|x|<a} \Delta(\ln(|x|)) = 2\pi; \tag{8.12}$$

$$\Delta(\ln(|x|)) = 2\pi\delta. \tag{8.13}$$

Last, if D is a sufficiently smooth domain in $\mathbb{R}^2$, we have

$$\int_D \Delta[\ln(|x-y|)]\psi(y)dy = \begin{cases} 0 & \text{if } x \notin D, \\ 2\pi\psi(x) & \text{if } x \in D - \partial D, \\ \pi\psi(x) & \text{if } x \in \partial D. \end{cases} \tag{8.14}$$

The proofs are similar to the ones we have given. □

ORIENTATION We are now concerned with the numerical solution of (8.1) in the case $f = 0$.

A possibility would be to search for φ in the form

$$\varphi(x) = \sum_{i=1}^{N} \frac{\lambda_i}{|x - x^i|}, \tag{8.15}$$

for a given set of N points x^i located on the boundary Γ.

Since such a function automatically satisfies Laplace's equation in the interior of Ω, it would suffice to find the scalars λ_i so as also to satisfy the boundary condition. But, because (8.15) is singular at the nodes x^i, this boundary condition can only be satisfied in a weak sense, for instance by writing:

$$\forall j \in \{1, ..., N\}, \quad \sum_{i=1}^{N} \lambda_i \int_\Gamma \frac{1}{|x - x^i||x - x^j|} d\gamma(x) = \int_\Gamma \frac{\varphi_\Gamma(x)}{|x - x^j|} d\gamma(x), \tag{8.16}$$

which provides a system of N equations in the N unknowns λ_i. This method actually works, but it has two drawbacks:

1. The integrals on the left hand side are singular, all the more so for $i = j$.
2. This method is expensive, because it requires a double loop over the indices i and j of the boundary nodes, as well as another loop to compute the integrals.

We shall now propose another much less expensive method, based upon an *integral representation* of the solutions of (8.1).

8.1.2 *Integral Representation of the Solutions of (8.1)*

Let us use again the notation of the previous section, and denote by Ω_i a smooth bounded domain in $\mathbb{R}^3$ with boundary Γ, whose complement in $\mathbb{R}^3$ is denoted by $\bar{\Omega}_e$. The normal to Γ oriented towards the exterior of Ω_i (resp. Ω_e) is denoted by n_i (resp. n_e); we set $n = n_i = -n_e$. By analogy, we shall denote by φ_i (resp. φ_e) any function defined over $\bar{\Omega}_i$ (resp. $\bar{\Omega}_e$) and we shall let φ be the function defined over $\Omega_i \cup \Omega_e$ by:

$$\varphi_{|\Omega_i} = \varphi_i, \quad \varphi_{|\Omega_e} = \varphi_e. \tag{8.17}$$

We have the following theorem:

Theorem 8.5 Let us assume $\varphi_i \in C^2(\Omega_i) \cap C^1(\bar{\Omega}_i)$, $\varphi_e \in C^2(\Omega_e) \cap C^1(\bar{\Omega}_e)$, and that the following hypotheses on the behaviour at infinity are satisfied: φ_e and $\nabla\varphi_e$ decay like $1/r$ and like $1/r^2$ respectively as $r = |x| \to +\infty$. Then, if φ is harmonic in $\Omega_i \cup \Omega_e$, it admits an integral representation defined by:

- if $x \in \Omega_i \cup \Omega_e$:

$$\varphi(x) = \frac{1}{4\pi}\int_\Gamma \left[\frac{\partial\varphi}{\partial n}\right](y)\frac{1}{|x-y|}\mathrm{d}\gamma(y) - \frac{1}{4\pi}\int_\Gamma [\varphi](y)\frac{\partial}{\partial n}\left(\frac{1}{|x-y|}\right)(y)\mathrm{d}\gamma(y), \quad (8.18)$$

with the notation

$$[\varphi] = \varphi_i - \varphi_e, \quad \left[\frac{\partial\varphi}{\partial n}\right] = \frac{\partial\varphi_i}{\partial n} - \frac{\partial\varphi_e}{\partial n}; \quad (8.19)$$

- if x is on the boundary Γ:

$$\frac{\varphi_i(x)+\varphi_e(x)}{2} = \frac{1}{4\pi}\int_\Gamma \left[\frac{\partial\varphi}{\partial n}\right](y)\frac{1}{|x-y|}\mathrm{d}\gamma(y) - \frac{1}{4\pi}\int_\Gamma [\varphi](y)\frac{\partial}{\partial n}\left(\frac{1}{|x-y|}\right)(y)\mathrm{d}\gamma(y). \quad (8.20)$$

To prove this result, we need to introduce some notation that we shall now make precise. Let us denote by B_R the open ball centred at the origin with a radius R sufficiently large that it contains $\bar{\Omega}_i$. For any point x, we denote by $B_\varepsilon(x)$ the open ball centred at x with radius $\varepsilon << 1$. The boundaries of these balls will be denoted by ∂B_R and $\partial B_\varepsilon(x)$ respectively; last, n_R and n_ε will denote the normals to these boundaries oriented towards the exterior of these balls. These normals, as well as the exterior normal $n = n_i = -n_e$ to Ω_i at each point of Γ, are functions of the integration variable on the boundary of the domain (y in the sequel); for clarity, we shall not make this dependence explicit. In the same way, the abbreviated notation $\frac{\partial}{\partial n}(1/|x-y|)$, y being the integration variable over Γ, means $(\nabla_y(1/|x-y|)\cdot n)(y)$, where ∇_y denotes the derivation operator with respect to the variable y.

PROOF OF THEOREM 8.5 Let us first consider a point x in Ω_i. By using Green's formula in $\Omega_i - \bar{B}_\varepsilon(x)$, we have, since the functions φ_i and $(y \to 1/|x-y|)$ are harmonic there:

$$\begin{aligned} 0 &= \int_{\Omega_i - \bar{B}_\varepsilon(x)} \left(\Delta\varphi_i(y)\frac{1}{|x-y|} - \varphi_i(y)\Delta\left(\frac{1}{|x-y|}\right)(y)\right)\mathrm{d}y \\ &= \int_\Gamma \frac{\partial\varphi_i}{\partial n}(y)\frac{1}{|x-y|}\mathrm{d}\gamma(y) - \int_\Gamma \varphi_i(y)\frac{\partial}{\partial n}\left(\frac{1}{|x-y|}\right)(y)\mathrm{d}\gamma(y) \\ &\quad - \int_{\partial B_\varepsilon(x)} \frac{\partial\varphi_i}{\partial n_\varepsilon}(y)\frac{1}{|x-y|}\mathrm{d}\gamma_\varepsilon(y) + \int_{\partial B_\varepsilon(x)} \varphi_i(y)\frac{\partial}{\partial n_\varepsilon}\left(\frac{1}{|x-y|}\right)(y)\mathrm{d}\gamma_\varepsilon(y). \end{aligned} \quad (8.21)$$

The third (resp. fourth) integral of the right hand side has a slightly more general, but similar expression to the second (resp. third) integral in (8.8); by a similar argument to that used in the proof of Proposition 8.2, when $\varepsilon \to 0$, we deduce it has 0 (resp. $-4\pi\varphi_i(x)$) as a limit. We then have:

$$4\pi\varphi_i(x) = \int_\Gamma \frac{\partial\varphi_i}{\partial n}(y)\frac{1}{|x-y|}\mathrm{d}\gamma(y) - \int_\Gamma \varphi_i(y)\frac{\partial}{\partial n}\left(\frac{1}{|x-y|}\right)(y)\mathrm{d}\gamma(y). \quad (8.22)$$

Let us then use Green's formula in $B_R \cap \Omega_e$ (since $x \in \Omega_i$, the function $(y \to |x-y|)$ does not vanish); we have, by virtue of the choice $n = -n_e$:

$$\begin{aligned}
0 &= \int_{B_R \cap \Omega_e} \left(\Delta\varphi_e(y) \frac{1}{|x-y|} - \varphi_e(y) \Delta\left(\frac{1}{|x-y|}\right)(y) \right) \mathrm{d}y \\
&= -\int_\Gamma \frac{\partial \varphi_e}{\partial n}(y) \frac{1}{|x-y|} \mathrm{d}\gamma(y) + \int_\Gamma \varphi_e(y) \frac{\partial}{\partial n}\left(\frac{1}{|x-y|}\right)(y) \mathrm{d}\gamma(y) \\
&\quad + \int_{\partial B_R} \frac{\partial \varphi_e}{\partial n_R}(y) \frac{1}{|x-y|} \mathrm{d}\gamma_R(y) - \int_{\partial B_R} \varphi_e(y) \frac{\partial}{\partial n_R}\left(\frac{1}{|x-y|}\right)(y) \mathrm{d}\gamma_R(y).
\end{aligned} \tag{8.23}$$

Now, the hypotheses on the behaviour at infinity of φ_e are sufficient to show that the third and fourth integrals in the right hand side of (8.23) tend to 0 as R goes to $+\infty$ (to see this, it suffices to perform, as in the computation of the terms in (8.8), a change of variables to spherical coordinates, the Jacobian being proportional to R^2). We thus deduce:

$$0 = -\int_\Gamma \frac{\partial \varphi_e}{\partial n}(y) \frac{1}{|x-y|} \mathrm{d}\gamma(y) + \int_\Gamma \varphi_e(y) \frac{\partial}{\partial n}\left(\frac{1}{|x-y|}\right)(y) \mathrm{d}\gamma(y). \tag{8.24}$$

If we then add equations (8.22) and (8.24) termwise, we obtain (8.18), for points x in Ω_i. The argument is identical for points in Ω_e, except we must integrate first over Ω_i, then over $B_R \cap \Omega_e - \bar{B}_\varepsilon(x)$.

It remains to study the case of points x that belong to Γ. Let us use Green's formula in $\Omega_i - \bar{B}_\varepsilon(x)$ once again, we obtain:

$$\begin{aligned}
0 &= \int_{\Gamma - \Gamma \cap B_\varepsilon(x)} \frac{\partial \varphi_i}{\partial n}(y) \frac{1}{|x-y|} \mathrm{d}\gamma(y) - \int_{\Gamma - \Gamma \cap B_\varepsilon(x)} \varphi_i(y) \frac{\partial}{\partial n}\left(\frac{1}{|x-y|}\right)(y) \mathrm{d}\gamma(y) \\
&\quad - \int_{\partial B_\varepsilon(x) \cap \Omega_i} \frac{\partial \varphi_i}{\partial n_\varepsilon}(y) \frac{1}{|x-y|} \mathrm{d}\gamma_\varepsilon(y) + \int_{\partial B_\varepsilon(x) \cap \Omega_i} \varphi_i(y) \frac{\partial}{\partial n_\varepsilon}\left(\frac{1}{|x-y|}\right)(y) \mathrm{d}\gamma_\varepsilon(y).
\end{aligned} \tag{8.25}$$

As above, we show that the third integral in the right hand side of (8.25) tends to 0 with ε. The fourth integral is computed as in the proof of Proposition 8.2, except that, because we now integrate (for ε infinitely small) over a half-sphere, this integral tends toward $-2\pi\varphi_i(x)$ when ε goes to 0. Let us notice that it is necessary, in the latter computation, to assume that the boundary Γ has no angular points; indeed, if there were an angular point x, since the tangents at this point make up an angle θ in Ω_i, the above limit would be $-\theta\varphi_i(x)$.

Now, in the limit $\varepsilon \to 0$, we have

$$2\pi\varphi_i(x) = \int_\Gamma \frac{\partial \varphi_i}{\partial n}(y) \frac{1}{|x-y|} \mathrm{d}\gamma(y) - \int_\Gamma \varphi_i(y) \frac{\partial}{\partial n}\left(\frac{1}{|x-y|}\right)(y) \mathrm{d}\gamma(y). \tag{8.26}$$

By a similar argument, by integrating over $B_R \cap \Omega_e - \bar{B}_\varepsilon(x)$, we obtain, in the limit $R \to +\infty$ and $\varepsilon \to 0$:

$$2\pi\varphi_e(x) = -\int_\Gamma \frac{\partial\varphi_e}{\partial n}(y)\frac{1}{|x-y|}\mathrm{d}\gamma(y) + \int_\Gamma \varphi_e(y)\frac{\partial}{\partial n}\left(\frac{1}{|x-y|}\right)(y)\mathrm{d}\gamma(y). \quad (8.27)$$

The expression (8.20) is then obtained by adding equations (8.26) and (8.27) termwise. □

Remark 8.6 The hypotheses regarding the behaviour at infinity of φ_e may actually be weakened: it suffices that φ_e/r and $\nabla\varphi_e$ be square integrable in the neighbourhood of infinity. □

Remark 8.7 This integral representation theorem can be generalised to two dimensions, with the same hypotheses, except for the behaviour at infinity, which can now only be of the type:

$$\text{if } r = |x| \to +\infty, \quad \varphi_e(x) = a\ln(r) + 0\left(\frac{1}{r}\right). \quad (8.28)$$

We obtain:

- if $x \in \Omega_i \cup \Omega_e$, then:

$$\varphi(x) = -\frac{1}{2\pi}\int_\Gamma \left[\frac{\partial\varphi}{\partial n}\right](y)\ln(|x-y|)\mathrm{d}\gamma(y) + \frac{1}{2\pi}\int_\Gamma [\varphi](y)\frac{\partial}{\partial n}(\ln(|x-y|))(y)\mathrm{d}\gamma(y)\ ; \quad (8.29)$$

- if x is on the boundary Γ, we have:

$$\frac{\varphi_i(x)+\varphi_e(x)}{2} = -\frac{1}{2\pi}\int_\Gamma \left[\frac{\partial\varphi}{\partial n}\right](y)\ln(|x-y|)\mathrm{d}\gamma(y) + \frac{1}{2\pi}\int_\Gamma [\varphi](y)\frac{\partial}{\partial n}(\ln(|x-y|))(y)\mathrm{d}\gamma(y). \quad (8.30)$$

Let us note that the conditions at infinity required by Theorem 8.5, i.e.

$$|\varphi_e(x)| \sim \frac{C}{r}, \quad |\nabla\varphi_e(x)| \sim \frac{C}{r^2}; \quad (8.31)$$

are here equivalent to a zero flux condition on Γ:

$$\int_\Gamma \frac{\partial\varphi_e}{\partial n}(y)\mathrm{d}\gamma(y) = 0. \quad (8.32)$$

This will be an important condition for the solution of the exterior Neumann problem for the Laplacian. It is specific to the two-demensional case. □

From Theorem 8.5, we immediately deduce the following two corollaries.

Corollary 8.8 Under the hypotheses of Theorem 8.5, if $[\varphi] = 0$, we have, denoting $q = [\frac{\partial \varphi}{\partial n}]$:

$$\forall x \in \mathbb{R}^3, \quad \varphi(x) = \frac{1}{4\pi} \int_\Gamma \frac{q(y)}{|x-y|} \mathrm{d}\gamma(y). \tag{8.33}$$

Corollary 8.9 Under the hypotheses of Theorem 8.5, if $[\frac{\partial \varphi}{\partial n}] = 0$, we have, setting $\mu = [\varphi] = \varphi_i - \varphi_e$:

$$\forall x \in \mathbb{R}^3 - \Gamma, \quad \varphi(x) = -\frac{1}{4\pi} \int_\Gamma \mu(y) \frac{\partial}{\partial n}\left(\frac{1}{|x-y|}\right)(y)\mathrm{d}\gamma(y), \tag{8.34}$$

$$\forall x \in \Gamma, \quad \varphi_i(x) = \frac{\mu(x)}{2} - \frac{1}{4\pi} \int_\Gamma \mu(y) \frac{\partial}{\partial n}\left(\frac{1}{|x-y|}\right)(y)\mathrm{d}\gamma(y). \tag{8.35}$$

The integral in (8.33) is called a *single layer potential*, whereas the one in (8.34) is a *double layer potential*; we shall come back to these names later. We shall now use these results to solve the Dirichlet problem (8.1), when $\Omega = \Omega_i$ or $\Omega = \Omega_e$, by first using the integral representation (8.33).

8.2 SOLUTION OF THE DIRICHLET PROBLEM BY A SINGLE LAYER POTENTIAL

Our goal being chiefly to describe a numerical method for solving the Dirichlet problem (8.1), we shall briefly summarise, without dwelling over the proofs, the theoretical results linked to our problem. For further results, the reader is referred to Nedelec (1977), Dautray and Lions (1990).

8.2.1 Summary of the Theoretical Aspects

To begin with, we shall assume that f is zero. From a continuous point of view, we have seen in the previous chapters that the interior Dirichlet problem is 'well posed' as soon as the data φ_Γ is sufficiently smooth; more precisely, if $\varphi_\Gamma \in H^{\frac{1}{2}}(\Gamma)$, problem (8.1) with $\Omega = \Omega_i$ has a unique solution φ_i in $H^1(\Omega_i)$.

One also shows (and we shall note it without proof) that under the same smoothness hypotheses on φ_Γ, the exterior Dirichlet problem admits a unique solution φ_e in the weighted space (we still write $r = |x|$):

$$W^1(\Omega_e) = \left\{ \psi, \frac{\psi}{\sqrt{1+r^2}} \in L^2(\Omega_e), \nabla\psi \in (L^2(\Omega_e))^3 \right\}. \tag{8.36}$$

By construction, the functions φ_i and φ_e coincide on Γ, so that, with notation as above, $[\varphi] = 0$. Thus, thanks to Corollary 8.8 (and to Remark 8.6 about the behaviour at infinity of φ_e), we may deduce an explicit expression for φ in the form (8.33). This expression involves an a priori unknown function q that represents the jump of the normal derivative of φ across the interface Γ between the two domains. This unknown function q is solution of an *integral equation*, obtained by writing (8.33) for points x on the boundary; indeed, on Γ, the function is perfectly known, since it is a datum of the problem: $\varphi_i = \varphi_e = \varphi_\Gamma$. Thus we have:

$$\forall x \in \Gamma, \quad \varphi_\Gamma(x) = \frac{1}{4\pi}\int_\Gamma \frac{q(y)}{|x-y|}\mathrm{d}\gamma(y). \tag{8.37}$$

If we could solve this equation in the unknown q, then the expression (8.33) gives us the value of φ at each point of $\mathbb{R}^3$, which concludes the (theoretical) problem of solving (8.1): we have an *integral representation of φ by a single layer potential.*

It remains to solve (8.37). The theoretical study shows that if φ_Γ is in the space $H^{\frac{1}{2}}(\Gamma)$, problem (8.37) admits a unique solution in the dual of the above space, denoted by $H^{-\frac{1}{2}}(\Gamma)$.

Because the solution space is not a classical function space, problem (8.37) will be solved in weak form, i.e. by writing its variational formulation:

$$\text{find } q \in H^{-\frac{1}{2}}(\Gamma) \text{ such that } \forall w \in H^{-\frac{1}{2}}(\Gamma), \text{ we have:} \tag{8.38}$$

$$\frac{1}{4\pi}\int_\Gamma\int_\Gamma \frac{q(y)w(x)}{|x-y|}\mathrm{d}\gamma(x)\mathrm{d}\gamma(y) = \int_\Gamma \varphi_\Gamma(x)w(x)\mathrm{d}\gamma(x).$$

It is precisely this formulation that we shall discretise in the next section.

Let us point out one last result that can be useful in practice, and whose proof, similar to that of Theorem 8.5, is given for instance in Kellog (1967).

Proposition 8.10 *Let φ be given by (8.33). Then, if $q \in H^{-\frac{1}{2}}(\Gamma)$, we have with the above notation,*

$$\forall x \in \Gamma, \quad \frac{\partial \varphi_i}{\partial n}(x) = \frac{q(x)}{2} + \frac{1}{4\pi}\int_\Gamma q(y)\frac{\partial}{\partial n_x}\left(\frac{1}{|x-y|}\right)\mathrm{d}\gamma(y), \tag{8.39}$$

$$\forall x \in \Gamma, \quad \frac{\partial \varphi_e}{\partial n}(x) = -\frac{q(x)}{2} + \frac{1}{4\pi}\int_\Gamma q(y)\frac{\partial}{\partial n_x}\left(\frac{1}{|x-y|}\right)\mathrm{d}\gamma(y),$$

where the notation n_x means that the derivative bears on the x variable (and not on the integration variable y).

Remark 8.11 The name 'single layer' comes from electrostatics: the expression (8.37) is indeed the one giving the density q of the electric charge at the surface of a conductor Γ held at a potential φ_Γ. □

Still from a theoretical point of view, let us end by pointing out the conditions that should be required from a possible right hand side f so that problem (8.1) still has a unique solution in the above function spaces: for the interior problem, it suffices that f be in $L^2(\Omega_i)$, and for the exterior problem, it suffices that f satisfies: $\sqrt{1+r^2}f \in L^2(\Omega_i)$.

8.2.2 Numerical Aspects

The scenario for solving (8.1) with an arbitrary right hand side f, has three stages:

STAGE 1: find φ_0 solution of $-\Delta\varphi_0 = f$ in the whole space (for instance), then set $\psi = \varphi - \varphi_0$; (8.1) is then equivalent to the solution of the following problem in the unknown ψ:

$$-\Delta\psi = 0 \text{ in } \Omega, \quad \psi|_\Gamma = \psi_\Gamma = \varphi_\Gamma - \varphi_0|_\Gamma. \tag{8.40}$$

STAGE 2: find q solution of the integral equation:

$$\forall x \in \Gamma, \quad \psi_\Gamma(x) = \frac{1}{4\pi}\int_\Gamma \frac{q(y)}{|x-y|}\,d\gamma(y). \tag{8.41}$$

STAGE 3: reconstruct φ thanks to (8.33), that is:

$$\forall x \in \mathbb{R}^3, \quad \varphi(x) = \varphi_0(x) + \frac{1}{4\pi}\int_\Gamma \frac{q(y)}{|x-y|}\,d\gamma(y). \tag{8.42}$$

Remark 8.12 If $f = 0$, the first stage disappears, so that to know φ at any point of Ω, it suffices to compute the function q defined over Γ by solving problem (8.41). We see one of the interests of the boundary integral method: solving a problem in three dimensions (seeking φ in the whole of Ω) is reduced to the solution of a problem in two dimensions (seeking q on Γ). In practice, it suffices to mesh the surface Γ, in contrast to what we would have had to do for a classical finite difference or finite element method. The computational cost will thus be *a priori* reduced.

Moreover, if the domain is an exterior domain, i.e. if $\Omega = \Omega_e$, it is clear that if we wish to solve (8.1) with a usual finite difference or finite element method, it is necessary, in practice, to bound the computational domain, for example by embedding Ω_e inside a large ball B_R. This leads to the following delicate problem: what boundary condition should we impose on the artificial boundary ∂B_R, so that the solution found in this way is identical (or at least almost identical), at all points in $\Omega_e \cap B_R$, to the restriction to this set of the solution of the initial problem? This problem does not arise with the boundary integral method, since the discretisation is carried out on the boundary Γ, which is bounded: this is a second advantage of this method.

What are its drawbacks? First, it is clear that the first argument fails as soon as $f \neq 0$, because, as we shall see later on, the function φ_0 from stage 1 is defined as a function of f by a volume integral. This means that, to compute it, we must, in theory, mesh the whole space! Fortunately, in practice it is possible to overcome this difficulty.

There remains, however, a drawback we cannot avoid: we shall see that the matrix of the linear system obtained from the discretisation of (8.41) is full, in contrast to what was the case for the other methods we quoted above. The solution of the linear system itself will be costly in both computer time and storage. Moreover, we shall see in Section 8.3, other situations where the matrix will not be symmetric, which limits the choice of possible solution methods. □

We shall now specify the numerical implementation of the various stages.

8.2.2.1 Implementation of stage 1

If the right hand side f is a function whose analytical expression is known, then it is possible to solve exactly the problem: $\Delta\varphi_0 = f$ in $\mathbb{R}^3$. Indeed, φ_0 is the convolution of the elementary solution of the operator $-\Delta$ with f, which gives us, using (8.6):

$$\forall x \in \mathbb{R}^3, \quad \varphi_0(x) = \frac{1}{4\pi}\int_{\mathbb{R}^3} \frac{1}{|x-y|} f(y)\mathrm{d}y. \tag{8.43}$$

So as to avoid having to mesh the whole space (!), it is possible to take the Fourier transform of the above relation, and then to solve this problem using the Fast Fourier Transform, or FFT.

Another possibility for the case $\Omega = \Omega_i$ is to choose a large rectangle R containing Ω_i and to solve the following Dirichlet problem in R by finite differences:

$$-\Delta\varphi_0 = f \text{ in } R, \quad \varphi_0|_\Gamma = 0. \tag{8.44}$$

If we do this, the values of φ_0 will only be (approximately) known at the nodes of the finite difference grid. When we need to know these values at other points later, we shall interpolate linearly from values taken at the closest grid nodes. If we wish to solve the linear system from the discretisation of (8.44) as fast as possible, we should use the multigrid method alluded to in Chapter 4: it is particularly efficient for a solution in a rectangle by the finite difference method.

8.2.2.2 Implementation of stage 2

With a view to discretising (8.41) by a 'boundary' element method, we construct an admissible triangulation of the surface Γ with nt triangles T_k, and we denote by h the size of the mesh thus defined and by Γ_h the computational domain (i.e. the union of all the triangles T_k).

A possible choice is to approximate the unknown q by its P^0 interpolant, denoted by q_h, i.e. there is one value q_h by triangle. The problem is then to solve:

$$\forall k \in \{1, \ldots, nt\}, \quad \frac{1}{4\pi}\int_{\Gamma_h} \frac{q_h(y)}{|x^k - y|}\mathrm{d}\gamma(y) = \psi_\Gamma(x^k), \tag{8.45}$$

if we denote by x^k the centre of mass of triangle T_k. This method is very popular with engineers [Brebbia (1978)], but it has two drawbacks:

- it leads to nonsymmetric systems;
- it is difficult to prove the existence of such a solution q_h.

This is why we prefer using the weak formulation (8.38) with φ_Γ replaced by ψ_Γ, or:

$$\text{find } q \in H^{-\frac{1}{2}}(\Gamma) \text{ such that } \forall w \in H^{-\frac{1}{2}}(\Gamma), \text{we have :} \qquad (8.46)$$
$$\frac{1}{4\pi}\int_\Gamma\int_\Gamma \frac{q(y)w(x)}{|x-y|}\mathrm{d}\gamma(x)\mathrm{d}\gamma(y) = \int_\Gamma \psi_\Gamma(x)w(x)\mathrm{d}\gamma(x).$$

Then we solve the following discrete problem

$$\text{find } q_h \in L_h \text{ such that } \forall w_h \in L_h, \text{ we have:} \qquad (8.47)$$
$$\frac{1}{4\pi}\int_{\Gamma_h}\int_{\Gamma_h} \frac{q_h(y)w_h(x)}{|x-y|}\mathrm{d}\gamma(x)\mathrm{d}\gamma(y) = \int_{\Gamma_h} \psi_\Gamma(x)w_h(x)\mathrm{d}\gamma(x),$$

where L_h is the space of functions constant on each triangle, a basis of which is made of the nt functions 1_{T_j} equal to 1 on triangle T_j and 0 on all other triangles. We then have, denoting by q_j the value of q_h on the triangle T_j,

$$q_h = \sum_{j=1}^{nt} q_j 1_{T_j},$$

where the q_j satisfy

$$\forall i \in \{1,\ldots,nt\}, \quad \frac{1}{4\pi}\sum_{j=1}^{nt}\int_{T_i}\int_{T_j}\frac{q_j}{|x-y|}\mathrm{d}y\mathrm{d}x = \int_{T_i}\psi_\Gamma(x)\mathrm{d}x. \qquad (8.48)$$

The equation above is obtained by taking for the function w_h in (8.47) each of the basis functions 1_{T_i} of the space L_h in turn. The discrete problem (8.47) thus reduces to the solution of the linear system with unknown $Q = (q_1, \ldots, q_{nt})^T$ and with matrix $A = ((A_{ij}))$

$$AQ = \Psi, \qquad (8.49)$$

with:

$$A_{ij} = \int_{T_i}\left(\int_{T_j}\frac{1}{|x-y|}\mathrm{d}y\right)\mathrm{d}x, \quad \Psi_i = 4\pi\int_{T_i}\psi_\Gamma(x)\mathrm{d}x. \qquad (8.50)$$

The matrix A is symmetric; moreover, if $z = (z_1, \ldots, z_{nt})$, we have

$$z^T A z = \int_\Gamma\int_\Gamma \frac{z_h(x)z_h(y)}{|x-y|}\mathrm{d}y\mathrm{d}x, \quad z_h = \sum_{i=1}^{nt} z_i 1_{T_i},$$

and this relation shows that A is positive definite. The solution of the system (8.49) is thus possible, but it is more complicated that that of system (8.45) since it requires computing double integrals. We shall treat this computation later, after having explained how to handle stage 3.

8.2.2.3 Implementation of stage 3

Once q_h has been computed, in order to reconstruct the approximation φ_h to φ, it suffices to write:

$$\varphi_h(x) = \varphi_{0h}(x) + \frac{1}{4\pi}\sum_{i=1}^{nt} q_i \int_{T_i} \frac{1}{|x-y|}\,dy, \tag{8.51}$$

where φ_{0h} is the approximation of φ_0 given by stage 1. If necessary, we can differentiate this equation at all points $x \notin \Gamma$, thus obtaining an approximation of the gradient of φ. For a point x in Γ, it suffices to use the results of Proposition 8.10.

We shall now describe the computation of the integrals involved in (8.51) as well as in the matrix and the right hand side of the linear system (8.49).

8.2.2.4 Integral computations

The computation of the right hand side is done using the quadrature formula (2.58), or, denoting the area of the triangle T_i by $|T_i|$ and its centre of mass by x_i:

$$\Psi_i \simeq 4\pi |T_i|\ \psi_\Gamma(x_i). \tag{8.52}$$

The same formula lets us compute the A_{ij} coefficients for indices i and j such that $T_i \cap T_j = \emptyset$, since the denominator in the expression for A_{ij} cannot vanish. We obtain:

$$\text{if } T_i \cap T_j = \emptyset, \quad A_{ij} \simeq \frac{|T_i|.|T_j|}{|x_i - x_j|}. \tag{8.53}$$

But this computation becomes very singular as soon as $i = j$ or T_i and T_j have a common edge or a common vertex.

We thus decide to use another method, based on a 'semi-analytical' computation of these integrals. First, the integral with respect to the variable x in the expression for A_{ij} is approximated by the midpoints quadrature formula, i.e. equation (2.60). This gives, denoting the midpoints of the edges of the triangle T by $x_k(T)$, $k \in \{1,2,3\}$:

$$A_{ij} \simeq \frac{|T_i|}{3}\sum_{k=1}^{3} A_j(x_k(T_i)), \quad A_j(x) = \int_{T_j} \frac{1}{|x-y|}\,dy. \tag{8.54}$$

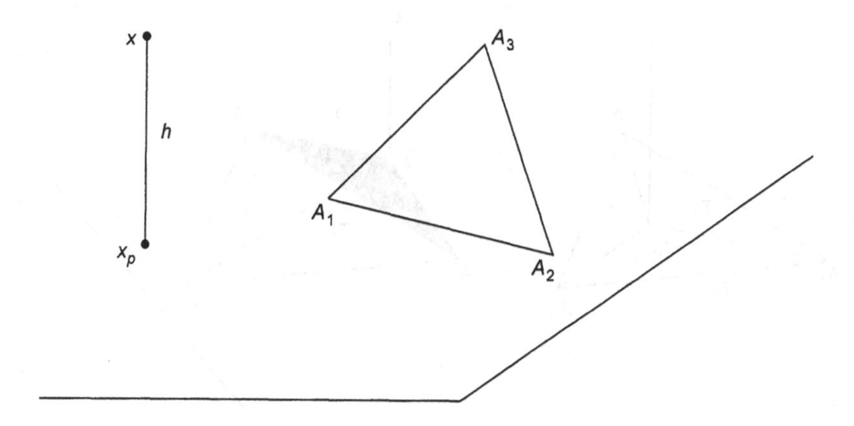

Figure 8.1 Notation for semi-analytical computation.

It thus suffices to compute integrals of the type:

$$I_T(x) = \int_T \frac{1}{|x-y|} \mathrm{d}y, \tag{8.55}$$

for points x located on edges of triangles of Γ_h. The name 'semi-analytical' comes from the fact that this second computation will be carried out partly analytically: this is what we shall now detail.

The notation is that of Figure 8.1: A_1, A_2, A_3 are the three vertices of the triangle T, x_p is the projection of point x on the plane P containing this triangle, and h is the distance between the points x and x_p. We shall begin by showing that it is sufficient to compute $I_T(x)$ for points x such that x_p is one of the vertices of T.

Let us join x_p to the three vertices A_1, A_2, A_3; this defines three new triangles denoted by T_1, T_2, T_3 with the convention: T_i is that of the three triangles that does not contain the vertex A_i (cf. Figure 8.2). The orientation of these three triangles is completely determined by that of T since each of these triangles has an edge in common with T. To illustrate this, let us refer to Figure 8.2: on the left, the case where x_p is internal to, or on the boundary of, the triangle T, the three triangles T_i are positively oriented (i.e. counterclockwise), $T = T_1 + T_2 + T_3$, and to this new local triangulation there corresponds a decomposition of $I_T(x)$ into three pieces:

$$I_T(x) = \sum_{i \in \{1,2,3\}} I_{T_i}(x), \quad I_{T_i}(x) = \int_{T_i} \frac{1}{|x-y|} \mathrm{d}y. \tag{8.56}$$

Thus, it suffices to compute the three integrals $I_{T_i}(x)$ by noting that now x_p is a vertex of T_i, for all $i \in \{1, 2, 3\}$.

However, if x_p is outside (or on the boundary of) the triangle T, as is the case on the right of Figure 8.2, one of the three triangles T_i has a negative area: in the example shown here, it is triangle T_2, the two other triangles having a positive area. We observe that formula (8.56) above remains true, as the computation of

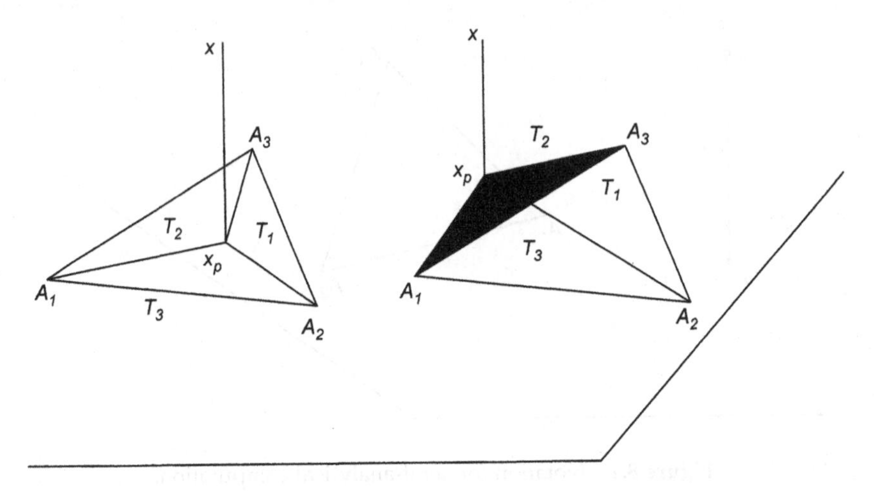

Figure 8.2 Illustration of the computation of three integrals.

the integral over $T_2 \cap T_1$ is globally zero, because of the different orientations of the triangles T_2 and T_1; the same is of course true for the computation over $T_2 \cap T_3$.

The formula (8.56), which is thus valid in all configurations, shows that, modulo changes in the triangulation around point x, the computation reduces to that of quantities of the type $I_T(x)$, with x_p a vertex of the triangle T.

We now carry out this computation, referring to the notation in Figure 8.3.

Let us put the coordinate origin at $x_p = A^1$, and use polar coordinates (r, θ) in the plane P, so that the line (A^2A^3) has $r = \rho / \cos\theta$ as an equation, where ρ is the distance between x_p and this line.

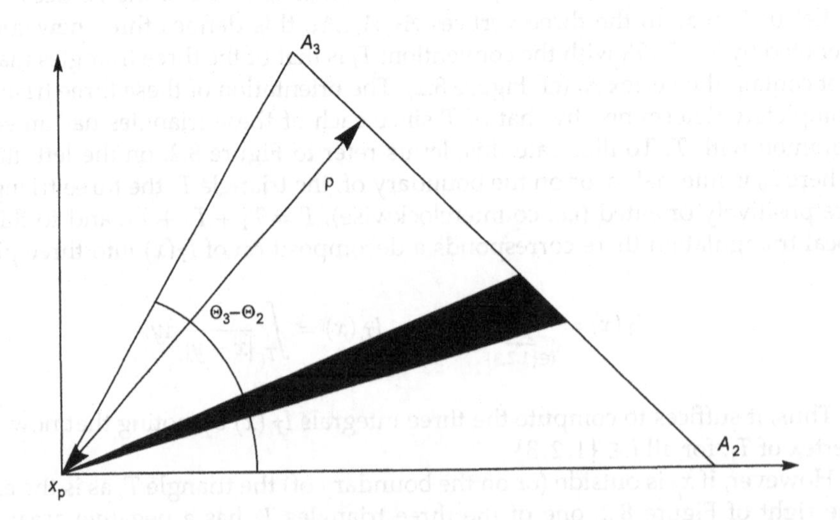

Figure 8.3 Notation for computation of quantities of type $I_T(x)$ with x_p a vertex of triangle T.

Then we have, still denoting by h the distance between the points x and x_p:

$$I_T(x) = \int_T \frac{1}{|x-y|}\mathrm{d}y = \int_T \frac{1}{\sqrt{h^2+(y-x_p)^2}}\mathrm{d}y$$
$$= \int_{\theta_2}^{\theta_3}\int_0^{\rho/\cos\theta} \frac{r\mathrm{d}r\mathrm{d}\theta}{\sqrt{h^2+r^2}} = \int_{\theta_2}^{\theta_3}\left(\sqrt{h^2+\frac{\rho^2}{\cos^2\theta}}-h\right)\mathrm{d}\theta. \tag{8.57}$$

It is possible, but painful, to carry out the analytical computation to the end, so we choose another solution. There are two different cases:

- If h is zero, which will always be the case for $i=j$, integration is exact:

$$I_T(x) = \int_{\theta_2}^{\theta_3} \frac{\rho\mathrm{d}\theta}{\cos\theta} = \rho\left[\ln\left(\frac{1+t}{1-t}\right)\right]_{t=\tan\left(\frac{\theta_2}{2}\right)}^{t=\tan\left(\frac{\theta_3}{2}\right)}; \tag{8.58}$$

- If, on the other hand, h is non-zero, the integral is nonsingular, and we can approximate it by using a Gaussian quadrature formula, (equation (2.54)), which gives us:

$$I_T(x) \simeq \left(\sqrt{h^2+\frac{\rho^2}{\cos^2\left(\frac{\theta_2+\theta_3}{2}\right)}}-h\right)(\theta_3-\theta_2). \tag{8.59}$$

To compute (8.51), we essentially distinguish two cases. If x is far from Γ, we can be content with a Gaussian quadrature formula:

$$\varphi_h(x) = \varphi_{0h}(x) + \frac{1}{4\pi}\frac{1}{3}\sum_{i=1}^{i=nt}\frac{q_i|T_i|}{|x-x^i|}. \tag{8.60}$$

In the other case, it will be necessary, for some triangles, to use an exact integration, like (8.58).

8.2.2.5 Accuracy of the method

We have made three different approximations:

- the boundary Γ has been approximated by a piecewise polynomial surface, defined by degree k (here $k=1$) polynomials;
- the function q has been approximated by a piecewise polynomial function q_h of degree m (here $m=0$);
- quadrature formulae with different orders (essentially 1 and 2) have been used to compute the integrals.

Without taking into account the latter approximation, the error can be written as [Nedelec (1976)]:

$$|(\varphi - \varphi_h)(x)| \leq \frac{c}{d(x,\Gamma) + d(x,\Gamma)^2}\{h^{k+1}\|q\|_0 + h^{m+2}\|q\|_{m+1} + \|\varphi_\Gamma - \tilde{\varphi}_{h\Gamma}\|_0 + h^{\frac{1}{2}}\|\varphi_\Gamma - \tilde{\varphi}_{h\Gamma}\|_{\frac{1}{2}}\}, \tag{8.61}$$

where $d(x,\Gamma)$ is the distance from x to Γ, h is the mesh size, and $\varphi_\Gamma - \tilde{\varphi_{h\Gamma}}$ is the error on the boundary data. Without going into further details (the norms, left unspecified here, are usual Sobolev norms), the estimate (8.61) is useful for two things. On the one hand, it shows that the error 'blows up' in the neighbourhood of the boundary Γ. On the other hand, it allows us to determine the optimal parameter pair (k, m), i.e. the one for which the errors in the geometry and the function approximation are of the same order of magnitude: the right choice is to take $k = m + 1$. With $k = 1, m = 0$, the method is in $O(h^2)$, except in the neighbourhood of the boundary.

For the derivatives, we have

$$|D^\alpha(\psi - \psi_h)|(x) \leq \frac{1}{d(x,\Gamma)^{|\alpha|}}M, \tag{8.62}$$

if we denote by M the right hand side of (8.61): far from the boundary, the method is in $O(h)$.

It is necessary to take quadrature formulae of order two in order to preserve these estimates. However, in practice, we can be satisfied with formulae involving only one Gauss point (when triangles are not neighbours). For the mathematical analysis of the error incurred by the use of Gaussian quadrature formulae the reader is referred to an article by Johnson and Scott (1989) and also to Guermond (1992).

8.3 OTHER PROBLEMS

8.3.1 Solution of the Dirichlet Problem by a Double Layer Potential

Let us study the case $f = 0$ and let us set about solving the interior Dirichlet problem, i.e.

$$-\Delta\varphi_i = 0 \quad \text{in } \Omega_i, \quad \varphi_i|_\Gamma = \varphi_\Gamma. \tag{8.63}$$

We assume in the sequel that Ω_i and Ω_e are connected. In our previous approach, we had actually 'extended' the equation outside Ω_i by solving the Dirichlet problem

$$-\Delta\varphi_e = 0 \quad \text{in } \Omega_e, \quad \varphi_e|_\Gamma = \varphi_i|_\Gamma, \tag{8.64}$$

i.e. by extending φ by continuity across the interface Γ.

We could also have chosen to match the normal derivatives of φ across Γ, i.e. to extend (8.63) by the following exterior Neumann problem:

$$-\Delta\varphi_e = 0 \quad \text{in } \Omega_e, \quad \frac{\partial \varphi_e}{\partial n}_{|\Gamma} = \frac{\partial \varphi_i}{\partial n}_{|\Gamma}. \tag{8.65}$$

Let us note that in three dimensions, the exterior Neumann problem admits a unique solution in the space $W^1(\Omega_e)$, as soon as the boundary data is in $H^{-\frac{1}{2}}(\Gamma)$, so that if φ_Γ is in the space $H^{\frac{1}{2}}(\Gamma)$, both of the above problems (8.63) and (8.65) have a unique solution respectively in the spaces $H^1(\Omega_i)$ and $W^1(\Omega_e)$.

Let us set $\mu = [\varphi] = (\varphi_i - \varphi_e)_{|\Gamma}$. Corollary 8.9 gives us the following integral representation of φ_i

$$\forall x \in \Omega_i, \quad \varphi_i(x) = -\frac{1}{4\pi}\int_\Gamma \mu(y)\frac{\partial}{\partial n}\left(\frac{1}{|x-y|}\right)(y)\mathrm{d}\gamma(y), \tag{8.66}$$

where μ is solution of the integral equation:

$$\forall x \in \Gamma, \quad \varphi_\Gamma(x) = \frac{\mu(x)}{2} - \frac{1}{4\pi}\int_\Gamma \mu(y)\frac{\partial}{\partial n}\left(\frac{1}{|x-y|}\right)(y)\mathrm{d}\gamma(y). \tag{8.67}$$

From a theoretical point of view, we can show that this problem admits a unique solution (if Ω_e is connected) in $H^{\frac{1}{2}}(\Gamma)$ as soon as the Dirichlet data φ_Γ is itself in the space $H^{\frac{1}{2}}(\Gamma)$.

Remark 8.13 The name 'double layer potential' for the expression (8.66) refers to the electric potentials generated by doublets. □

The numerical solution of problem (8.67) starts, as before, from its variational formulation

find $\mu \in H^{\frac{1}{2}}(\Gamma)$ such that $\forall w \in H^{\frac{1}{2}}(\Gamma)$, we have :

$$\frac{1}{2}\int_\Gamma \mu(x)w(x)\mathrm{d}\gamma(x) - \frac{1}{4\pi}\int_\Gamma\int_\Gamma \mu(y)w(x)\frac{\partial}{\partial n}\left(\frac{1}{|x-y|}\right)(y)\mathrm{d}\gamma(y)\mathrm{d}\gamma(x)$$
$$= \int_\Gamma \varphi_\Gamma(x)w(x)\mathrm{d}\gamma(x), \tag{8.68}$$

that is then discretised by P^0 or P^1 finite elements, (i.e. μ and w are approximated by piecewise constant or piecewise linear functions on each triangle). For instance, the matrix of the linear system derived from approximating problem (8.68) using P^0 finite elements has entries given by:

$$A_{ij} = |T_i|\delta_{ij} - \frac{1}{2\pi}\int_{T_i}\left(\int_{T_j}\frac{\partial}{\partial n}\left(\frac{1}{|x-y|}\right)(y)\mathrm{d}\gamma(y)\right)\mathrm{d}\gamma(x). \tag{8.69}$$

The linear system has a smaller condition number, but the integrals are more difficult to compute because of the stronger singularity (in the neighbourhood of $x = y$).

Remark 8.14 The solution of the exterior Dirichlet problem by a double layer potential only raises a few theoretical problems linked to the fact that the interior Neumann problem (obtained by 'extension') only has solutions if the *Neumann data has a vanishing integral* over Γ; this condition means, for φ_Γ, that:

$$\int_\Gamma \varphi_\Gamma(x)\frac{\partial u}{\partial n}(x)\mathrm{d}\gamma(x) = 0, \tag{8.70}$$

where $u \in W^1(\Omega_e)$ solves the problem:

$$-\Delta u = f \text{ in } \Omega_e, \quad u_{|\Gamma} = 1. \tag{8.71}$$

We then have the integral representation (8.34) of φ_e, except that μ is now a solution of

$$\forall x \in \Gamma, \quad \varphi_\Gamma(x) = -\frac{\mu(x)}{2} - \frac{1}{4\pi}\int_\Gamma \mu(y)\frac{\partial}{\partial n}\left(\frac{1}{|x-y|}\right)(y)\mathrm{d}\gamma(y). \tag{8.72}$$

This problem has a unique solution, up to an additive constant, in $H^{\frac{1}{2}}(\Gamma)$ (still assuming $\varphi_\Gamma \in H^{\frac{1}{2}}(\Gamma)$). In particular, if μ is a solution, so is $\mu + 1$, and we recover in this way the usual solid angle relations:

$$\int_\Gamma \frac{\partial}{\partial n}\left(\frac{1}{|x-y|}\right)(y)\mathrm{d}\gamma(y) = \begin{cases} 0 & \text{if} \in \Omega_e, \\ -4\pi \text{ if } x \in \Omega_i, \\ -2\pi \text{ if } x \in \Gamma. \end{cases} \tag{8.73}$$

□

8.3.2 *Solution of the Neumann Problem*

Let us consider, for instance, the exterior Neumann problem

$$-\Delta\varphi_e = 0 \quad \text{in } \Omega_e, \quad \frac{\partial \varphi_e}{\partial n}_{|\Gamma} = g \text{ in } \Gamma. \tag{8.74}$$

We have seen that this problem admits a unique solution in $W^1(\Omega_e)$ (if we assume that Ω_e is connected), if the data g is in $H^{-\frac{1}{2}}(\Gamma)$.

If we extend φ_e to Ω_i by a function φ_i that is a solution of the Dirichlet problem

$$-\Delta\varphi_i = 0 \quad \text{in } \Omega_i, \quad \varphi_{i|_\Gamma} = \varphi_{e|_\Gamma}, \tag{8.75}$$

we obtain an integral representation of φ by a single layer potential: in Ω_e, φ_e is written in the form (8.33), where $q = [\partial\varphi/\partial n]$ is, by virtue of (8.39), solution of:

$$\forall x \in \Gamma, \quad g(x) = -\frac{q(x)}{2} + \frac{1}{4\pi}\int_\Gamma q(y)\frac{\partial}{\partial n_x}\left(\frac{1}{|x-y|}\right)\mathrm{d}\gamma(y). \tag{8.76}$$

This problem admits a unique solution in $H^{-\frac{1}{2}}(\Gamma)$.

Let us now study the integral representation of the exterior Neumann problem by double layer potential. φ_e is given by (8.34), with μ solving formally the problem

$$‘g(x) = -\frac{1}{4\pi}\int_\Gamma \mu(y)\frac{\partial^2}{\partial n_x \partial n_y}\left(\frac{1}{|x-y|}\right)\mathrm{d}\gamma(y)’. \tag{8.77}$$

We have placed this problem between quotes, because it turns out that the kernel in (8.77) is not integrable, as it has a singularity in $|x-y|^{-3}$ (it is this very singular behaviour of the various kernels that gives the usual name 'singularity method'). The idea is then to turn around the difficulty by writing a variational formulation associated with this integral equation. In Dautray and Lions (1990), Nedelec shows that if we assume

$$\int_\Gamma g(x)\mathrm{d}\gamma(x) = 0,$$

this formulation can be written in the form

$$\begin{aligned} \frac{1}{4\pi}\int_\Gamma\int_\Gamma (\mu(y)-\mu(x))(w(y)-w(x))\frac{\partial^2}{\partial n_x \partial n_y}\left(\frac{1}{|x-y|}\right)(y)\mathrm{d}\gamma(y)\mathrm{d}\gamma(x) \\ = \int_\Gamma (gw)(x)\mathrm{d}\gamma(x), \end{aligned} \tag{8.78}$$

for all w in $H^{\frac{1}{2}}(\Gamma)$ defined up to an additive constant. It is then shown that this problem admits a unique solution, up to an additive constant, in the space $H^{\frac{1}{2}}(\Gamma)$. Moreover, there exists another formulation, equivalent the one above, giving integrals that are easier to compute; it is written as

$$\frac{1}{8\pi}\int_\Gamma\int_\Gamma (\nabla_\Gamma \times \mu)(y)\cdot(\nabla_\Gamma \times w)(x)\frac{1}{|x-y|}\mathrm{d}\gamma(x)\mathrm{d}\gamma(y) = \int_\Gamma (gw)(x)\mathrm{d}\gamma(x), \tag{8.79}$$

with the notation: $\nabla_\Gamma \times \varphi = \nabla\varphi \times n$. Because of the terms with derivatives, it is necessary to discretise μ and w by P^1 finite elements over a triangulation of Γ.

Last, let us note that to solve the exterior Neumann problem (8.74), it is also possible to extend φ_e by $\varphi_i = 0$ inside Ω_i. The jump $\mu = [\varphi] = \varphi_i - \varphi_e = -\varphi_e$ of φ across the interface Γ is an unknown function, whereas the jump in the normal derivative of φ is perfectly known, since we have:

$$\left[\frac{\partial\varphi}{\partial n}\right] = 0 - \frac{\partial\varphi_e}{\partial n}_{|\Gamma} = -g.$$

We then obtain an integral representation of φ_e by single and double layer potentials

$$\text{if } x \in \Omega_e, \quad \varphi_e(x) = -\frac{1}{4\pi}\int_\Gamma g(y)\frac{1}{|x-y|}\,\mathrm{d}\gamma(y) - \frac{1}{4\pi}\int_\Gamma \mu(y)\frac{\partial}{\partial n}\left(\frac{1}{|x-y|}\right)(y)\,\mathrm{d}\gamma(y), \tag{8.80}$$

where the function μ is a solution of (8.20), and this gives here:

$$\text{if } x \in \Gamma, \quad \frac{\mu(x)}{2} = \frac{1}{4\pi}\int_\Gamma g(y)\frac{1}{|x-y|}\,\mathrm{d}\gamma(y) + \frac{1}{4\pi}\int_\Gamma \mu(y)\frac{\partial}{\partial n}\left(\frac{1}{|x-y|}\right)(y)\,\mathrm{d}\gamma(y). \tag{8.81}$$

8.3.3 Another Example: Helmholtz Equation

If we look for periodic solutions of the wave equation ($\partial^2 v/\partial t^2 - \Delta v = 0$) in the form $v(x,t) = \exp(i\omega t)\,u(x)$, we are led to solve the equation

$$\forall x \in \Omega_e \text{ or } \forall x \in \Omega_i, \quad (\Delta + k^2)u(x) = 0, \tag{8.82}$$

whose solutions are complex valued. This equation is called the *Helmholtz equation*. We associate it with Neumann or Dirichlet boundary conditions.

Let us consider for instance the exterior Neumann problem:

$$(\Delta + k^2)u(x) = 0 \text{ in } \Omega_e, \quad \frac{\partial u}{\partial n}_{|\Gamma} = g \text{ on } \Gamma. \tag{8.83}$$

Let us assume g is in the space $H^{-\frac{1}{2}}(\Gamma)$. Then problem (8.83) admits a unique complex valued solution u that satisfies

$$ru \text{ bounded as } r = |x| \to +\infty, \tag{8.84}$$

$$r^2\left(\frac{\partial u}{\partial n} + iku\right) \text{ bounded as } r = |x| \to +\infty, \tag{8.85}$$

if we denote by n the exterior normal to the ball centred at the origin with radius r; this solution is in the space $H^1_{loc}(\Omega_e)$. The condition (8.85) is called the *radiation condition* or the *Sommerfeld condition*; it prescribes a behaviour at infinity for the solutions that rules out plane waves and ensures the uniqueness of the solution.

Among all the elementary solutions of the Helmholtz operator, only

$$E(x) = \frac{\exp(-ikr)}{4\pi r}, \quad r = |x|, \tag{8.86}$$

satisfies the radiation condition (8.85) and behaves as $1/r$ at infinity. This allows us to obtain the following integral representation theorem [Dautray and Lions (1990)]:

Theorem 8.15 Let $u_i \in C^\infty(\bar{\Omega}_i)$ be a solution of (8.82) in Ω_i and $u_e \in C^\infty(\bar{\Omega}_e)$ be a solution of (8.82) in Ω_e satisfying the conditions (8.84)–(8.85). We let u be the function defined over $\Omega_i \cup \Omega_e$ by $u_{|\Omega_e} = u_e$, $u_{|\Omega_i} = u_i$. Then:

- if $x \in \Omega_i \cup \Omega_e$,

$$u(x) = \frac{1}{4\pi}\int_\Gamma \left[\frac{\partial u}{\partial n}\right](y)\frac{\exp(-ik|x-y|)}{|x-y|}\mathrm{d}\gamma(y) \tag{8.87}$$
$$-\frac{1}{4\pi}\int_\Gamma [u(y)]\frac{\partial}{\partial n}\left(\frac{\exp(-ik|x-y|)}{|x-y|}\right)(y)\mathrm{d}\gamma(y),$$

still with the notation:

$$[u] = u_i - u_e, \qquad \left[\frac{\partial u}{\partial n}\right] = \frac{\partial u_i}{\partial n} - \frac{\partial u_e}{\partial n}\ ; \tag{8.88}$$

- if x is on the boundary Γ:

$$\frac{u_i(x) + u_e(x)}{2} = \frac{1}{4\pi}\int_\Gamma \left[\frac{\partial u}{\partial n}\right](y)\frac{\exp(-ik|x-y|)}{|x-y|}\mathrm{d}\gamma(y)$$
$$-\frac{1}{4\pi}\int_\Gamma [u(y)]\frac{\partial}{\partial n}\left(\frac{\exp(-ik|x-y|)}{|x-y|}\right)(y)\mathrm{d}\gamma(y). \tag{8.89}$$

This integral representation theorem then enables us to solve the interior or exterior Dirichlet or Neumann problems associated with the Helmholtz operator. We refer the reader to Dautray and Lions (1990) for the developments and proofs of the results briefly alluded to in this section.

Beyond the references already given in this chapter, let us point out some general works about boundary integral methods: Nedelec and Planchard (1973), Hanouzet (1971) for weighted spaces, Hsiao and Wendland (1977) and Le Roux (1977) for the two-demensional case, Vladimirov (1971), for the 'mathematical physics' aspects, and last Colton and Kress (1983) for the Helmholtz equation.

9 Some Algorithms for Parallel Computing

Summary

After introducing multiprocessing, we describe two mathematical approaches useful in parallel computing.

9.1 ARCHITECTURES AND PERFORMANCES

9.1.1 Vectorisation

One can no longer ignore the structure of computers when one is interested in their performance, as algorithm choice and programming have a deep influence on the computational speed, either because compilers are not able to exploit fully the resources of the computer, or because programmer intervention is unavoidable to optimise the code.

Vector computers appeared in the early 80's. They can perform the same operation on a large number of data, say 64, all the more rapidly as those data are stored in an appropriate way. In the beginning, compilers were not very effective at vectorising Fortran programs, and they had to be partially rewritten according to the following criteria [Levesque and Williamson (1989)]. Knowing that vectorisation is chiefly in DO loops starting with the innermost ones, we must avoid:

- simultaneous references to several elements of an array (for instance if a(i) requires knowledge of a(i-1));
- indirect addressing (like a(b(i)));
- subroutine calls and evaluation of transcendental functions (example: sin(a(i)));
- input and output;
- assigned GOTOs, GOTOs to exit from loops and nested IF statements.

Now that compilers have vastly improved, optimising vectorisation becomes rather easy, as long as the problem allows it, because vector computers give indications on those parts of the code where vectorisation could not be carried out. As a rule, finite element methods make almost constant use of DO loops with indirect addressing, as in

```
          DO 1 k=1,nt
            DO 1 i1=1,3
            ... q(me(k,i1)) ...
  1       CONTINUE
```

One way to avoid these *scatter/gather* (read to and write from memory) is to program the matrix and right hand side assembly by do loops over the edges instead of do loops over elements.

9.1.2 Hierarchical Memories

Even workstations are provided with hierarchical memories, of which the fastest (cache memory) are used for storing the data and instruction to handle next. If the most expensive parts of a computation hold in the cache, then the yield is very good. One can try to rewrite programs to take this factor into account.

9.1.3 Parallelisation

The above considerations result in modifications to the programs rather than to the algorithms. For computers that have several processing units, the situation is quite different, and results in a reworking of the algorithms to make full use of the computer architecture. In this chapter, we give several useful methods to parallelise programs for solving partial differential equations.

9.2 PARALLELISM

9.2.1 Multiprocessors

It is estimated that the computing power of microprocessors doubles every two years, and this will remain true at least until 2005. Since the end of the 80's, the microprocessors found in workstations, or even personal computers, have computing speeds that are close to those of supercomputers. The difference is only a question of architecture, of transfer speed, etc. It is currently cheaper to build supercomputers from several off-the-shelf microprocessors rather than to try and produce a faster computing unit. Moreover, workstations are usually networked and are available at night for large computations. Several libraries for message passing, like PVM (Parallel Virtual Machine), or MPI, are available under Unix, to help workstations cooperate. Thus, a cluster of workstations can be seen by the user as a single parallel machine.

One distinguishes between several types of parallel computers:

- SIMD: *Single Instruction Multiple Data* all computing units do the same thing at the same time.
- MIMD: *Mulitiple Instructions Multiple Data:* all computing units are independent.
- Fine grain (say more than 256 processors) or coarse grain machine.
- Shared memory (all the processors access all the memory) or distributed memory machine (a communication request must be sent by the processor over the network).

In this chapter we shall restrict ourselves to coarse grain MIMD machines, with a distributed memory. For this kind of machine (workstation networks belong to this class), PVM and MPI are good programming models.

It would take too long to describe PVM in this chapter [Geist *et al.* (1994)]. All we need to know is that this library has SEND and RECEIVE commands to send and receive blocks of data from one processor to another. Each transmission takes time proportional to $a+bT$, where b is the number of data packets of size T transmitted. The ratio between communication time and computation time determines what kind of algorithm to use.

Parallelising a sequential program can, up to some limits, be carried out by parallelising loops or blocks; this is necessary when the code has already been written, but the best performances are obtained by changing the solution algorithm. A domain decomposition method, like the Schwarz method, can be applied as an 'outer loop' to an already written program; this means that we treat the program as a black box having, as input, boundary conditions on the interfaces of the subdomains, and as output, the value of the solution at other interfaces. To be actually confident we are making full use of the machine architecture, the real method is to change the algorithm. We shall describe here three algorithms used in parallel computing for solving partial differential equations like those met in the previous chapters. All three are based on a subdivision of the domain of definition of the partial differential equation into subdomains: domain decomposition.

9.2.2 *Mesh Partitioning*

It is not easy to partition a triangulated domain Ω into two or more subdomains Ω_m. The problem lies with the number of vertices, because we need, for the LU factorisation, the number of vertices to be of the same order in every processor, and that the bandwidth μ be minimal in all subdomains. We recall the definition of the geometric bandwidth:

$$\mu_m = \max_{i,j,k}\{|i-j| : q^i \in T_k, q^j \in T_k, T_k \subset \Omega_m\}. \tag{9.1}$$

A specific program will have to be dedicated to this task.

9.2.3 *Load Balancing*

If all the processors have the same computing speed, load balancing can be performed *a priori* by affecting the same number of vertices to each processor. We must then know how to partition the domain triangulation into several equal parts. If the processors do not have the same computing speed, putting the same number of points in each subdomain will not be optimal. On the other hand, if some processors must manage other tasks, we must be able to quickly repartition the domain so as to balance the tasks better.

A method applicable in all cases is to divide the domain Ω into $q = np$ subdomains, where p is the number of processors. When the processors are identical and are not busy, we shall assign n subdomains to each processor. When they are busy with other tasks, or if they have different speeds, we shall assign to each processor a number of subdomains larger or smaller than n, according to the processor characteristics.

EXERCISE On a rectangle divided uniformly in $l \times L$ quadrangles, themselves split into two triangles, assess, by estimating the number of multiplications and divisions, for which values of l and L it becomes interesting to apply the subdomain method, even on a sequential processor (because the bandwidth is strictly smaller than the bandwidth over the whole domain). □

9.3 *SUBDOMAIN PARALLELISATION*

9.3.1 *The Problem*

Let us consider the following model problem: find $u(x,t)$ solution of the heat equation

$$\partial_t u - \Delta u = f, \quad u|_{t=0} = u^0 \text{ in } \Omega,$$
$$u = 0 \text{ on } \Gamma = \partial\Omega. \tag{9.2}$$

An explicit scheme in time gives

$$u^{m+1} = u^m + (\Delta u^m + f^m)\delta t. \tag{9.3}$$

and the same scheme implicit in time gives

$$u^{m+1} - \Delta u^{m+1}\delta t = u^m + f^m \delta t. \tag{9.4}$$

9.3.2 *Parallelising the Explicit Scheme*

After discretising in space by a finite element method and mass lumping (cf. Remark 2.21), we obtain

$$u_i^{m+1} = u_i^m - \frac{1}{\sigma_i}\int_\Omega (\nabla u_h^m \cdot \nabla w^i - f^m w^i)(x)\delta t dx, \tag{9.5}$$

where σ_i is the area of the triangles that have point q^i as a vertex, and where u_h^m is the approximation of u^m given by:

$$u_h^m(x) = \sum_i u_i^m w^i(x). \tag{9.6}$$

Consider a machine with p processors. We subdivide the domain Ω into q subdomains ($q = np$, $n \geq 1$). To each processor, we associate q/p subdomains, and we store all the data corresponding to these subdomains on the corresponding processor. Data exchanges are limited to the subdomain interfaces, so as to be able to compute Δu on these interfaces. Thus, for two subdomains Ω_1 and Ω_2, since

$$\int_\Omega (\nabla u_h^m \cdot \nabla w^i)(x)\mathrm{d}x = \int_{\Omega_1} (\nabla u_h^m \cdot \nabla w^i)(x)\mathrm{d}x + \int_{\Omega_2} (\nabla u_h^m \cdot \nabla w^i)(x)\mathrm{d}x, \tag{9.7}$$

each integral can be computed in each processor if q^i is on the interface; the sum will be carried out on both processors, so as to preserve the symmetry of the computation. Thus, it is more advisable to store information pertaining to the interface in all the processors handling subdomains that have points on this interface. This duplicates information, but is necessary to preserve symmetry.

To make the method faster, we can also write (9.5), (9.6) in matrix form and store the matrix once and for all

$$U^{m+1} = U^m + AU^m + F^m \tag{9.8},$$

with:

$$U_j = u_j, \quad F_i = \frac{\delta t}{\sigma_i}\int_\Omega f\, w^i \mathrm{d}x, \quad A_{ij} = \frac{\delta t}{\sigma_i}\int_\Omega \nabla w^i \nabla w^j \mathrm{d}x. \tag{9.9}$$

We split U into three blocks: $U = (u^1, u^2, u^3)^T$. Here u^1 gathers all unknowns in Ω_1, u_2 those of Ω_2 and u_3 those on the interface $\overline{\Omega_1} \cap \overline{\Omega_2}$. For parallelising, we could assign all operations on u^i to processor i and carry out operations on u^3 in a third processor, or better, assign half of the operations to one processor and the other half to the other processor (see below). With this decomposition, the matrix multiplication has the following structure:

$$\begin{pmatrix} A^{11} & 0 & A^{13} \\ 0 & A^{22} & A^{23} \\ A^{31} & A^{32} & A^{33} \end{pmatrix} \begin{pmatrix} u^1 \\ u^2 \\ u^3 \end{pmatrix}. \tag{9.10}$$

In (9.8), all operations can be carried out independently in each processor, except for the matrix-vector product AU^m, which we shall now study in detail.

9.3.3 *Parallelising the Matrix-vector Product*

Parallelisation is made easier by the fact that $A^{12} = A^{21} = 0$ and $A^{33} = A_1^{33} + A_2^{33}$ where A_l^{33} involves only data available on processor l. Here

$$A^{33}_{lij} = \frac{\delta t}{\sigma_i} \int_{\Omega_l} \nabla w^i \nabla w^j \mathrm{d}x \,. \tag{9.11}$$

Algorithm
1 Duplicate u^3 in both processors.
2 Compute on processor $i, i = 1, 2$:

$$v^i = \begin{pmatrix} A^{ii} & A^{i3} \end{pmatrix} \begin{pmatrix} u^i \\ u^3 \end{pmatrix}, v^3_i = A^{3i} u^i \text{ and } v^3_{3i} = A^{33}_i u^3.$$

3 Send v^3_1 and v^3_{31} (resp. v^3_2 and v^3_{32}) to processor 2 (resp. 1).
4 In both processors do $v^3 = v^3_1 + v^3_2 + v^3_{31} + v^3_{32}$.

Note that step 4 implies that step 1 will not be needed subsequently.

Remark 9.1 For this method, operations in each processor are in $O(N^d)$ when u has size N^d and the matrix is sparse, where d is the dimension of Ω. Duplicated computations and communications are in $O(N^{d-1})$, so the ratio is in $O(N)$. □

Remark 9.2 The linear system matrix A is an integral over Ω, which itself is a union of triangles T_k; thus $A = \sum_k A_k$. If we do not want to 'assemble' this matrix, we can use this property by computing the matrix-vector product triangle by triangle, that is subdomain by subdomain, without ever computing A, only A_k. □

9.3.4 *Parallelising the Implicit Scheme*

For an implicit algorithm like the Euler scheme

$$u^{m+1} - \delta t \Delta u^{m+1} = u^m + f^m \delta t, \tag{9.12}$$

the same strategy requires the solution of a linear system at each time step. We no longer need to lump the mass matrix; the linear system still has the form (9.9) with:

$$A_{ij} = \int_\Omega (w^i w^j + \delta t \nabla w^i \nabla w^j) \mathrm{d}x. \tag{9.13}$$

Then (9.12) can be written in matrix form as

$$AU^{m+1} = BU^m + F^m, \quad F^m_i = \delta t \int_\Omega f^m w^i \mathrm{d}x, \tag{9.14}$$

with:

$$B_{ij} = \int_\Omega w^i w^j \mathrm{d}x. \tag{9.15}$$

The product BU^m is as before and for an optimal performance we must solve the linear system in parallel on p processors, for example by the conjugate gradient method.

9.3.5 Parallelising the Conjugate Gradient Method

We assume that A is positive semi-definite. Let us recall the algorithm for solving $Ax = b$ preconditioned by a symmetric positive definite matrix C of the same size as A (refer to chapter 4 for a more thorough discussion of this algorithm).

Algorithm [solution of $\mathbf{Ax = b}$]
Initialisation: Assume $b \in \text{Range}(A)$; chose a computing accuracy $0 < \epsilon << 1$ and a starting point x^0 such that $Cx^0 \in \textit{Range}(A)$, ($x^0 = 0$ for example); set

$$n = 0, \quad g^0 = h^0 = C^{-1}(b - Ax^0), r^0 = b - Ax^0 \tag{9.16}$$

While $||g^n|| < \epsilon$ do

$$\begin{aligned} d &= Ah^n, \\ \rho &= \frac{r^n \cdot h^n}{h^n \cdot d}, \\ r^{n+1} &= r^n - \rho d, \\ x^{n+1} &= x^n + \rho h^n, \\ g^{n+1} &= g^n - \rho C^{-1} d, \\ (gr)^{n+1} &= g^{n+1} \cdot r^{n+1}, \\ \gamma &= \frac{(gr)^{n+1}}{(gr)^n}, \\ h^{n+1} &= g^{n+1} + \gamma h^n. \end{aligned} \tag{9.17}$$

Computation of Ah^n will be carried out in the same way as in Section 9.3.3, that is by decomposing the operations over the processors associated with the subdomains and by duplicating the data on the interfaces. What is left is essentially inner products with vectors distributed over several processors; we leave it to the reader to check that he can parallelise this operation. Last, the solution of the systems $C^{-1}d$ should also be done in parallel, a non trivial task if C is not a diagonal matrix.

9.3.6 The Schwarz Method

Solution algorithm for linear systems on parallel machines are still under development. In a different direction, we can also rethink the continuous problem. To solve an elliptic PDE, Schwarz (1869) has proposed the following method (called the additive Schwarz method). We subdivide the computational domain Ω into several (here two) overlapping subdomains: the overlap $\Omega_1 \cap \Omega_2$ has boundary Γ_{12} (resp Γ_{21}) that is internal to Ω_1 (resp. Ω_2), and apply the following iterative algorithm.

Algorithm [symmetric Schwarz with overlapping]
while $|u_1^k - u_2^k|_{\Omega_1 \cap \Omega_2} > \epsilon$,

1 Compute u_1^{k+1} solution of the PDE in Ω_1 with $u_1^{k+1} = u_2^k$ on Γ_{12};
2 Compute u_2^{k+1} solution of the PDE in Ω_2 with $u_2^{k+1} = u_1^k$ on Γ_{21};
3 In $\Omega_1 \cap \Omega_2$ set $u^{k+1} = \frac{1}{2}[u_1^{k+1} + u_2^{k+1}]$.

As the overlap can be difficult to manage, we might prefer a non-overlapping 'Dirichlet–Neumann' algorithm, based on the same idea, the common thread being the reduction of the jump of the normal derivative across the interfaces by an fixed point algorithm (choose Ω_1, Ω_2 with $\overline{\Omega}_1 \cap \overline{\Omega}_2 = \Gamma_{12} = \Gamma_{21}$, choose $0 < \theta < 1$, set $\lambda^0 = 0$):

Algorithm [non-overlapping Schwarz Dirichlet–Neumann]
While $|u_1^k - u_2^k|_{\Gamma_{12}} > \epsilon$,

1 Compute u_1^{k+1} solution of the PDE in Ω_1 with $u_1^{k+1} = \lambda^k$ on Γ_{12};
2 Compute u_2^{k+1} solution of the PDE in Ω_2 with $u_2^{k+1} = \lambda^k$ on Γ_{21};
3 Set $\lambda^{k+1} = \lambda^k + \theta[\frac{\partial u^{k+1}}{\partial n}]$.

NOTATION In the sequel $[f]|_\Sigma$ denotes the jump of f across a curve Σ, that is the right value minus the left value, right and left being defined with respect to the orientation of Σ.

Convergence
Convergence of the general method has been shown by Lions (1988). In one space dimension, and in the framework of the Dirichlet problem for the Laplacian, for instance, we have an analytical solution, which makes a simpler proof possible. Let us then consider the problem

$$-\frac{d^2u}{dx^2} = f, \text{ in }]0, L[, u(0) = 0, u(L) = 0,$$

and the symmetric Schwarz algorithm on two intervals $]0, d+h[$ and $]d, L[$ overlapping over a length h, with $0 < d+h < L, d > 0, h > 0$:

$$-\frac{d^2u^{k+1}}{dx^2} = f, \text{ in }]0,\ d+h[,\ u^{k+1}(0) = 0,\ u^{k+1}(d+h) = v^k(d+h).$$

$$-\frac{d^2v^{k+1}}{dx^2} = f, \text{ in }]d, L[,\ v^{k+1}(d) = u^k(d),\ v^{k+1}(L) = 0.$$

The solutions are

$$u(x) = \tilde{u}(x) + \frac{x}{d+h}(u(d+h) - \tilde{u}(d+h)), v(x) = \tilde{v}(x) + \frac{x-L}{d-L}(v(d) - \tilde{v}(d)),$$

where:

$$\tilde{u}(x) = -\int_0^x \int_0^t f(s)\mathrm{d}s\mathrm{d}t, \tilde{v}(x) = -\int_L^x \int_L^t f(s)\mathrm{d}s\mathrm{d}t.$$

If we substitute the values of $u(d+h)$ and $v(d)$, we obtain:

$$\begin{aligned} u^{k+1}(x) &= \tilde{u}(x) + \frac{x}{d+h}(v^k(d+h) - \tilde{u}(d+h)) \\ &= \tilde{u}(x) + \frac{x}{d+h}\left[\tilde{v}(d+h) + \frac{d+h-L}{d-L}(u^{k-1}(d) - \tilde{v}(d)) - \tilde{u}(d+h)\right]. \end{aligned}$$

If we take $x = d$ we obtain a recurrence relation of the form

$$u^{k+1}(d) - u^* = \frac{d}{d+h}\frac{d+h-L}{d-L}(u^{k-1}(d) - u^*),$$

with:

$$u^* = \left[\tilde{u}(d) + \frac{d}{d+h}(\tilde{v}(d+h) - \frac{d+h-L}{d-L}\tilde{v}(d) - \tilde{u}(d+h))\right] \Big/ \left(1 - \frac{d}{d+h}\frac{d+h-L}{d-L}\right).$$

Thus we have

$$u^{2p}(d) - u^* = \left(\frac{d+h-L}{d-L}\right)^p (u^0(d) - u^*) = \left(1 - \frac{h}{L-d}\right)^p (u^0(d) - u^*),$$

that is the speed of convergence is polynomial, and is all the faster when $h/(L-d)$ is close to 1.

If both processors are identical, then to balance the load, we must take $L - d = d + h$ which gives a convergence speed equal to $2h/(L+h)$.

Remark 9.3 There remains to show that convergence on the interface implies convergence at any point. We leave that to the reader. □

Similarly, in one space dimension, convergence of the symmetric Dirichlet–Neumann algorithm on two intervals $]0, d[\times]d, L[$ can be analysed from the analytical solution; more precisely:

$$u^{k+1}(x) = \tilde{u}(x) + \frac{x}{d}(\lambda^k - \tilde{u}(d)), \quad v^{k+1}(x) = \tilde{v}(x) + \frac{x-L}{d-L}(\lambda^k - \tilde{v}(d)).$$

Let us choose $n = x/|x|$; we have ($d > 0$):

$$\frac{\partial v^{k+1}}{\partial n}(d) - \frac{\partial u^{k+1}}{\partial n}(d) = \frac{\mathrm{d}\tilde{v}}{\mathrm{d}x}(d) + \frac{1}{d-L}(\lambda^k - \tilde{v}(d)) - \frac{\mathrm{d}\tilde{u}}{\mathrm{d}x}(d) - \frac{1}{d}(\lambda^k - \tilde{u}(d)).$$

We thus have the following recursion:

$$\lambda^{k+1} = \lambda^k - \theta\left(\frac{L}{L-d}\lambda^k - \frac{\tilde{u}(d)}{d} - \frac{\tilde{v}(d)}{L-d}\right) - \frac{\mathrm{d}}{\mathrm{d}x}(\tilde{u} - \tilde{v})(d).$$

If we denote, as above, the solution of the recursion by

$$\lambda^* = \frac{L-d}{L}\left[\frac{\tilde{u}(d)}{d} + \frac{\tilde{v}(d)}{L-d} - \frac{1}{\theta}\frac{\mathrm{d}}{\mathrm{d}x}(\tilde{u} - \tilde{v})(d)\right],$$

we rewrite the recursion in the following form:

$$\lambda^{k+1} - \lambda^* = \frac{L(1-\theta) - d}{L-d}(\lambda^k - \lambda^*).$$

The algorithm will always converge, for any d, if for example, $0 < \theta \leq 1$. Load balancing implies the choice $d = L/2$ and the convergence speed is then $4(1-2\theta)/L$. The optimal value is $\theta = 1/2$ and the method converges in one iteration, which is natural since we are in only one space dimension and the problem is linear.

9.4 THE SCHUR COMPLEMENT METHOD

9.4.1 Principle

Let us consider again the linear algebra problem with $A \in R^{n\times n}, b \in \mathbb{R}^n$:

$$\text{find } x \in \mathbb{R}^n \quad \text{solving } Ax = b \tag{9.18}$$

Let us partition A into four blocks: $A^{11} \in \mathbb{R}^{l\times l}$, $A^{12} \in \mathbb{R}^{l\times(n-l)}$, $A^{21} \in \mathbb{R}^{(n-l)\times l}$ and $A^{22} \in \mathbb{R}^{(n-l)\times(n-l)}$ such that:

$$A = \begin{pmatrix} A^{11} & A^{12} \\ A^{21} & A^{22} \end{pmatrix}. \tag{9.19}$$

Let b^1 be the vector formed with the l first entries of b and b^2 the remainder, and similarly for x. Equation (9.18) is then written as

$$\begin{pmatrix} A^{11} & A^{12} \\ A^{21} & A^{22} \end{pmatrix}\begin{pmatrix} x^1 \\ x^2 \end{pmatrix} = \begin{pmatrix} b^1 \\ b^2 \end{pmatrix}, \tag{9.20}$$

or:

$$A^{11}x^1 + A^{12}x^2 = b^1,$$
$$A^{21}x^1 + A^{22}x^2 = b^2.$$

If $(A^{11})^{-1}$ exists, then

$$x^1 = (A^{11})^{-1}(b^1 - A^{12}x^2), \tag{9.21}$$

and

$$(A^{22} - A^{21}(A^{11})^{-1}A^{12})x^2 = b^2 - A^{21}(A^{11})^{-1}b^1. \tag{9.22}$$

The matrix $A^{22} - A^{21}(A^{11})^{-1}A^{12}$ is called the *Schur complement* of A.

9.4.2 Schur Complement and Parallel Computing

To simplify notation, we only consider here a three processors machine: two slaves and a master processor. Let us recall that solving equation (9.2) by an implicit scheme in time leads to system (9.10) $Ax = b$ of the form:

$$\begin{pmatrix} A^{11} & 0 & A^{13} \\ 0 & A^{22} & A^{23} \\ A^{31} & A^{32} & A^{33} \end{pmatrix} \begin{pmatrix} x^1 \\ x^2 \\ x^3 \end{pmatrix} = \begin{pmatrix} b^1 \\ b^2 \\ b^3 \end{pmatrix}. \tag{9.23}$$

Let us assume A^{11} and A^{22} are invertible. Let us express x^1 and x^2 as functions of x^3 by using the first two equations, giving:

$$\begin{aligned} x^1 &= (A^{11})^{-1}(b^1 - A^{13}x^3), \\ x^2 &= (A^{22})^{-1}(b^2 - A^{23}x^3). \end{aligned} \tag{9.24}$$

The last equation becomes

$$Sx^3 = c, \tag{9.25}$$

because (9.24) substituted in (9.23) gives:

$$\begin{aligned} (A^{33} - A^{31}(A^{11})^{-1}A^{13} - A^{32}(A^{22})^{-1}A^{23})x^3 \\ = b^3 - A^{31}(A^{11})^{-1}b^1 - A^{32}(A^{22})^{-1}b^2. \end{aligned} \tag{9.26}$$

Thus we have:

$$\begin{aligned} S &= A^{33} - A^{31}(A^{11})^{-1}A^{13} - A^{32}(A^{22})^{-1}A^{23}, \\ c &= b^3 - A^{31}(A^{11})^{-1}b^1 - A^{32}(A^{22})^{-1}b^2. \end{aligned} \tag{9.27}$$

Note that computing S is both difficult and costly. Equation (9.25) will then be solved by a preconditioned conjugate gradient method, or any other iterative method; let us, for instance, give the details of the algorithm when the system is solved by a fixed point method

$$\tilde{S}(x^{3^{m+1}} - x^{3^m}) = c - Sx^{3^m}, \tag{9.28}$$

where $\tilde{S}$ is the preconditioner for S. This algorithm converges when $\tilde{S}$ is spectrally close to S (the eigenvalues $\lambda(\tilde{S})$ of $\tilde{S}$ are close to the eigenvalues $\lambda(S)$ of S), since if x^{3^*} solves (9.25), we can rewrite (9.28) as

$$\tilde{S}(x^{3^{m+1}} - x^{3^*}) = (\tilde{S} - S)(x^{3^m} - x^{3^*}),$$

which implies:

$$||x^{3^{m+1}} - x^{3^*}|| < \frac{\lambda_{max}(\tilde{S} - S)}{\lambda_{min}(\tilde{S})} ||x^{3^m} - x^{3^*}||.$$

Let us organise the computation in the following way: let us factorise A^{ii} and solve the systems $A^{ii}v^i = b^i$ by Gaussian elimination on each processor i ($i = 1,2$).

Algorithm

0 Choose x^{3^0}.

For $m = 0$ to mMax do

1 for each processor $i = 1,2$

1.1 Solve by Gaussian elimination $A^{ii}x^{i^m} = b^i - A^{i3}x^{3^m}$

1.2 Compute $y^{i^m} = A^{3i}x^{i^m}$

2 On the master processor, solve

$$\tilde{S}x^{3^{m+1}} = b^3 - A^{33}x^{3^m} - y^1 - y^2 + \tilde{S}x^{3^m}$$

end loop over m .

Remark 9.4 This computation can also be carried out on only two processors, as in Section 9.3.3. But parallelising step 2 is not simple. □

9.4.3 *Properties of S*

Proposition 9.5 *If A as well as the diagonal blocks A^{ii} are non-singular, then this is also true of S.*

PROOF To simplify notation, let us consider the case where $S = A^{22} - A^{21}(A^{11})^{-1}A^{12}$. Take z such that $Sz = 0$. Let $y = -(A^{11})^{-1}A^{12}z$, we then have

$$0 = A^{22}z + A^{21}y, \quad A^{11}y + A^{12}z = 0, \tag{9.29}$$

that is $A\binom{y}{z} = 0$, and this implies, in particular, that $z = 0$ since A is invertible. □

Proposition 9.6 *If A is positive definite, then so is S.*

PROOF First it is clear that if A is symmetric, then so is S. Let us again consider the case of two blocks instead of three, so that $S = A^{22} - A^{21}(A^{11})^{-1}A^{12}$. The assumption A positive definite gives

$$x^{1^T}A^{11}x^1 + 2x^{1^T}A^{12}x^2 + x^{2^T}A^{22}x^2 > 0 \text{ for any } x \neq 0. \tag{9.30}$$

Let us take $x^1 = -(A^{11})^{-1}A^{12}x^2$. We then have

$$x^{1^T}A^{11}x^1 = -x^{1^T}A^{12}x^2 = x^{2^T}A^{21}(A^{11})^{-1}A^{12}x^2, \tag{9.31}$$

so that (9.30) becomes

$$x^{2^T}A^{22}x^2 - x^{2^T}A^{21}(A^{11})^{-1}A^{12}x^2 > 0 \text{ for any } x^2 \neq 0, \tag{9.32}$$

i.e. $x^{2^T}Sx^2 > 0$ for any $x^2 \neq 0$.

The exercise in Appendix 9.A aims at showing, on an example, that the condition number of the Schur matrix is in $O(h^{-1})$. This is to be compared to the condition number of the Laplacian, which is $O(h^{-2})$. Even if the Schur linear systems are better conditioned, it is still necessary to precondition iterative methods to solve them, all the more so as each iteration requires a large amount of work because of the presence of $(A^{ii})^{-1}$.

9.4.4 Choice of $\tilde{S}$

When $S = A^{33} - A^{31}(A^{11})^{-1}A^{13} - A^{32}(A^{22})^{-1}A^{23}$, one of the possible choices for $\tilde{S}$ is

$$\tilde{S} = A^{33} - A^{31}(\text{diag}(A^{11}))^{-1}A^{13} - A^{32}(\text{diag}(A^{22}))^{-1}A^{23}, \tag{9.33}$$

where diag(A) is a diagonal matrix whose diagonal is that of A.

A more sophisticated method, called the *Neumann–Neumann preconditioner*, and for which we can show that $\tilde{S}^{-1}S$ has a condition number in $O(1)$ stems from the following Remark (cf. Chan and Mathew [1994] or Smith *et al.* [1996] for further details).

In the case of problem (9.2), or any other constant coefficients PDE, we have seen (cf. (9.11)) that the block of the linear system matrix corresponding to the interface between the two subdomains has the form: $A^{33} = A_1^{33} + A_2^{33}$. Thus $S = S^1 + S^2$ with

$$S^i = A_i^{33} - A^{3i}(A^{ii})^{-1}A^{i3}, i = 1, 2.$$

When Ω_2 and Ω_1 are symmetric with respect to Σ, we have $S^1 = S^2 = S/2$. But it is easy to see that, up to a sign, Sv is none other than the trace on Σ of the PDE solution in Ω_i with a Neumann condition equal to v on Σ. For example, in the case of (9.4) (cf. (9.2)) with a Dirichlet boundary condition on $\partial\Omega$,

$$\int_{\Omega_i}\left(\frac{1}{\delta t}u_h w^j + \nabla u_h \nabla w^j\right)\mathrm{d}x = \int_{\Sigma} v w^j, \forall j, \text{vertex } q^j \in \partial\Omega_i - \Sigma$$
$$\Leftrightarrow \quad u_h(q^j) = (S^i v)_j, \forall q^j \in \Sigma.$$

This leads to the idea of defining $\tilde{S}v$, in the general case, as half the sum of the traces on Σ of the solutions of the above equation.

9.5 A LAGRANGIAN METHOD

9.5.1 Principle of the Method

Let us consider the following Dirichlet problem:

$$-\Delta u = f \text{ in } \Omega, \quad u|_{\Gamma} = 0. \tag{9.34}$$

Let us consider again a non-overlapping partition of the domain:

$$\overline{\Omega} = \overline{\Omega_1} \cup \overline{\Omega_2}, \quad \Sigma = \overline{\Omega_1} \cap \overline{\Omega_2}, \text{ with } \Omega_1 \cap \Omega_2 = \emptyset. \tag{9.35}$$

We replace it by two problems over the Ω_i with a Neumann condition on the interface Σ

$$-\Delta u^i = f \text{ in } \Omega_i, \quad u^i|_{\Gamma} = 0, \quad \frac{\partial u^i}{\partial n}|_{\Sigma} = \lambda, \tag{9.36}$$

where n is a normal to Σ (for instance the outer normal to Ω_1) and where λ will be determined so as to have:

$$u^1 = u^2 \text{ on } \Sigma. \tag{9.37}$$

In variational form, this is written:
Find $\{u^1, u^2, \lambda\} \in H^1(\Omega_1) \times H^1(\Omega_2) \times L^2(\Sigma)$ such that

$$\begin{aligned} &\int_{\Omega_i} \nabla u^i \cdot \nabla w \,\mathrm{d}x + (-1)^i \int_{\Sigma} \lambda w = \int_{\Omega_i} f w \,\mathrm{d}x, \forall w \in H_0^1(\Omega), i = 1, 2, \\ &\int_{\Sigma} (u^2 - u^1)\mu = 0, \forall \mu \in L^2(\Sigma). \end{aligned} \tag{9.38}$$

We must remember that with this method u takes two values on the interface Σ, a right value and a left value, though they are meant to become equal.

9.5.2 Discretisation

We approximate the problem with finite elements of degree 1 for u and degree 0 for λ:

Find $u_h \in H_{0h}$ and $\lambda_h \in L_h$ such that

$$\int_{\Omega_{hi}} \nabla u_h^i \cdot \nabla w_h dx + (-1)^i \int_{\Sigma_h} \lambda_h w_h = \int_{\Omega_{hi}} f w_h dx, \forall w \in H_{0h},$$
$$\int_{\Sigma_h} (u_h^2 - u_h^1)\mu_h = 0, \quad \forall \mu_h \in L_h, \tag{9.39}$$

where H_h is the space of piecewise linear continuous functions on the triangulation Ω_h of Ω, H_{0h} is the subspace of H_h of functions vanishing on the boundary Γ and L_h is the space of piecewise constant functions on the discrete interface Σ_h. Here we also compute a right value and a left value of the solution on Σ_h, but these will be equal to the solution.

Let us rewrite (9.39) as a linear system

$$\begin{pmatrix} A^{11} & 0 & B^1 \\ 0 & A^{22} & B^2 \\ B^{1T} & B^{2T} & 0 \end{pmatrix} \begin{pmatrix} U^1 \\ U^2 \\ \Lambda \end{pmatrix} = \begin{pmatrix} b^1 \\ b^2 \\ 0 \end{pmatrix}. \tag{9.40}$$

The notation is as follows: w^i is the function in H_h that takes the value 1 at the ith vertex and 0 at all other vertices; $\bar{w}^i$ is the function of $L^2(\Sigma)$ that takes the value 1 on the ith edge and 0 on the other edges. The vector U^i with components u_j^i and the vector Λ with components λ_k are such that

$$u_h^i(x) = \sum_j u_j^i w^j(x) \text{ if } x \in \Omega^i, i = 1,2; \lambda_h(x) = \sum_k \lambda_k \bar{w}^k(x) \text{ if } x \in \Sigma, \tag{9.41}$$

$$A_{ij}^{kk} = \int_{\Omega_{hk}} \nabla w^i \nabla w^j dx, \quad B_{ij}^k = (-1)^k \int_{\Sigma_h} w^i \bar{w}^j, k = 1,2.$$

We could also solve this system by the Schur method.

Convergence can be shown as soon as the spaces H_h and L_h are 'compatible', meaning that they satisfy the 'inf-sup' condition (cf. for instance Brezzi and Fortin (1991)).

9.6 MORTAR ELEMENTS

9.6.1 Generalities

The domain decomposition method of Section 9.5 depends on the triangulation of the domain. The question that arises is: is it not possible to make the partition of the domain independent of the triangulation? That is, we want to partition first, then triangulate each subdomain. This triangulation could even be carried out in parallel.

If we allow the triangulations not to fit together on the interfaces, the problem becomes much simpler, since the meshing operations in the subdomains are now independent.

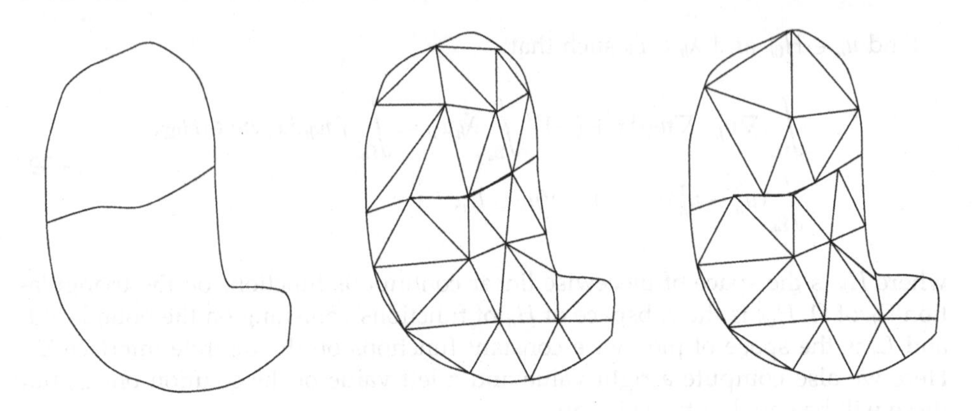

Figure 9.1 Two subdomains with matching (left) and non matching (right) triangulations

Figure 9.1 shows a domain partitioned into two subdomains, each one with its own triangulation. These triangulations may (centre picture) or may not (right picture) fit together on the interface between the two subdomains.

When a vertex on one side does not coincide with a vertex on the other side, we have to deal with a *non-conforming* finite element method. Such a method has been proposed by Bernardi, Maday and Patera (1994); we shall now explain it.

Let us consider again the Dirichlet problem on the domain Ω:

$$-\Delta\phi = f \text{ in } \Omega, \quad \phi|_{\partial\Omega} = 0, \tag{9.42}$$

with:

$$\overline{\Omega} = \overline{\Omega}_1 \cup \overline{\Omega}_2, \quad \Sigma = \overline{\Omega_1} \cap \overline{\Omega_2}. \tag{9.43}$$

Its variational formulation is

$$\int_{\Omega} (\nabla\phi \cdot \nabla w)(x)\mathrm{d}x = \int_{\Omega} (f\,w)(x)\mathrm{d}x, \quad \forall w \in H_0^1(\Omega), \tag{9.44}$$

whence

$$\int_{\Omega_1\cup\Omega_2} \nabla\phi \cdot \nabla w\mathrm{d}x = \int_{\Omega_1\cup\Omega_2} f\,w\mathrm{d}x, \quad \forall w \in H, \tag{9.45}$$

where

$$H = \{w \in H^1(\Omega_1 \cup \Omega_2) : [w]_\Sigma = 0, \quad w|_{\partial\Omega} = 0\}, \tag{9.46}$$

$[w]$ denoting the jump of w across the interface Σ between Ω_1 and Ω_2.

9.6.2 Discretisation

The discrete problem will thus be: for any $\phi_h \in H_h$, we have

$$\int_{\Omega_1 \cup \Omega_2} (\nabla \phi_h . \nabla w_h)(x) dx = \int_{\Omega_1 \cup \Omega_2} (f\, w_h)(x) dx \quad \forall w_h \in H_h, \tag{9.47}$$

where

$$H_h = \{w_h \in H_h(\Omega_1) \cup H_h(\Omega_2) : \int_\Sigma [w_h] v_h d\gamma = 0, \forall v_h \in W_h, w_h|_{\partial\Omega} = 0\}. \tag{9.48}$$

$H_h(\Omega_i)$ is the finite element space made up of piecewise polynomial and continuous functions on the triangulation of Ω_i. Let us also consider W_h, the space of piecewise polynomial and continuous functions on the interface Σ. Theory suggests taking a discretisation $\Sigma_h = \cup_{0 \ldots N-1} [z_j, z_{j+1}]$ of Σ made up of line segments and to consider the space of piecewise polynomials and continuous functions on Σ_h with strong Neumann conditions at both ends:

$$\begin{aligned} W_h = \{w_h \in C^0(\Sigma_h) : w_h|_{[z_j, z_{j+1}]} \in P^k, \forall j \in \{0, .., N-1\}, \\ w_h(z_0) = w_h(z_1), w_h(z_{N-1}) = w_h(z_N)\}. \end{aligned} \tag{9.49}$$

Thus, when k=1, the dimension of W_h is $N-1$ and the basis functions are indexed from 1 to $N-1$.

The error estimate has been obtained by Bernardi *et al.* (1994). When linear elements are used in Ω_1 and Ω_2, taking $k = 1$ in (9.49) is a good choice. We then have:

$$||\phi_h - \phi||_{H^1(\Omega_1 \cup \Omega_2)} \le Ch. \tag{9.50}$$

9.6.3 Numerical Solution

This method allows for *unstructured* meshes. On the other hand, the price to be paid is the solution of a saddle point linear system:

$$\begin{pmatrix} A^{11} & 0 & B^{1T} \\ 0 & A^{22} & B^{2T} \\ B^1 & B^2 & 0 \end{pmatrix} \begin{pmatrix} u^1 \\ u^2 \\ \lambda \end{pmatrix} = \begin{pmatrix} f^1 \\ f^2 \\ 0 \end{pmatrix}. \tag{9.51}$$

Indeed, let M be the number of vertices in Ω_1 (including those in Σ) and let $\{w^j\}_{j \le M}$ be the hat functions on the domain Ω. We have

$$\phi_h(x) = I_{\Omega_1}(x) \sum_{j \le M} u_j w^j(x) + I_{\Omega_2}(x) \sum_{j > M} u_j w^j(x), \tag{9.52}$$

where $I_{\Omega_i}, i = 1, 2$ is the characteristic function of Ω_i. Weak continuity (the constraint in (9.48)) requires that, for all $k \in \{1, .., N-1\}$:

$$\sum_{j\leq M} u_j \int_\Sigma ([w^j I_{\Omega_1}]\omega^k)(x)\mathrm{d}\gamma(x) + \sum_{j>M} u_j \int_\Sigma ([w^j I_{\Omega_2}]\omega^k)(x)\mathrm{d}\gamma(x) = 0. \qquad (9.53)$$

In matrix form, we have:

$$Bu = 0, \text{ with } B_{kj} = \int_\Sigma \left(\left[w^j I_{\Omega_{l(j)}}\right]\omega^k\right)(x)\mathrm{d}x, \qquad (9.54)$$

where $l(j) = 1$ if $j \leq M$ and $l(j) = 2$ otherwise. The expression (9.47) becomes

$$\sum_{j\leq M} u_j \int_{\Omega_1} \nabla w^j \cdot \nabla w_h \mathrm{d}x + \sum_{j>M} u_j \int_{\Omega_2} \nabla w^j \cdot \nabla w_h \mathrm{d}x = \int_\Omega f\, w_h \mathrm{d}x, \qquad (9.55)$$

for any w_h such that

$$w_h|_{\partial\Omega} = 0, \text{and} \int_\Sigma [w_h]\omega^k \mathrm{d}\gamma = 0, \quad \mathrm{k} = 1, ..., \mathrm{N}-1. \qquad (9.56)$$

Let us expand w_h:

$$w_h(x) = I_{\Omega_1}(x) \sum_{j\leq M} w_j w^j(x) + I_{\Omega_2}(x) \sum_{j>M} w_j w^j(x). \qquad (9.57)$$

Let $w = (\overline{w^1}, \overline{w^2})^T$ be the column vector with entries w_j, with the M first components in $\overline{w^1}$, (9.55) becomes

$$\overline{w^1}^T A^{11} u^1 + \overline{w^2}^T A^{22} u^2 = \overline{w^1} f^1 + \overline{w^2} f^2, \qquad (9.58)$$

with

$$A^{ii}_{kj} = \int_{\Omega_i} (\nabla w^k \cdot \nabla w^j)(x)\mathrm{d}x,$$
$$f^i = ((f^i_j)), \quad f^i_j = \int_{\Omega_i} (f\, w^j)(x)\mathrm{d}x, \qquad (9.59)$$

and the constraint:

$$B^1\overline{w^1} + B^2\overline{w^2} = 0,$$
$$B^i = ((B^i_{kj})), \quad B^i_{kj} = \int_\Sigma [w^j 1_{\Omega_i}]\omega^k. \qquad (9.60)$$

To remove the constraint (9.60), we use Lagrange multipliers $\{\lambda_k\}_1^{N-1}$ and we rewrite (9.59) and (9.60) in the form ($\lambda^T = (\lambda_1, ..., \lambda_{N-1})$)

$$\overline{w^1}^T A^{11} u^1 + \overline{w^2}^T A^{22} u^2 + \lambda^T (B^1\overline{w^1} + B^2\overline{w^2}) = \overline{w^1}\, f^1 + \overline{w^2}\, f^2, \qquad (9.61)$$

for any $\overline{w^i}$ and any λ. Writing this relation for any $\overline{w^i}$, for a given λ, we obtain:

$$A^{ii}u^i + B^{i^T}\lambda = f^i, \; i = 1, 2. \tag{9.62}$$

Adding the constraint on u

$$B^1u^1 + B^2u^2 = 0,$$

gives the system (9.51).

9.6.4 Iterative Solution of (9.51)

We shall apply the Schur complement method by eliminating u:

$$\begin{aligned}(B^1(A^{11})^{-1}B^{1^T} &+ B^2(A^{22})^{-1}B^{2^T})\lambda \\ &= B^1(A^{11})^{-1}f^1 + B^2(A^{22})^{-1}f^2.\end{aligned} \tag{9.63}$$

The matrix of this linear system is positive semi-definite. The problem can be solved by the preconditioned conjugate gradient method. Several preconditioners exist in the literature, but this choice, though essential, remains difficult.

Remark 9.7 As long as the subdomains have part of their boundary in common with the boundary of the global domain, the corresponding matrix block A^{ii} has a well defined inverse. If this is not the case, this block corresponds to a Neumann problem, and the inverse is no longer well defined. A numerical solution is to replace the operator $-\Delta$ by $\epsilon I - \Delta$ for a small ϵ. □

9.7 PERSPECTIVES

The methods we have given are still the subject of current research. The reader will find more details in the DDM (Domain Decomposition Methods) proceedings(*). Moreover, computer architectures are still evolving. Thus the market is currently more oriented towards shared memory machines. However, we can predict that it will not be possible to go beyond a few tens of processors with this kind of architecture, until connection buses become faster. But, once again, with shared memory, it is unclear whether or not the subdomains method is optimal.

Finally, debugging, load balancing techniques, and programming methodologies are still in a rudimentary state.

APPENDIX 9.A

EXERCISE Condition Number of the Schur Matrix

Let $\Omega =]-l, L[\times]0, 2\pi[$ be a rectangle. We apply the Schur complement method to a Dirichlet problem for the Laplacian:

(*) see also http://www.ddm.org

$$-\Delta u = f, \text{ in } \Omega, \quad u|_{\partial\Omega} = 0.$$

We assume that the Fourier expansion of f in y does not contain any terms in $\cos(y)$. The domain is divided in two by the line $\Sigma = \{(0, y), 0 < y < 2\pi\}$.

1. We are interested in solutions of the PDE with $f = 0$ in $\Omega\backslash\Sigma$ that are equal to $v(y)$ on Σ. Show that the solution to this problem is

$$u(x, y) = \sum_k v^k \sin(ky) \frac{\sinh(k(l + x))}{\sinh(kl)} \quad \text{for } x < 0,$$

$$u(x, y) = \sum_k v^k \sin(ky) \frac{\sinh(k(L - x))}{\sinh(kL)} \quad \text{for } x > 0,$$

   if $v(y) = \sum_k v^k \sin(ky)$ is the Fourier expansion of v.

2. Show that the Schur complement operator is the one that associates to $v = u|_\Sigma$ the jump of the normal derivative of u on Σ, that is

$$v \to -\sum_k k v^k \sin(ky) \left(\frac{1}{\tanh(kL)} + \frac{1}{\tanh(kl)} \right).$$

3. On the discrete level, if the grid is uniform and we use a five point difference scheme, we have:

$$-\frac{1}{h^2}(u_{i+1,j} + u_{i-1,j} + u_{i,j+1} + u_{i,j-1} - 4u_{i,j}) = f_{i,j},$$

   where $u_{i,j} \approx u(ih, jh)$, $i, j = 1, \ldots, M - 1$, $2\pi = Mh$. Show that functions of the form

$$u_h(x, y) = u^k e^{k_1(x+l)+ik_2y},$$

   with $k = \{k_1, k_2\}$ and $k_1 = \pm\frac{2}{h}\sinh^{-1}[\sin(\pm\frac{k_2 h}{2})]$ satisfy the scheme with $f = 0$.

4. For k_2 fixed, we note that the equation linking k_1 to k_2 always has a solution and that if k_1 is a solution, then so is $-k_1$; finally, changing k_2 into $-k_2$ does not change the equation. Show that

$$u_h(x, y) = \sum_{k_2=1}^{M-1} v^{k_2} \frac{\sinh(k_1(x + l))}{\sinh(k_1 l)} \sin(k_2 y), \text{ if } x < 0,$$

$$u_h(x, y) = \sum_{k_2=1}^{M-1} v^{k_2} \frac{\sinh(k_1(L - x))}{\sinh(k_1 L)} \sin(k_2 y), \text{ if } x \geq 0,$$

   is a general solution of the scheme, vanishing on the boundary and equal to v on Σ.

5. Writing the scheme for points of Σ gives the image of v by the Schur complement operator C:

$$v = \sum_{k_2=1}^{M-1} v^{k_2} \sin(k_2 y), y = jh, j = 1, ..., M-1,$$

$$Cv = -\sum_{k_2=1}^{M-1} \frac{v^{k_2}}{h^2} \left[\left(\frac{\sinh(k_1(l-h))}{\sinh(k_1 l)} + \frac{\sinh(k_1(L-h))}{\sinh(k_1 L)} - 4 \right) \sin(k_2 y) \right.$$
$$\left. -(\sin(k_2(y+h)) + \sin(k_2(y-h))) \right].$$

We conclude that the eigenvalues of this operator are

$$\frac{1}{h^2} \left[\frac{\sinh(k_1(l-h))}{\sinh(k_1 l)} + \frac{\sinh(k_1(L-h))}{\sinh(k_1 L)} - 2 + 4\sin^2 \frac{k_2 h}{2} \right].$$

6. Taking $k_2 = O(1)$ then $k_2 = M - 1 = O(\frac{1}{h})$, show that the condition number of C is $O(h^{-1})$.

Bibliography

AXELSSON O., (1994) *Iterative solution methods*, Cambridge University Press.

AXELSSON O. and BARKER V., (1984) *Finite element solution of boundary value problems : theory and computation*, Computer science and applied mathematics, Academic press.

BANK R.E., (1983) The efficient implementation of local mesh refinement algorithms, *Adaptive computational methods for partial differential equations*, I. Babuska ed., SIAM, Philadelphia, Pa., 74–81.

BANK R.E., DUPONT T. and YSERENTANT H., (1988) The hierarchical basis multigrid method, *Num. Math.*, **52**, 427–458.

BERNARDI C., MADAY Y. and PATERA A.T., (1994) New nonconforming approach to domain decomposition : the mortar element method, in *Collège de France*, Seminar **11**, H. Brézis and J.L. Lions, ed., 13–51.

BESPALOV A., KUZNETSOV Y., PIRONNEAU O. and VALLET M.G., (1992) Fictitious domains with separable preconditioners versus unstructured adapted meshes, *IMPACT of Computing in Science and Engineering*, Vol. 4, *n^o* 3, 217–249.

BRAMBLE J., PASCIAK J. and XU J., (1990) Parallel multi-level preconditioners, *Math. Comp.*, **55**, 1–22.

BREBBIA C.A., (1978) *The boundary element method for engineers*, Pentech Press, London.

BRENNER S.C. and SCOTT L.R., (1994) *The mathematical theory of finite element methods*, Texts in Applied Mathematics, Springer.

BREZIS H., (1987) *Analyse fonctionnelle*, Collection Mathématiques appliquées pour la maîtrise, Masson.

BREZZI F. and FORTIN M., (1991) *Mixed and hybrid finite element methods*, Springer series in computational mathematics, Springer.

BROWN P.N. and SAAD Y., (1990) Hybrid Krylov methods for nonlinear systems of equations, *SIAM J. Sci. Statist. Comput.*, **11**, *n^o* 3, 450–481.

CHAN T. and MATHEW T., (1994) Domain decomposition algorithms, in *Acta Numerica*, 61–143.

CIARLET P.G., (1978) *The finite element method for elliptic problems*, North Holland, Amsterdam.

CIARLET P.G., (1989) *Introduction to numerical linear algebra and optimisation*, Cambridge University Press.

COLTON D. and KRESS R., (1983) *Integral equation methods in scatterring theory*, John Wiley and Sons, New York.

CROUZEIX M. and MIGNOT A.L., (1984) *Analyse numérique des équations différentielles*, Collection Mathématiques appliquées pour la maîtrise, Masson.

DAUTRAY R. and LIONS J.L., (1990) *Mathematical analysis and numerical methods for science and technology*, Springer.

EUVRARD D., (1990) *Résolution numérique des équations aux dérivées partielles de la physique, de la mécanique et des sciences de l'ingénieur*, Masson.

EYMARD R. and GALLOUËT T (1993) Convergence d'un schéma de type éléments finis-volumes finis pour un système forméd'une équation elliptique et d'une équation hyberbolique. RAIRO Modél. *Math. Anal. Numér.* **27**, no. 7, 843-861.

EYMARD R., GALLOUËT T. and HERBIN R., (1998) An introduction to finite volume methods, in preparation for *Handbook of numerical analysis*, CIARLET P.G. and LIONS J.L. (eds).

GEIST A., BEGUELIN A., DONGARRA J., JIANG W., MANCHEK R. and SUNDERAM V., (1994) *PVM : Parallel Virtual Machine. A user's guide and tutorial for networked parallel computing*, MIT Press.

GIRAULT V. and RAVIART P.A., (1986) *Finite element methods for the Navier–Stokes Equations*, Springer Verlag, Heidelberg.

GODUNOV S.K. and RYABENKII V.S., (1987) *Difference schemes*, Studies in mathematics and its applications, Vol. 19, North Holland, Amsterdam.

GOLUB G.H. and VAN LOAN C., (1996) *Matrix computations*, Third edition, Johns Hopkins University Press.

GOLUB G.H. and MEURANT G.A., (1982) *Résolution numérique des grands systèmes linéaires* Eyrolles, Paris.

GRISVARD P., (1992) *Singularities in boundary value problems*, Collection Recherche en mathématiques appliquées, RMA 22, Masson.

GUERMOND J.L., (1992) Numerical quadratures for layer potentials over curved domains in $\mathbb{R}^3$, *SIAM J. Numer. Anal.*, **29**, *n°* 5, 1347–1369.

HANOUZET B., (1971) Espaces de Sobolev avec poids. Application au problème de Dirichlet dans un demi-espace, *Rend. del Sem. Nat. della Univ. di Padova*, XLVI, 227–272.

HSIAO G.C. and WENDLAND W.L., (1977) A finite element method for some integral equations of the first kind, *J. of Math. Anal. and Appl.*, Vol 3, **58**, 449–481.

JAMESON A., BAKER J. and WEATHERHILL N., (1986) Calculation of the Inviscid Transonic flow over a complete aircraft, *AIAA* paper 86–0103.

JOHNSON C.G.L. and SCOTT L.R., (1989) An analysis of quadrature errors in second-kind boundary integral methods , *SIAM J. Numer. Anal.*, **26**, *n°* 6, 1356–1382.

JOLY P., (1990) *Mise en œuvre de la méthode des éléments finis*, Collection mathématiques et applications, Ellipses.

KELLOGG O.D., (1967) *Foundations of potential theory*, Springer Verlag, Berlin.

KNUTH D., (1973) *The art of computer programming*, Vol. 1–3. Addison-Wesley.

LASCAUX P., (1976) *Lectures on numerical methods for time dependent equations. Applications to fluid flow problems*, Tata Institute of Fundamental Research, Bombay.

LASCAUX P. and THÉODOR J., (1988) *Analyse numérique matricielle appliquée à l'art de l'ingénieur*, 2 tomes. Masson.

LE ROUX M.N., (1977) Méthode d'éléments finis pour la résolution de problèmes extérieurs en dimension deux, in *R.A.I.R.O.*, Vol 11, **1**, 27–60.

LEVESQUE J. and WILLIAMSON R., (1989) *A guide book to Fortran on supercomputers*, Academic Press.

LIONS P.L., (1988) On the Schwarz alternating method I., in *First international symposium on domain decomposition methods for partial differential equations* (Paris, 1987), SIAM, Philadelphia, PA, 1–42.

LUCQUIN B. and PIRONNEAU O., (1997) *Introduction au calcul scientifique, exercices*, To appear, Collection Mathématiques appliquées pour la maîtrise.

MORTON K. W. (1996) Finite volume method, *Numerical analysis 1995*, Pitman Res. Notes Math. Ser., 344, 123–139.

NECAS J., (1967) *Les méthodes directes en théorie des équations elliptiques*, Masson.

NEDELEC J.C., (1976) Curved finite element methods for the solution of singular integral equations on surfaces in $\mathbb{R}^3$, *Comput. Meth. Appl. Mech. Engrg.*, **8**, 61–80.

NEDELEC J.C., (1977) *Approximation des équations intégrales en mécanique et en physique, cours de l'école d'été d'analyse numérique*, CEA-EDF-INRIA.

NEDELEC J.C. and PLANCHARD J., (1973) Une méthode variationnelle d'éléments finis pour la résolution numérique d'un problème extérieur dans $\mathbb{R}^3$, *R.A.I.R.O.*, R3, 105–129.

NICOLAÏDES R.A., (1992) Analysis and convergence of the MAC scheme, *SIAM J. Numer. Ana.*, **29**, *n°* 6, 1579–1591

PIRONNEAU O., (1989) *Finite element methods for fluids*, John Wiley and Sons.

POLAK E., (1971) *Computational methods in optimization*, Academic Press.

RAVIART P.A. and THOMAS J.M., (1983) *Introduction à l'analyse numérique des équations aux dérivées partielles*, Collection Mathématiques appliquées pour la maîtrise, Masson.

RICHTMYER R.B. and MORTON K.W., (1967) *Difference methods for initial-value problems*, Collection Interscience tracts in pure and applied mathematics, Number 4, Interscience publishers, John Wiley and Sons.

RISLER J.J., (1991) *Méthodes mathématiques pour la CAO*, Collection Recherche en mathématiques appliquées, RMA 18, Masson.

SAAD Y. AND SCHULTZ M. (1986) GMRES : A GENERALIZED MINIMUM RESIDUAL ALGORITHM FOR SOLVING NONSYMMETRIC LINEAR SYSTEMS, *SIAM J. SCI. STAT. COMP.*, 7, 856–869.

SCHATZMAN M., (1991) *Analyse numérique, Cours et exercices pour la licence*, Inter-Editions.

SCHWARZ H.A., (1869) Über einige Abbildungsaufgaben, *Ges. Math. Abh.*, **11**, 65–83.

SMITH B., BJORSTAD P. and GROPP W., (1996) *Domain decomposition*, Cambridge University Press.

STRANG G., (1986) *Introduction to applied mathematics*, Wellesley, Cambridge Press.

VLADIMIROV V.S., (1971) *Equations of mathematical physics*, Marcel Dekker, Inc., New York.

ZIENKIEWICZ O.C., (1971) *The Finite element method in engineering science*, Mc Graw-Hill, London.

Index